NEW GENERATION SMALL TELESCOPES

Edited by:
Donald S. Hayes
Russell M. Genet
David R. Genet

FAIRBORN PRESS

Mesa, Arizona

© 1987 Fairborn Observatory

Published in the U. S. A.
by the Fairborn Press
P. O. Box 7531
Mesa, AZ 85206

ISBN 0-944389-00-7

Library of Congress Catalog Card Number: 87-82772

Printed by Arcata Graphics Fairfield

Participants of the eighth annual Fairborn Observatory/IAPPP symposium.

CONTENTS

	CONTRIBUTORS	i
	PREFACE	v
1.	SECTION 1 INTRODUCTION: NEW GENERATION SMALL TELESCOPES Russell M. Genet	1
2.	NEW GENERATION SMALL TELESCOPES David L. Crawford	7
3.	REMOTE-ACCESS AUTOMATIC PHOTOMETRY Donald S. Hayes, Russell M. Genet, Louis J. Boyd and David L. Crawford	19
4.	A 0.75-METER TELESCOPE DESIGNED FOR AUTOMATIC PHOTOMETRY Louis J. Boyd, Russell M. Genet and David R. Genet	27
5.	INTENSIFIED CCDs FOR AUTOMATIC ACQUISITION AND CENTERING Louis J. Boyd and Russell M. Genet	35
6.	CONTRIBUTIONS OF THE SLEEPING ASTRONOMER Norman L. Markworth and John Mullikin	41
7.	THE DETECTABILITY OF SUPERNOVAE AGAINST ELLIPTICAL GALACTIC DISKS Eric C. Pearce	51
8.	AUTOMATED K-LINE PHOTOMETRY OF STARS WITH ACTIVE CHROMOSPHERES Sallie L. Baliunas, Louis J. Boyd, Russell M. Genet, and Donald S. Hayes	65
9.	APPLICABILITY OF SPACE INSTRUMENTATION RELIABILITY CONCEPTS TO UNATTENDED AUTOMATIC TELESCOPES G. Eichhorn, R. B. Piercey, J. McKisson, F. Giovane and D. W. Ely	73
10.	THE APT SERVICE: A BRIEF PROGRESS REPORT Russell M. Genet and Louis J. Boyd	85

11. BROADBAND PHOTOMETRY OF BRIGHT STARS: THE FIRST YEAR OF APT'S AT THE F. L. WHIPPLE OBSERVATORY 97
Sallie L. Baliunas, Robert A. Donahue, Jennifer G. Loeser, Edward F. Guinan, Russell M. Genet and Louis J. Boyd,

12. PLANNING FOR A CONSORTIUM APT—A PROGRESS REPORT 117
Robert J. Dukes, Jr.

13. DELTA SCUTI STARS AS APT SUBJECTS 125
Robert J. Dukes, Jr.

14. THE CASE FOR PHOTOMETRY OF THE NEARBY STARS 133
Arthur R. Upgren

15. SECTION 2a
INTRODUCTION: AUTOMATED SPECTROPHOTOMETRY 139
Donald S. Hayes

16. SYNTHETIC PHOTOMETRY I: MODELING A PHOTOMETRIC SYSTEM 145
David L. Crawford

17. SOME THOUGHTS ABOUT THE NEXT GENERATION OF SPECTROPHOTOMETRIC INSTRUMENTS 157
Saul J. Adelman

18. SPECTROPHOTOMETRY AND PHOTOMETRY OF A—TYPE FIELD HORIZONTAL—BRANCH STARS 165
A. G. Davis Philip

19. MUSINGS ON A SYNTHETIC PHOTOMETER 183
Nathaniel M. White

20. A PRELIMINARY DESIGN FOR A SPECTROPHOTOMETER TO BE USED ON A FULLY—AUTOMATED TELESCOPE 185
Donald S. Hayes

21. AN ALGORITHM FOR CALCULATING ATMOSPHERIC EXTINCTION FOR SPECTROPHOTOMETRY AND NARROW—BAND PHOTOMETRY 193
Donald S. Hayes and Paul C. Schmidtke

22.	SECTION 2b INTRODUCTION: AUTOMATED SPECTROSCOPY Michael A. Seeds	201
23.	THE FEASIBILITY OF AN AUTOMATIC SPECTROGRAPHIC/PHOTOMETRIC TELESCOPE Michael A. Seeds	203
24.	METHODS OF IMAGE STORAGE FOR AUTOMATED TELESCOPES Gary B. Karshner	211
25.	INTERMEZZO AMATEURS AND THEIR PLACE IN PROFESSIONAL SCIENCE Robert A. Stebbins	217
26.	SECTION 3 INTRODUCTION: COMPUTER-CONTROLLED TELESCOPES Russell M. Genet	227
27.	BRAESIDE OBSERVATORY-A NEW BEGINNING Robert E. Fried	229
28.	BIG COTTONWOOD OBSERVATORY-THE FIRST YEAR Jerrold L. Foote	233
29.	PHOTOELECTRIC PHOTOMETRY AT LIMBER OBSERVATORY David McDavid	237
30.	THE SFA 41-INCH MICROCOMPUTER-CONTROLLED TELESCOPE James B. Rafert and Norman L. Markworth	247
31.	THE FIT 25-INCH PHOTOMETRIC REFLECTOR James B. Rafert and Terry D. Oswalt	257
32.	THE SFA TELESCOPE CONTROLLER/DATA LOGGER Norman L. Markworth and James B. Rafert	267

33. COUPLING IMAGE INTENSIFIERS TO TELEVISION
SENSORS 279
Peter L. Manly

34. CHIPS SET SUBSTITUTE FOR THE MVL 100 295
Jeffrey L. Hopkins

35. A LOW-COST AUTOMATIC TELESCOPE AND
PHOTOPOLARIMETER 307
F. Giovane, D. Ely, A. Weisenberger, G. Eichhorn,
J. Rilum and B. Soderberg

36. A VME-BUS-BASED MICROPROCESSOR SYSTEM
FOR AUTOMATIC TELESCOPE AND PHOTOMETER
OPERATION 317
G. Eichhorn and F. Giovane

37. A HIGH-SPEED PHOTOMETER INTERFACE FOR
THE MICROVAX Q-BUS 325
Mark Trueblood

38. SECTION 4
INTRODUCTION: PHOTOELECTRIC PHOTOMETRY 341
Donald S. Hayes

39. A SHORT TUTORIAL ON STROMGREN FOUR-COLOR
PHOTOMETRY 345
David L. Crawford

40. A QUALITY-CONTROL CHECK ON SCALE-FACTOR
TRANSFORMATION COEFFICIENTS 351
Edward C. Olson

41. PHOTOELECTRIC DATA REDUCTION FOR EVERYONE 355
Norman L. Markworth and James B. Rafert

42. OBSERVATIONS OF THE W SERPENTIS STARS: A
PROGRESS REPORT 361
Norman L. Markworth and James B. Rafert

43. SZ HERCULIS: NEW PHOTOELECTRIC OBSERVATIONS
AND PERIOD STUDY 369
E. J. Michaels

44. TIMES OF ECLIPSING MINIMA OF BINARY STARS 379
Frank Bradshaw Wood and Kwan-Yu Chen

45.	ECLIPSING BINARIES IN CONFLICT WITH GENERAL RELATIVITY Edward F. Guinan and Frank P. Maloney	383
46.	PHOTOELECTRIC PHOTOMETRY RESEARCH AT THE HIGH–SCHOOL LEVEL Kenneth W. Zeigler	403
47.	ULTRAVIOLET PHOTOMETRY WITH THE IUE SATELLITE Joel A. Eaton	411
	INDEX	433

CONTRIBUTORS

Saul J. Adelman
    Department of Physics, The Citadel

Sallie L. Baliunas
    Harvard–Smithsonian Center for Astrophysics

Louis J. Boyd
    Fairborn Observatory

Kwan–Yu Chen
    Rosemary Hill Observatory, University of Florida

David L. Crawford
    Kitt Peak National Observatory
    National Optical Astronomy Observatories

Robert A. Donahue
    Harvard–Smithsonian Center for Astrophysics

Robert J. Dukes, Jr.
    Physics Department, College of Charleston

Joel A. Eaton
    Astronomy Department, Indiana University

G. Eichhorn
    Space Astronomy Laboratory, University of Florida

D. W. Ely
    Space Astronomy Laboratory, University of Florida

Robert E. Fried
    Braeside Observatory

Jerrold L. Foote
    Big Cottonwood Observatory

Russell M. Genet
    Fairborn Observatory

David R. Genet
    AutoScope

F. Giovane
    Space Astronomy Laboratory, University of Florida

Edward F. Guinan
    Department of Astronomy and Astrophysics, Villanova University

Donald S. Hayes
    Fairborn Observatory

Jeffrey L. Hopkins
    Hopkins Phoenix Observatory

Gary B. Karshner
    Physics Department, Gettysburg College

Jennifer G. Loeser
    Harvard—Smithsonian Center for Astrophysics

Peter L. Manly
    Saguaro Astronomy Club, Metro Phoenix

Norman L. Markworth
    Department of Physics and Astronomy
    Stephen F. Austin State University

David McDavid
    Division of Earth & Physical Sciences
    The University of Texas at San Antonio

J. McKisson
    Space Astronomy Laboratory, University of Florida

E. J. Michaels
    Dept. of Physics and Astronomy
    Stephen F. Austin State University

John Mullikin
    Department of Physics and Astronomy
    Stephen F. Austin State University

Edward C. Olson
    Department of Astronomy
    University of Illinois, Champaign—Urbana

Terry D. Oswalt
    Department of Physics an Space Sciences,
    Florida Institute of Technology

CONTRIBUTORS

Eric C. Pearce
    Digitized Astronomy, Department of Physics
    New Mexico Institute of Mining and Technology

A. G. Davis Philip
    Van Vleck Observatory and Union College

R. B. Piercey
    Space Astronomy Laboratory, University of Florida

James B. Rafert
    Department of Physics and Space Sciences,
    Florida Institute of Technology

J. Rilum
    Optical, Materials and Devices Lab.,
    Dept. of Elect. Eng., Univ. of Southern California

Paul C. Schmidtke
    Department of Physics, Arizona State University

Michael A. Seeds
    Pennsylvania Astronomy Research Consortium
    Franklin and Marshall College

B. Soderberg
    Royal Institute of Technology, Stockholm

Robert A. Stebbins
    The University of Calgary

Mark Trueblood
    Winer Mobile Observatory

Arthur R. Upgren
    Van Vleck Observatory, Wesleyan University

A. Weisenberger
    Space Astronomy Laboratory, University of Florida

Nathaniel M. White
    Lowell Observatory

Frank Bradshaw Wood
    Rosemary Hill Observatory, University of Florida

Kenneth W. Zeigler
    Gila Astronomical Research Institute

PREFACE

The Eighth Annual Fairborn–IAPPP Symposium was held on February 12–14, 1987, at the Saguaro Lake Ranch, Mesa, Arizona. The Symposium was co–chaired by Russell M. Genet, Douglas S. Hall, and Donald S. Hayes. These three also served as chairs of sessions of the meeting; other session chairs were David L. Crawford, Norman L. Markworth, and Michael A Seeds. The main sessions of the meeting corresponded in title to the sections in these proceedings, except that the session entitled: "Automatic Photoelectric Telescopes" was re–named "Computer–controlled Telescopes" for the proceedings. The papers contained in these proceedings have been re–arranged, so that the order and organization of the papers does not correspond exactly to the meeting. Additional informal workshops were held on the last day; they were "The Four–College APT Consortium," chaired by Robert J. Dukes, Jr., "The Pennsylvania Astronomy Research Consortium,", chaired by Harold L. Nations, and a "General Photometry Workshop," chaired by James B. Rafert. No manuscripts resulted from these sessions.

On two evenings during the Symposium after–dinner speeches were given, much to our enjoyment. The Thursday evening speech was a witty and informative memoir by Robert C. Wolpert, entitled: The Confessions of a Photoelectric Photometry Has–Been." On Friday evening we were treated to a fascinating discourse by Robert A. Stebbins, of the Sociology Department of the University of Calgary. It was entitled: "Amateurs and Their Place in Professional Science," and a written version may be found in these proceedings.

The Saguaro Lake Ranch is located on the Salt River, 40 miles east of Phoenix. Sunny weather and stunningly beautiful surroundings encouraged horseback riding, hiking and rafting on the river in the afternoons after the sessions adjourned for the day. They also formed the backdrop for informal seminars during the meeting. We are most appreciative of the warm hospitality of the manager of the ranch, Steve Durand, and the assistant manager, Kelly Milliron, and their ever–helpful staff.

The editors would like to thank Douglas S. Hall for his share in organizing the meeting and chairing the first day. Doug also was co–host, representing the IAPPP (International Amateur–Professional Photoelectric Photometry).

The editors would also like to thank Michael A. Seeds for organizing and chairing the session on Automatic Spectroscopy and for writing the introduction to the corresponding section in these proceedings.

Many of the manuscripts were submitted on floppy disk, in various formats. Converting to IBM–PC–compatible format was challenging in some cases, but the result was worth it. Those manuscripts which were submitted in typed/printed form had to be entered into the computer by hand, and the editors wish to thank Karen A. Genet for her excellent work on this tedious task. We also would like to thank John Richard whose copy editing added appreciably to the quality of this volume. The final manuscript of the proceedings was edited and printed in its entirety with personal computers, a process which has resulted in vastly higher quality and much less effort than the old method of editing the so-called "camera—ready" manuscripts prepared by the authors (and their secretaries).

Finally, the editors would like to thank the participants for a lively, informative and stimulating Symposium. We hope the readers enjoy the Proceedings as much as we enjoyed the Meeting.

Donald. S. Hayes
Russell M. Genet
David R. Genet

SECTION 1—INTRODUCTION: NEW GENERATION SMALL TELESCOPES

Russell M. Genet

Fairborn Observatory

First editor Donald S. Hayes asked that I write a personal introduction to the first section of the book—something he knew I would be delighted to do. This symposium was originally billed as Automatic Photoelectric Telescopes II. The Photoelectric in the title began to look less appropriate as a number of papers were offered in the area of spectrophotometry and spectroscopy. Automated Photometry, Spectrophotometry, and Spectroscopy was considered, but then a paper was offered on automated area imaging (for supernovae searches).

While the rest of us had been looking at the individual trees in the forest, and some of us had been working hard on particular branches on a single tree, David L. Crawford had been looking at the entire forest, in fact several forests. He noted a number of similarities between what was being done at the forefront of technology with the largest telescopes, the 10-meter Keck, the 15-meter NNTT, etc., and what was being done with the small automatic photoelectric telescopes (APTs). In some areas, such as optics, the large telescopes were in the lead and the small

New Generation Small Telescopes, ed. D. S. Hayes, R. M. Genet, & D. R. Genet.
© 1987 Fairborn Observatory.

telescopes could benefit and learn from the large. In other areas, such as full automation, the small telescopes were in the lead, and the large could learn from the small.

Crawford termed the forests on the mountain New Generation Telescopes (NGTs). Individual forests on the mountain he noted as New Generation Large Telescopes (NGLTs), New Generation Small Telescopes (NGSTs), and also an intermediate size, NGITs. He noted that, of course, NGSTs were fully automatic, had no eyepieces, and that while the first of the NGSTs were just capable of photometry (a subspecies of NGSTs called APTs), that later ones would certainly be able to do spectrophotometry, spectroscopy, and even direct imaging.

Thus his insightful taxonomic analysis resolved the issue of what to name the symposium and New Generation Small Telescopes (NGSTs) became the official name. Of course Crawford needed to explain this new overview, which he ably did, a written version of which is given as the first chapter of these proceedings.

Two potential long-term problems arose at the Automatic Photoelectric Telescope Service as we considered life beyond a couple of APTs doing the most elementary differential photometry. We had originally thought of the APT Service as only making differential measurements on long-term synoptic programs. Measurements would be made once every clear night for years, even decades, with the APTs patiently observing in the same fixed, simple manner year after year. To some extent this has indeed happened. RS CVn binaries put on the program over three years ago by Douglas S. Hall are still on the program. Solar-type stars put on the first APT on Mt. Hopkins by Sallie L. Baliunas are still on the program and probably will be one or two decades from now. The value of some observations increases with longevity.

However, there were a number of astronomers who failed to adhere to our idealistic view of the simple future of the APT Service. They wanted to observe the same star (or couple of stars) hours, nights, or even months on end. Robert J. Dukes, in this section, suggests in some detail why APTs would be ideal to observe d Scuti stars—the same star—for weeks on end. Arthur R. Upgren, as explained later in this section, wanted to observe a large number of stars nearby in space just a few times each, and, worse yet, he wanted to observe them using the all-sky approach to photometry. Adelman, Dukes, Grayzeck, McCook, and Smith went even further, suggesting that a consortium on a single telescope would like to mix ordinary differential once-a-night photometry with continuous observations and all-sky photometry.

While the APT Service could see that these requirements were legitimate, it was not at all obvious at first how they might be handled. Already it could be seen that, even with simple differential once-a-night photometry, channeling all the requests for observations through the APT Service and all the results back through the service would gradually become a monumental task. Astronomers wanted this star pulled off for a while and replaced with that star. The standard reduction procedure or format was not quite what another astronomer wanted. Could the data be reduced more often than once per quarter; some of the results were urgent! Could a very important star be placed on for just one night in support of IUE observations?

Our problem was eventually given a name. We called it the middleman problem. The APT Service staff was standing between the astronomers and their APTs. While this was acceptable in the simple initial days of the APT Service, it was clear that it would not be acceptable in the future. When the College of Charleston, The Citadel, the University of Nevada, Las Vegas, and Villinova University formed a consortium, as described by Robert Dukes in this section, we knew that the middleman problem would have to be solved. The solution was Remote Access Automatic Photometry (RAAP).

The basic concept of RAAP is simple enough. The astronomers at their home institutions compose an observing program request for one or more nights. They do this on their own PCs using software supplied to them that simplifies this process. Completed observational requests are sent via modem over an ordinary phone line to a PC at the APT Service. The transaction is PC to PC and takes just a few minutes. The next morning or the next week, whenever they feel it is time, the astronomers retrieve the observational results in the same manner, from PC to PC over the phone lines using a modem. They then use software supplied to them to reduce their data.

The advantage of RAAP is that it puts the astronomers in full charge of their own APT. As long as they make types of observations for which request generation and data reduction software exists, then they do not have to be involved in software in any way (except analysis). If, however, they would like to make a modification or even develop a completely new type of observing, they are free to do so as long as the format (protocol) for requesting observations and receiving results is adhered to. This protocol is currently being developed. It will be in plain ASCII text, and thus computer independent, and will allow the full capabilities of APTs to be exercised.

Getting back to the forests, one would expect that the remote access concept would apply not only to small photometric telescopes, but to other instruments including large telescopes. Thus the more generic term might be Remote Access Automatic Telescopes (RAATs).

Other requests by astronomers have included a desire to go to fainter stars, and to cover more stars per hour with the automatic systems. Could the already high efficiency of APTs be increased even further? Thought was given to this challenge by a number of engineers and astronomers, and it was suggested that there were three major paths to improvement: (1) larger aperture APTs; (2) APTs that literally moved faster; and (3) a faster way of finding and centering stars.

Larger aperture makes observations of bright stars more efficient by reducing the scintillation noise, while on faint stars the efficiency is achieved, of course, by gathering more photons per second. What is the optimum aperture–size for an automatic photometric telescope? It might, at first, seem that this question would not have a single answer, but in fact there is, for the majority of stellar photometry applications, a single answer.

There are a number of APT costs that are not a function of aperture size. These include the control system, the photometer, cables, software, etc. Also, to some extent once an APT is installed, its cost of operation is not strongly dependent on size. There are other costs that are clearly a strong function of aperture, such as the optics and the mount (and the housing also). Now if one wanted to observe bright stars efficiently, such that the cost per observation were minimized over a ten-year period, then a 10-inch or even a 16-inch telescope would be non-optimally small. If one wanted to observe stars fainter than 8th magnitude, such as stars from 8th to 12th magnitude, then the 10- to 16-inch size would be decidedly non-optimally small. However, as the aperture size is made larger a point is reached where the costs of the optics make a sudden jump upwards. With current technology this point is somewhere between about 20 and 30 inches. These are the largest sized mirrors that 1 5/8ths-inch slumped-pyrex will support with good optical quality. Sizes much larger than this require either special technology (cellular, hex tubes, spinning oven, etc.) or if they utilize the old thick mirror technology which cause telescope mounts to regress back to the "battleship technology" approach.

Thus a consensus of sorts has been reached that with today's technology, for photometry of stars in the range of 1st to 12th magnitude, and for the lowest long run cost per observation, that an aperture of about 20- to 30-inches, 0.50- to 0.75-meters, is nearly optimal. Aperture sizes much larger or much smaller than this create systems that would cost more per observation in the long run (to the same level of accuracy, of course).

This conclusion only applies to telescopes designed specifically for automated photometry. General purpose telescopes are necessarily more expensive than specialized ones, and as the aperture gets larger, the disparity in cost between a general purpose and specialized automatic telescopes increases rapidly. At the 30-inch size, the general purpose telescope is more expensive by a factor of two. The reasons for this disparity are discussed in some detail in the paper in this section on the 0.75-meter telescope designed specifically for fully automatic operation. General purpose telescopes have to accommodate humans and changes of instruments. To do this they must be raised well off the floor, must usually be cantilevered over the astronomer, and must provide convenient eyepiece and instrument access at almost all telescope positions. The result is a physically much larger telescope that is much more expensive to build, more difficult to maintain, and one that takes up lots more room. There are some NGST requirements, besides compactness and low cost, that the general purpose telescopes may not meet well. These include very low backlash (almost zero) without preloading, very low moments of inertia, and low effective cross sections to minimize wind gust effects. Also, as pointed out by Gunter Eichhorn in a chapter in this section, specialized automatic telescopes need to have very high reliability.

NEW GENERATION SMALL TELESCOPES 5

Not only can efficiency be improved by increasing aperture to the optimal size, but it can be improved by making the telescopes move faster. The use of a new type of DC servo that has the high speed range and power of DC motors, but from the viewpoint of the control computer looks like a simple stepper, allows high performance to be achieved with simple control systems. Automatic telescopes can, of course, move at speeds that would be difficult for a human operator!

The final approach to improving efficiency which was discussed at the symposium was elimination of the spiral search and iterative centering procedures. Currently, these take about 25% of the observing time. The use of an area detector would allow the program star to be almost instantly found and centered, reducing the time spent to 1% or so of the total time—a negligible amount. Louis J. Boyd visualizes an approach that would use low—cost off—the—shelf commercial, uncooled CCD cameras and intensifiers and simple electronics and software. Norman L. Markworth envisions an approach that would have considerable built—in "smarts." A side benefit of these approaches would be highly accurate "electronic offsetting." This would allow very faint or nebulous objects to be accurately observed. A recognition system already working that is very sophisticated is that on the automated supernovae search system as described by Eric C. Pearce.

Larger aperture APTs allow one to consider, instead of fainter stars, narrower bandwidths. One very narrow—bandwidth project being given serious consideration is fully automated observations of solar—type stars in the K—line of Calcium II.

While it is exciting to think about the future potentials of NGSTs, what they are doing now and have already accomplished is the baseline from which we must work. One of the chapters in this section describes progress made at the APT Service over the past year, and another describes the observations made by the Smithsonian Institution of a variety of cool stars, and a few not so cool ones. Douglas S. Hall and others reported at last year's symposium on observations of RS CVn binaries and other types of stars by APTs.

This completes our tour of David Crawford's NGST forest. For the more detailed tour please read the individual chapters in this section. For advanced tree species read the sections on spectrophotometry and spectroscopy. What does David Crawford have in store for us next year? He let slip, over a second beer, that many of the NGST trees in the forest should be interconnected to form a Global Network of Automatic Telescopes (GNAT). Then a remote access request from a home institution would not trigger observations on just a single NGST—observations to be interrupted by the inconvenient rising of the Sun—but would cause observations to be made literally round the world, one NGST after another. Robert Dukes would then be able to get unparalleled d Scuti coverage, not to mention what could be done in stellar seismology. This would, of course, require some international cooperation—I trade some of my NGST time for some of yours—but astronomy has always led the way in international cooperation, and this would expand such cooperation to the modern electronic age. For a more detailed tour of the GNAT concept, however, you'll have to come to next year's 9th Annual Symposium, which will be held at the Lazy K Bar Guest Ranch in Tucson and on Mt. Hopkins from February 18 — 21, 1988.

NEW GENERATION SMALL TELESCOPES

David L. Crawford

Kitt Peak National Observatory
National Optical Astronomy Observatories*

ABSTRACT: Besides the current wave of new generation large telescopes, glamorous expensive instruments for frontier astronomical research, there is a coming wave of new generation small telescopes. This paper describes those things that make a telescope "new generation," whether large or small, and identifies those items that apply especially to small telescopes. From this inspection, we identify three general sizes: large, intermediate, and small. A simple example of a telescope cost scaling law is shown. Finally, we describe some of the presently operating new generation small telescopes, and make a few predictions about the future.

---

*Operated by the Association of Universities for Research in Astronomy, Inc. under contract with the National Science Foundation.

New Generation Small Telescopes, ed. D. S. Hayes, R. M. Genet, & D. R. Genet.
© 1987 Fairborn Observatory.

I. INTRODUCTION

Looking at the title of this paper raises two immediate questions: what do we mean by "new generation," and what do we mean by "small"? Most of the paper will address answers to those two questions.

I have attended meetings on telescopes where someone called an 8m aperture telescope "small." Note that a telescope that large has never yet been built. For a refracting type telescope, a 1m aperture is very large, yet for a reflecting type telescope, 1m is small. So the answer to the question of "What's small?" is quite relative. It depends on the context, the user, and the bias. In this paper, we will use "small" to mean a telescope of approximately 1m in aperture or smaller. As we indicate below, that is the size where the issues of mirror fabrication and, especially, mirror support are not a major issue.

As to "new generation," there is no question that many aspects of technology have improved greatly in recent years. The use of computers in design and control is an especially important change over the last wave of telescopes, such as the 4m class at KPNO, CTIO, ESO, CFHT, and AAO. There are other such changes, and we will discuss them in the next sections, where we also compare the present wave of New Generation Telescopes (NGTs) to previous Old Generation Telescopes (OGTs).

I will not discuss here any of the scientific justification, for the large or the small NGTs. That is for other papers.

II. THE PRESENT NEW GENERATION LARGE TELESCOPE PROJECTS

Let me here describe very briefly six of the current large telescopes being planned or built. I do this primarily to illustrate some of the new generation aspects inherent in the designs. I will then come back and list in the following section those things that make a telescope an NGT.

a) The NNTT: (National New Technology Telescope), a 15m for NOAO.

Design studies are well along at National Optical Astronomy Observatories (NOAO) for a 15m aperture multiple mirror telescope. It will use four light−weight, ribbed−pyrex 7.5m diameter mirrors (possibly 8.0m mirrors, thus making the effective aperture 16m rather than 15m). The light collecting power of these four mirrors is equivalent to one mirror of 15m diameter. The four mirrors are rather close together, on the same mounting, but the separation from edge to edge will still allow the angular resolution to be that of a 21m single mirror.

The telescope can operate with a combined focus, light from all the mirrors being brought to one focus and one instrument located there. It can also be operated as four 7.5m telescopes, pointed at approximately the same location in the sky, using four separate instruments at the four individual foci. An application for the latter would be to collect spectra on many individual galaxies in a cluster of galaxies; each instrument could be a multiple object spectrograph (MOS) allowing, say, 40 spectra to be taken at a time. Hence, one could obtain spectra on 160 individual galaxies, each with the power of a 7.5m telescope. One could then come back and use the combined focus for higher resolution spectroscopy on the most interesting individual galaxies.

b) The European Southern Observatory's (ESO) Very Large Telescope (VLT), a 16m linear array of four 8m individual telescopes.

ESO wants to emphasize the angular resolution potential of a new generation large telescope. Hence, they plan to use four individual 8m aperture telescopes as an array. The light—collecting power of the system is equivalent to a 16m aperture telescope, but the angular resolution is much greater. They may even use smaller telescopes as part of the system to further increase the efficiency of the array.

The telescope will be located in northern Chile, near La Silla, the current ESO site, or perhaps a bit further north, where observing conditions may be even better than at La Silla.

c) "The Giant Binocular," an 11.3—m equivalent aperture telescope, being planned by the University of Arizona in collaboration with Ohio State University and the University of Chicago.

The telescope system would use two 7.5m or 8m aperture mirrors, identical to those to be used in NOAO's NNTT. The two mirrors would be on the same mounting, hence the cute name. The light collecting power is that of an 11.3m telescope, but it has greater angular resolution potential. The system appears to be an interesting combination of large aperture and larger—yet angular resolution. It would be located on Mt. Graham, a new site for astronomy, about 60 miles east of Tucson, if objections of some environmentalists can be overcome and Forest Service approval obtained. Mirrors for this telescope, and for the NNTT, would be made at the University of Arizona, in a new facility located under the football stadium. The facility is essentially finished, and smaller mirrors are now being made on the program that will eventually turn out a significant number of 7.5m or 8m mirrors.

d) The Keck Telescope, a 10m aperture new generation telescope for Cal Tech and the University of California system.

This telescope would be of the "segmented mirror" design, using 36 approximately 1m mirrors, figured with spherical surfaces, but in a stressed condition, so that when the stress is released, the mirror surface becomes the correct segment of a single 10m mirror. The mirrors are cut into a hexagonal shape, fitted closely together, with their position accurately sensed and then adjusted to the correct position. The tolerances are very tight, of course (a millionth of an inch or so), but quite possible to realize. The result is a "single" mirror of 10m aperture, not the "multiple mirror" approach of the telescope projects mentioned above.

The project is funded, and fabrication has begun on all parts of the telescope system. The telescope will be located on Mauna Kea, the excellent location on the island of Hawaii.

e) There are several projects planning an NGT of 8m aperture, using one of the 7.5m or 8m mirrors that can be produced by the University of Arizona facility (or other similar ones). These telescopes are larger than any now in existence, and should be excellent new instruments for frontier astronomy.

f) In addition, there are several projects planning or building NGTs in the size range 2.5m to 4m. I will mention the Australian National telescope, the Nordic 2.5m Telescope, the University of Washington, et al. 3.5m Telescope, and the MOST project (described below). There are others as well, as in this Intermediate size range, the costs are within the range of a single university or small consortium to fund and to operate.

III. WHAT MAKES A TELESCOPE "NEW GENERATION"?

Let us list now some of those aspects of the above designs that distinguish these NGTs from the previous OGTs. All of these aspects allow improved performance at lower cost, as of course they must; otherwise they would not be of interest and probably not "new generation." They improve the Science Per Dollar (Sci/$) ratio, and by a large factor. Such is the "Bottom Line" of an NGT, whether large or small.

a). Thin, light–weight primary mirror. OGTs have relatively thick primary mirrors; the 4m at KPNO is typical: it is 24 inches thick, and solid. Such thickness was used to gain added stiffness, as mirror flexure is a critical item. Now we can make thinner mirrors (or ones with very thin ribs), so that both the weight and the thermal inertia are quite low. Due to improved understanding of mirror flexure and the ability to calculate accurately such flexure for any geometry of the mirror, we feel we can now use the low weight, thin primaries, without compromising surface quality.

In fact, some of the designs actually use the flexibility of the primary to advantage, by stressing it (bending it) to compensate for flexure due to gravity, wind gusts, or whatever. This is "active optics" applied to a single mirror. The Keck Telescope uses active optics by sensing and displacing the 36 individual components of the segmented primary mirror.

A significant additional advantage of the light–weight primaries is their relatively low thermal inertia. The glass adjusts quickly to changes in ambient air temperature. Hence adverse thermal turbulence is much less of a problem than in OGTs. One must do all possible not to compromise the excellent atmospheric seeing at the best observing sites. Current telescopes often do, because of this problem (high thermal inertia of the primary) as well as many others, such as the thermal inertia of the telescope mounting. We will see below that this latter item is better in the NGTs too.

Remarkable improvements in Sci/$ are possible, due to the lower weight (and hence lower weight of the support system and telescope mounting) and the lower thermal inertia (better images: better science).

b). *Active optics*, as discussed above. Since we can now adjust the primary (and secondaries too, for that matter), we have the potential of improving the image. One hopes to compensate for the bending effects of the optics due to gravity, temperature differences, wind gusts, or whatever. Also, if one can accurately sense the incoming wave front disturbances (due to thermal turbulence in the atmosphere), then one can perhaps do something to correct for such disturbances, improving the image. Such "active optics" may mean that we can have excellent images a large fraction of the observing time (at least, never having to suffer through degraded images due to our own man–made causes). The power of large telescopes is strongly dependent on the quality of the image at the focus. This is an area of NGTs where striking improvements in Sci/$ are possible.

c) *Fast focal ratios.* Due to the improvement in optical design, fabrication, and testing made possible by increased understanding and the use of computers to assist in design and testing, we believe that we can move to the use of considerably faster primary focal ratios. All the new large NGTs have focal ratios of f/2 or faster. There is even serious discussion of the use of a focal ratio of f/1. All of this means that the telescope tube can be shorter and lighter. The overall mounting can also be lighter, and lower cost. Hence, the thermal inertia of the system is also much lower, and less adverse thermal turbulence in the building will result. The telescope housing (building and dome) can be smaller. For example, the Keck Telescope dome is no larger than the KPNO or CFHT dome, even though it is a 10m telescope compared to the OGT 4m size.

d) *Image size.* Because of the quality of the best observing sites, one must and is specifying a smaller image size. One must take advantage of the site quality. The optics, mounting, and telescope surroundings must not be allowed to degrade the image significantly. Great care must be taken concerning the design so as not to introduce adverse mechanical or thermal effects that will increase image size.

None of this is easy! Not only are some of these NGTs much larger than any telescope built so far, but the tolerances are also much tighter than any existing telescopes. The sharper images, the faster focal ratios, the lighter components, and the thermal specifications all mean difficult tolerances, much tighter than with OGTs. Fortunately, we think all of these are possible. Such tight tolerances and specifications are part of all the new telescope designs. Time will tell if our optimism is correct.

e) The faster focal ratios mean shorter tubes, hence, lower weight, added stiffness, smaller domes. All this offers lower cost for a given aperture, a key aspect of an NGT.

f) All of the above mean lower weight for the telescope system, hence lower cost. To first order, the cost goes as the weight, at least down to a certain critical weight where the added complexity begins to raise the cost significantly. One would like to be at or near that minimum weight / minimum cost per aperture, of course.

g) All of the above mean a much smaller dome and housing for the telescope system. In an OGT, a large share of the overall project cost goes into the building and dome. In the NGTs that share should be much lower; hence, more aperture per dollar and better Sci/$, at any telescope size. In addition, the building and dome may well be simpler than in an OGT. One wants to keep heat sources to a minimum, hence such things may well be located in a separate (low cost) housing some distance away from the telescope itself. As we will see below, computer control also means a simpler housing for the telescope should be possible.

h) The extensive use of computers in all phases of the telescope system will mean a very versatile control system, lots more bang for the buck. All the NGTs, of any size, use full computer control of the telescope, instrumentation, and data handling (and for the large NGTs, also for the active optical control). The new generation of instrumentation, whether used on an OGT or an NGT, means that the amount of data to be handled and analyzed is remarkably larger than just a few years ago. It is unthinkable to be able to cope with these vast amounts of data without full use of computers. Certainly the larger telescopes with complex instrumentation will require larger and more complex computer systems, but the small NGTs will make full use of computer systems as well.

i) Mentioned in some of the above items was the quality of the best of the observing sites. It makes no sense to put a large, and very expensive, NGT at a site that is not first rank. That is clear. But it also means that intermediate size NGTs, and even small NGTs, should also be at excellent sites. The amount and quality of the data produced will be much greater at such sites. With the full use of computer control, NGTs can be operated remotely (or automatically in the case of small NGTs) for maximum cost effectiveness. For a low cost installation, such as a small NGT, there may be arguments for immediate access: instrument testing, student access, "show and tells," or whatever. For the larger, more costly telescopes, however, it is essential to locate them at the best observing sites. I personally believe that this rationality holds also for almost all of the small NGTs, whether used for research or for educational purposes. (See the separate papers in this conference and elsewhere.)

j) Scheduling can be done differently with NGTs than with OGTs. The full use of computers means that the potential is there for more rational scheduling of the telescope. Programs that need the lowest water vapor content can be switched to when such conditions occur. The same is true for partial clouds, less-than-the-best seeing, etc. All of this can well be called "queue scheduling." A number of observing programs are in the queue. When each gets done depends on sky conditions, priority of other programs, time of night or month, or whatever one chooses to put in the computer algorithm. Thus, if all goes well, and why shouldn't it?, one will be maximizing the Sci done, and hence the Sci/$. It is a goal to shoot for.

As noted above, automatic observing is also possible, probably with the small NGTs. Here one need only communicate with the telescope occasionally, loading the new program or sub-program, and downloading the data so far obtained. The observer need not (ever) be at the telescope, but can be in the cloudy east or midwest while the telescope is in Arizona, for example. The observer spends the time preparing programs, analyzing data, doing science, while the telescope spends time observing, under clearer skies (whether cold or not). Each does what each does best. Hence, better Sci/$.

k) Finally, in summary, the Bottom Line. All of the above mean better, much better, Science Per Dollar. That's the advantage of the NGTs over the OGTs.

In principle, one may well come to the state where the most money goes into data handling and analysis, the next most into the instrumentation, and the least into the telescope system. That should certainly be possible, and common, for the small NGTs, if not of this generation of NGTs, then at least of the next one.

There are two other aspects or views of NGTs that I would like to mention before going on to the section on NGSTs:

One is the issue of *Versatile vs. Special Purpose*. It may well be, especially for small and intermediate NGTs, that the cost is low enough that it makes sense to have several special-purpose NGTs rather that one versatile (or general purpose) one. There is no question that one can achieve even lower cost using the NGT aspects mentioned above, and others, by designing a special purpose telescope. Generality costs extra money, sometimes a lot extra. One example I might mention without discussion is the proposed MOST project at KPNO. It is a Multiple Object Spectroscopic Telescope. It is a 4m aperture NGT incorporating all the aspects above, but it is not very versatile. What it does, it does very well. And what it does is of very wide-spread interest and application. So it really is versatile on the science it can be used for, but not versatile in its design. Seems like a real scientific winner to me. It attempts to maximize the Sci/$.

The second issue is **complimentary**. I think this is, and should be, one of the main issues of present day astronomy, science, and even life. I will give two examples only, without discussion (it is not the topic of this paper); it is easy to think of many others:

1. Ground—based optical telescopes **complement** space telescopes, which **complement** radio telescopes, which **complement** theory, *etc*. None replace the others. All desperately require the best from the others.

2. Small telescopes complement large telescopes. The latter require the former, for backup, for standards, for added insight, for follow up, *etc*. Not all exciting, interesting, useful, important astronomy is done only with the largest telescope, in the infrared, at the furthest reaches of the universe.

IV. WHAT APPLIES TO THE NEW GENERATION SMALL TELESCOPES?

As noted in the Introduction, I think that one can perhaps divide the NGTs into three categories when thinking of aperture:

1. Small, where issues of mirror fabrication and support are not a major issue. They are important, of course, but easily handled. Many of the aspects of the optics go with a fairly high power of the mirror size.

2. Intermediate, where the issue of mirror support is more of a problem. Solutions exist, though, without having to resort to active optics. The primary mirror and its support, so as to produce the (new) tight specs on image quality, are very important, but the design is not at the cutting edge of the new technology.

3. Large, where the issue of mirror fabrication, design, and support are critical, right at the frontier of the new technology needed for successful NGTs of this size.

I would propose:
Small means: One meter or less
Intermediate: Two to four meters
Large: Eight meters and up

These ranges seem to include all the sizes now being discussed.

With this in mind, what are the aspects that apply to NGSTs?

NEW GENERATION SMALL TELESCOPES 15

1. Few new developments are needed. Technology already exists. Optics ought not to be any problem, even at f/2.

2. Simple systems, can be alt–az, alt–alt, or equatorial. Must be low weight and small with very simple housings.

3. One can use existing, simple personal computers as is. Same for most other electronics.

4. Queue scheduling is a key concept. Dynamite for NGSTs.

5. As the cost is low, one can afford to do maintenance when one can; instant response is not required, as with a 15m.

6. Data handling, while large compared to an OGT small telescope, is not large compared to the large NGTs. Existing computers, modems, and software can easily cope. Few new developments, even for software, are needed. One uses what already exists.

7. While it is possible, one would not imagine that real–time remote observing would be used. It's too expensive, and will be only used on the larger NGTs. One would and should imagine extensive automatic remote observing though.

8. What instrumentation? Currently, photoelectric photometers are used. In the not too distant future, one would expect simple spectrometers, small–chip direct CCDs, and fiber feeds to spectrometers, or whatever, sitting on the ground, in or out of the telescope housing (preferably out, in a separate building?). Since the cost of the telescope is so low, it would be easy to imagine it being lower than the cost of an instrument at the end of a fiber feed, and the instrument being of lower cost than the needs for data handling and analysis: truly a new generation of astro–economics.

9. Since the cost is so low, one can easily imagine many such small NGTs being built and used. If rationality exists, then even more cost savings can result because of standardization and reproducibility. Making them all the same will have obvious cost (and other) advantages. The same holds for instruments on the telescopes, for software, *etc*.

10. As with large NGTs, cost and use–sharing between several universities or institutions is certainly possible, and often desirable.

V. EXISTING NGSTs

Several of these new generation small telescopes already exist. The three now operating on Mt. Hopkins have been discussed at length, and are reviewed again at this meeting. Other papers here also deal with such telescopes and with their use. I am sure there are others. If you know of them, please do let me know details (and send photos). I fully intend to be an evangelist for these NGSTs, as I have become for dark skies.

VI. A SIMPLE EXAMPLE OF A SIMPLE SCALING LAW FOR TELESCOPE COSTS

I would like to show here one table illustrating a "scaling law" look at telescope costs. A truer scaling law approach would use a more involved study, including several terms with different exponents, as not all parts of a telescope system scale the same. Even a simple law can add some insight to telescope costs, however.

Table I gives the telescope aperture in meters, and several powers thereof. A simple calculator can easily do the calculation (y^x button). The cost follows: Cost = Constant × Aperture to the specific exponent. I have shown only one column of cost estimates, where the cost of a NGST of 1m aperture is set at $200,000. Note that this is lower, by quite a large factor, for an OGT of 1m aperture. The key to NGT success is the large shift in the zero point of a scaling law curve.

Note that the estimated costs, even with this simple scaling law, are not far from those quoted for the new wave of NGTs, from the small ones to the largest. Lengthy discussion can easily follow from inspection of such a table. That is another paper. Several other authors have already pointed the way.

The scaling law assumes that all items of the telescope scale with the same power, or at least do so in the average. This is not a realistic assumption, and so is a 0th—order approach only. For example, for the smallest telescopes, some things must scale with a lower power (such as the computer system?), and thus be higher in cost than the law estimates. Some items must scale with powers higher than 3.0. These are ones that must receive great attention in any design study of large NGTs. The problem is: how to get the exponent down for the items that scale with large powers? Many considerations show this to be difficult.

TABLE I: A SIMPLE SCALING LAW ESTIMATE OF TELESCOPE COSTS

Aperture	2.00	2.25	2.40	2.50	2.60	2.75	Cost with 2.6 law
16	256	512	776	1024	1331	2048	$270 M
15	225	443	665	871	1142	1715	230
11.3	128	235	338	430	548	790	110
10	100	178	251	316	398	560	80
8	64	108	147	180	223	304	45
4	16	23	28	32	37	42	7.5
3.5	12	17	20	23	26	31	5.2
2.5	6.2	7.9	9.0	9.9	10.8	12	2.2
2.0	4.0	4.8	5.3	5.7	6.0	6.7	1.2
1.5	2.2	2.5	2.6	2.8	2.9	3.0	.58
1.0	1	1	1	1	1	1	.200
0.5	0.25	0.21	0.19	0.18	0.16	0.15	.032

The cost of the 1m has been set at $200,000 = Zero Point of Law.

Even such a simple law shows what one has to shoot for, though, and how important it is to have as low a scaling power as possible, and as low a zero-point value as possible too. Any "error" in the exponent has a very large impact on the cost of a large NGT.

No instrumentation is included in the cost. This is a non-negligible part of the overall system cost.

VII. SUMMARY

I hope I have whetted the appetite of the reader—of the present and future user of small telescopes. I personally think that we are entering a new generation of such telescopes. I think that few people so far see the advantages of the new technology as applied to small new generation telescopes. Such advantages are as great for small telescopes as for large. But the small ones will be so inexpensive, and offer such power in their Sci/$ potential, that it is easy to imagine hundreds of them being put into use, nearly all at first class observing sites. The impact on astronomy will be enormous. For the first time, adequate telescope time with an excellent telescope at a first class site will be available to all astronomers for basic research, for work on standards, for variable star research, for all sky photometry, for surveys, for catalog production, for synoptic research, for quick response to new things, for supernova or comet searches, *etc.* The mouth waters with the potentials!

All of this means that the firm scaffolding of research will be in better shape than at any time in history. Small telescope research is fundamental stuff: the basics, the material that adds greatly to our understanding of the fascinating and exciting universe in which we live. It is critical to progress in advancing astronomy. It is inexpensive, and it has a fantastic Sci/$ ratio. It is complementary to research on large telescopes and to space astronomy. I believe that the new generation small telescopes will revolutionize how we do research, and offer fantastic potential improvements in education as well. Anyone, anywhere, can use them, regularly, with high efficiency and low cost, to great effect in advancing astronomy through the exciting and productive decades to come.

REMOTE-ACCESS AUTOMATIC PHOTOMETRY

Donald S. Hayes, Russell M. Genet, Louis J. Boyd

Fairborn Observatory

David L. Crawford

Kitt Peak National Observatory
National Optical Astronomy Observatories[*]

ABSTRACT. Remote access to automatic photoelectric telescopes over phone lines *via* modems would allow loading observing programs for automatic photometry and would return photometric observations and data giving the telescope status. Compared to the current APT Service operation, this remote access would give observers more direct control over their observing programs and shorten the time scale for changes in the programs and for retrieval of the data.

---

[*] Operated by the Association of Universities for Research in Astronomy, Inc. under contract with the National Science Foundation

New Generation Small Telescopes, ed. D. S. Hayes, R. M. Genet, & D. R. Genet.
© 1987 Fairborn Observatory.

I. AUTOMATIC PHOTOELECTRIC TELESCOPES

Automatic photoelectric telescopes, or APTs (Genet 1986; Boyd, Genet and Hall 1986), are pioneering examples of Next Generation Small Telescopes (NGSTs), a concept discussed in the previous chapter by Crawford (1987). Three APTs are operated by the APT Service on Mt. Hopkins, Arizona, the site of the Fred L. Whipple Observatory of the Smithsonian Institution, and also the site of the Multiple−Mirror Telescope Observatory. The Automatic Photoelectric Telescope Service (APT Service; Boyd, Genet and Baliunas 1986) is a joint operation of the Fairborn Observatory (equipment and operations) and the Smithsonian Astrophysical Observatory (site and facilities); it operates two ten−inch and one 16−inch aperture telescopes.

The unique characteristic of these APTs is that they make photometric observations while being completely unattended by human operators for weeks at a time. The APTs do differential photometry of variable stars for over a dozen astronomers in the United States, Canada, and Europe. One of the telescopes has operated in the fully−automatic mode for over three years. Many publications have resulted (partial list by Boyd, Genet and Hall 1985a; others have appeared since then), and a number of new variable stars have been discovered (Boyd, Genet and Hall 1985b; Hall, Kirkpatrick and Seufert 1986; Hall, et al. 1986). Further expansion in the number and aperture of the telescopes operated by the APT Service is in progress.

In the current mode of operation, an astronomer mails an observing program to the APT Service headquarters in Mesa, Arizona. A floppy disk is prepared by an APT Service staff member and taken to the mountain on the next trip, where it is loaded into the telescope control computer. This program remains in effect for about three months. After each night's observations, the site computer composes a "morning report," which summarizes the previous night's performance and the telescope status. It automatically telephones a computer in Phoenix and transfers the information so that an APT Service staff member can monitor the operation of the system. The morning report does not include any photometric data. About every two or three weeks, an APT Service staff member goes to the mountain to retrieve the data, which has been written to floppy disk. At the headquarters in Mesa, the data for each astronomer is separated from the others and mailed out quarterly. Thus, the system is very effective for rather long−term observing programs. It **cannot** accommodate changes on short time scales, nor can the results of a night's observations be obtained quickly.

II. THE RAAP CONCEPT

In the case of Remote−Access Automatic Photometry (RAAP), the telescopes would still operate in the fully automatic mode. What is different is that a new observing program for the telescope could be entered by the investigator *via* direct communication with a Remote Access Control Computer (RACC), in contrast to the process of mailing the information to the APT Service, it being loaded onto a floppy disk, and the disk being driven to the mountain to be loaded into the computer. Likewise, the data would be downloaded by direct communication between the astronomer and the RACC, rather than being retrieved by APT Service staff from the mountain and then mailed on floppy disk to the astronomer.

The RAAP concept does not include "on−line," or "real−time" operation, since it would be inefficient, the continuous communication costs would be excessive, the system integrity could be compromised, and the development effort required would be large. Thus, the RAAP concept is to be distinguished from "remote observing," as it has been realized previously at other observatories. In such cases, the remote observer was in continuous communication with the telescope, the auxiliary equipment and a telescope operator (or a subset of these) for the entire night. This worked well for certain types of observations, but it was expensive to operate. A significant advantage of the RAAP approach is that the system would be invulnerable to noisy phone lines and interrupted connections, not only because the communications software would have error checking and retransmission, but because the telescope would continue to observe according to the programs already loaded.

The "observer" using RAAP could be located at any distance from the telescope. The observing program would be put together "off−line" at the "home" location and saved to disk. The new observing program would be entered through menu−driven software that would require all necessary information to be entered. The software would then format the program correctly and verify it as observable; this software would be provided by the APT Service. The program would then be sent efficiently as a file over phone lines to a file on a hard disk in the RACC, located at the observing site. The file would only be accepted if the communications software verified that the transmission was error free. The telescope control system would access this file in the early evening before the roof was opened. Thus, the observing program for the night would be determined before the night began; changes could be made as late as that afternoon. Often, however, the observing program would be loaded into the RACC and left unchanged for days or weeks. Commercially available communication software would be used as the kernel of the system.

A block diagram of the system is shown in Figure 1; note that the remote computers would not have to be of the same type, as long as they could run the software (a common language, Turbo Pascal, would be used). Passwords would be used to restrict access only to authorized observers, and would further restrict their access to program and data files for the particular telescope executing their programs. All communication would be *via* files on the hard disk in the RACC; there would be no direct access to the telescope control system. In the case of consortia, a single investigator would be designated to enter observing programs; any member of the consortium could download data.

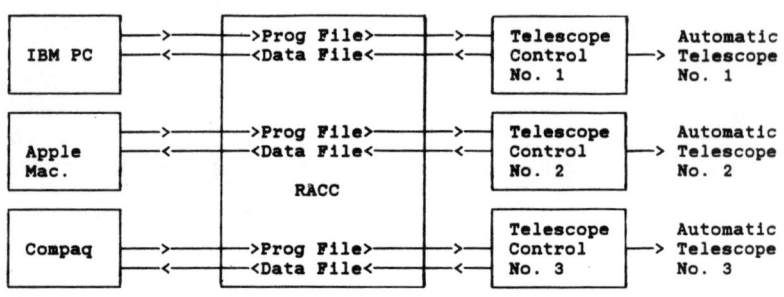

Figure 1. An example of a RAAP System. The computers on the left would be located remotely at the observer's home institution. The communication with the Remote Access Control Computer (RACC) would be over ordinary phone lines via modems.

The observing programs would be of two main types. One would simply be the current "first in the west" (FIW) mode. In this case, differential photometry of a variable star is performed. The observing program is a star list, including the variables and their check and comparison stars. It is not a detailed script for the night; the telescope control computer has some "intelligence" built in so that it can select which star to observe at a given time from the list. Some telescopes would operate with FIW programs nearly all of the time; others would retain FIW programs as "default" programs which would be used if so directed, or if for some reason the second type (see below) could not be executed.

The second type of program would, in effect, determine a detailed script for the night. It would specify a time interval within which each observation would take place. Included in this category would be programs to observe one star for an extended time, as in the case of the eclipse of a binary. Also included would be all—sky photometry. It would be possible to mix observing modes in a single night's program. The detailed script would be executed unless certain predetermined conditions occurred, in which case the default FIW program would be executed. Last-hour changes in the program could be made by remote access in the late afternoon. Data could be downloaded the next morning, but for routine observing programs it could be "saved up" for weeks.

An important extension of the APT concept would be the operation of a telescope at a remote site including ones in the southern hemisphere. Any remote operation must contend with the high cost of communication; the remote-access approach described here would be practicable because the communication sessions would be kept short. The local support staff would only have to deal with keeping the system running, as the home-based observers would supervise the observing programs and the collection of data.

III. NEW DEVELOPMENTS REQUIRED

In order to realize the RAAP concept, the software would have to be developed. It would include software needed at the observer's location which would compose the scripts and FIW star lists and write them to disk in the correct format for transmission. It would also include the software which would translate data files into reduceable data. Some reduction software could also be provided by the APT Service, but reduction programs would probably be developed by many observers. Commercial communications software would be used. On the mountain, software for communication between the RACC and the telescope control system would be required.

We envision three development phases. The first phase would include the development of the software described above; the communications link would go from the Mesa, Arizona office of Fairborn Observatory to the observing site 150 miles away. The second phase would continue the development of the system by working with a single remote observer. The third phase would result in the refinement of the service by making it available to a number of observers at different institutions. A single telescope could easily be shared by several institutions.

Although we intend to develop the RAAP concept for use in the APT Service, the RAAP capability would be useful for other observatories. The Fairborn Observatory has made a practice of publishing detailed descriptions of its efforts (see, *e.g.*, Trueblood and Genet 1985), and descriptions of the software developed for the RAAP concept would be published fully.

IV. SUMMARY

1) Remote access automatic photometry would add additional versatility and control to the photometric research programs of each college or university subscribing to the APT Service.

2) Since only a single phone connection would be needed, and only for limited periods of time (a few minutes for weeks of observing), the cost of communication would be low.

3) Since each telescope could be shared by a number of institutions, the cost of the telescope and its operation and maintenance would be low for each one.

4) Because of the capability for loading detailed scripts for each night into the system, any mode of observing, whether it be differential photometry, all-sky photometry, or the continuous monitoring of a single object, could be accommodated.

5) The primary capability needed to implement this concept, that of small fully automatic telescopes operating at a photometrically excellent site, is already developed and in operation. This means that the implementation of this concept requires developing only an increment to an existing system.

6) The institutions subscribing to the APT Service, whether sharing a telescope or using their own, would not need to be directly concerned with the development of software or hardware, although they would be welcome to develop software for reduction and analysis for non-standard programs. They would not be concerned with the maintenance of the telescope, building, or site. These activities would be performed by the APT Service.

The development of this concept would add a new dimension of versatility and control to photometric research programs in colleges and universities throughout the country. The cost would be low enough to permit even the smallest institutions to participate. Their access to a research-grade telescope would not depend upon their climate nor would it depend upon expensive travel. Nor would it depend upon being able to get away from campus for extended periods of time. The data obtained would be of the highest quality, so "front-line" research would be possible.

When fully developed and with a description published in detail, a RAAP system could be reproduced and used by astronomers at any institution to observe at any remote observatory.

REFERENCES

Boyd, L. J., Genet, R. M. and Hall, D. S. 1985a *I.A.P.P.P. Comm.* No. 19, 1.
Boyd, L. J., Genet, R. M. and Hall, D. S. 1985b *Sky and Tel.* 70, 16.
Boyd, L. J., Genet, R. M. and Hall, D. S. 1986 *Publ. Astron. Soc. Pacific* 98, 618.
Boyd, L. J., Genet, R. M. and Baliunas, S. L. 1986 in *Automatic Photoelectric Telescopes*, ed. D. S. Hall, R. M. Genet and B. L. Thurston, (Mesa: Fairborn Press), p. 15.
Crawford, D. L. 1987 in *New Generation Small Telescopes*, ed. D. S. Hayes, R. M. Genet, and D. R. Genet, (Mesa: Fairborn Press), p.7.
Genet, R. M. 1986 in *Automatic Photoelectric Telescopes*, ed. D. S. Hall, R. M. Genet and B. L. Thurston, (Mesa: Fairborn Press), p. 1.
Hall, D. S., Kirkpatrick, J. D. and Seufert, E. R. 1986 in *Automatic Photoelectric Telescopes*, ed. D. S. Hall, R. M. Genet and B. L. Thurston, (Mesa: Fairborn Press), p. 32.
Hall, D. S., Kirkpatrick, J. D., Seufert, E. R. and Henry, G. W. 1986 in *Automatic Photoelectric Telescopes*, ed. D. S. Hall, R. M. Genet and B. L. Thurston, (Mesa: Fairborn Press), p. 43.
Trueblood, M. and Genet, R. 1985 *Microcomputer Control of Telescopes* (Richmond: Willmann–Bell).

A 0.75−METER TELESCOPE DESIGNED FOR AUTOMATIC OPERATION

Russell M. Genet and Louis J. Boyd

Fairborn Observatory

David R. Genet

Fairborn Press

I. INTRODUCTION

It is possible to design small telescopes specifically for fully automatic or remote operation, or conversely, one can adapt manually operated small telescopes to this task. While both are possible, it became clear to us that conventional telescopes are usually not designed with remote operation or automation in mind, and a penalty has to be paid for their use—especially for larger sizes. That this is the case should not be surprising, as a manual telescope must accommodate human beings.

New Generation Small Telescopes, ed. D. S. Hayes, R. M. Genet, & D. R. Genet.
© 1987 Fairborn Observatory.

This means that the telescope must be raised off the floor to allow convenient human access to the eyepiece in operation. Usually the telescope must be cantilevered out over the top of the astronomer so that the space below the telescope will be clear for the astronomer to move about. The instruments must be mounted in a way that they can be easily changed, and so the observer can manipulate the instrument controls. These are all design constraints not faced by an automatic telescope (AT).

Figure 1. Overview of the 0.75−meter telescope. The telescope consists of four major subassemblies plus the mirrors. The major subassemblies are: (1) Base/RA Drive; (2) Horseshoe/Back Support/Dec Drive; (3) Primary Mirror Cell/Dec Ring; and (4) Superstructure/Secondary Mirror Cell. Photo by R. Genet from a 1/4 scale model by D. Genet.

On the other hand, there are design constraints that are peculiar to automatic telescopes that are not always met by manual telescopes. Examples of special requirements important for automation include: (1) having extremely low backlash (essentially none), preferably without any preloading; (2) a very low moment of inertia, allowing the rapid movements needed for acquisition and centering; and (3) a low aerodynamic cross section to minimize the effects of wind gusts on operation.

II. DESIGN

a) Requirements

The optics need to fit in a compact structure (to reduce structural demands and costs), and thus the primary f–ratio must be fast. A physically small system generally costs less to build and house. Also, a physically small system has a lower moment of inertia, making it easy to move about quickly, and it has a low wind cross section—decreasing resistance to the effects of wind gusts. Similarly the optics must be light in weight, not only to reduce inertia and lower the cost of the mount, but to simplify procedures for installation and removal of the optics. Light–weight optics also have a low thermal inertia.

There is no requirement for an eyepiece on a remotely operated or other electronic instrument. Stars are sensed with a CCD camera. As the instruments are permanently attached and totally automatic, there is no requirement for human observational or operational access to the telescopes—only occasional maintenance access.

The drive system for an AT needs to be very stiff, accelerate rapidly, and have essentially zero backlash (preferably without preloading). Efficiency of operation requires fast movement between stars. Stiffness is required for rapid settling after a fast move (and to minimize adverse wind gust effects), while zero backlash is required for accurate and efficient computer control and offsets.

Finally, a goal of the development and operation of automatic telescopes is to improve the cost–effectiveness of observational astronomy. Thus it is a vital requirement that the ATs be low in cost. Because ATs do not need to be manually operated machines or interface regularly with humans, they have the potential for costing much less.

b) Design Considerations

i) Optics

To keep the system compact, the primary f–ratio needs to be low. As one goes towards low f–ratios, however, the cost of the optics increases, and the requirements for precise alignment also increases. Currently a primary f—ratio of 2.0 represents the "knee of the curve."

The primary mirror diameter was set after trading off a number of parameters. One was coverage of stars with reasonable integration times. Another was keeping the AT of manageable size and cost. Finally there was the question of the state—of—the—art of light—weight f/2.0 mirrors. We concluded that a 0.75—meter mirror would be appropriate, although design work has been initiated on 0.5— and 1.0—meter versions of this telescope.

The base—line optics for the 0.75—meter design consists of a 30—inch diameter, f/2.0 primary mirror of 1 5/8—in thick, slumped pyrex, with a 4.5—inch central hole. The secondary is 8—inches in diameter, 1—inch thick, with an overall effective f—ratio of 8.0. The configuration and figure of the mirrors are classical Cassegrain.

ii) Mount

Both equatorial and non—equatorial mounts were considered. A complication of non—equatorial mounts is rotation of the field. While not a problem in photometry per se, it is a problem in CCD imaging. In addition, the computation of the drive rates in RA and Declination are more complex. Furthermore, our approach of injecting the stellar rate via software setable hardware (which makes the sky stand still from a software point of view) could not be used with a non—equatorial mount, and one would be forced to go either to an additional processor to make the translations or use an interrupt—driven system. While either can be done, they would increase cost and complexity. While justified in the case of large telescopes (where the structural cost can be reduced significantly), they are not justified in the case of small telescopes, especially those at moderate latitudes where the horseshoe configuration mount is as or more compact than either an alt—az mount or an alt—alt mount.

Several options were considered for focus adjustment, including movement of the secondary mirror, movement of the photometer, and adjustment of a lens or diaphragm within the instrument. We chose to move the instrument because this was relatively easy to do at a point where there was not a strong weight/inertia penalty, and because it kept the primary and secondary mirrors at their design separation—thus maintaining image quality.

The telescope's mechanical structure consists of four subassemblies, each of which is described below.

A. Base/RA Drive. The base mates the telescope to the floor (which can be an isolated floor section if desired). It also provides for adjustments of the entire mount in azimuth and elevation. The base is triangular in shape, welded from standard 3—inch square, 1/8—inch wall, steel tubing. The base holds, at its northern end (in the northern hemisphere), the two rollers on which the RA horseshoe rides, the bearing supports for these rollers, the reduction drive, and the DC servo motor. The south end of the mount base holds the south polar bearing.

Figure 2. The Base/RA Drive subassembly. This supports the Horseshoe and rotates it in RA. It also provides for altitude and azimuth adjustments to bring the entire telescope into polar alignment.

B. Horseshoe/Back Support/Dec Drive. As mentioned earlier, a horseshoe design is the most compact for middle latitudes. This design is not often used on small telescopes because it would make telescopes so compact that humans could not get at the eyepieces or instruments – not a consideration here. The horseshoe in cross section is "I–beam" in shape with a wide web and narrow flanges. The flanges are welded onto the top and bottom of the web. The bottom of the bottom flange is the surface that rolls on the rollers, and it is turned on a large lathe in Phoenix. The RA horseshoe also contains the Dec bearings. They are mounted somewhat higher than the center of rotation of the horseshoe, such that the "missing weight" (or first moment, to be more precise) on the open end of the horseshoe, the bottom RA back support arm, and Dec drive are just compensated for by the higher position of the Dec bearings and hence the main telescope assembly.

The RA back support runs between the RA Horseshoe and the south polar bearing. At its north end, the RA back support is welded at three places to the RA horseshoe, while at its south end, the three back support members (each of 3–inch square steel tubing) come together on a 12–inch diameter plate. On the other side of this plate is the south polar stub axle.

Figure 3. The Horseshoe/Back Support/Dec Drive subassembly. The horseshoe is "I" beam in cross section. The declination bearings are mounted on the horseshoe, offset somewhat above the center to compensate for the "missing mass" of the open end of the horseshoe and "4th" back support.

Figure 4. Mirror Cell/Dec Ring. Shown with the mirror laying face down and the photometer removed.

The declination drive is located on the bottom back support near the horseshoe. It is spring loaded against the Dec ring to maintain friction contact at all telescope positions.

Figure 5. Superstructure/Secondary Mirror Cell. Light, yet unusually rigid, this subassembly was designed to meet the low wind loading and fast motion requirements of remotely operated/automatic telescopes.

C. Primary Mirror Cell/Dec Ring. This single, welded structure not only holds the primary mirror, but it is the attachment point for the the superstructure (struts, primary baffle, secondary mirror, etc.), the instrument package, the Dec drive ring, and the Dec stub axles. The mirror is held in place from behind with a conventional 9−point suspension, and radially from the center. The Dec ring consists of a semicircular ring on the back of the primary mirror cell.

D. Superstructure/Secondary Mirror Cell. The superstructure consists of four struts (a quadrapod) that supports the secondary mirror cell, with cross braces to stiffen the structure. The secondary mirror is hand adjustable in tilt and centering.

c) Construction and Cost Considerations

The design emphasizes the use of standard, easily obtained components and materials, and minimizes precision machining and assembly. The friction contact surfaces in RA and Dec are precision machined, but little else requires fine tolerances.

That this telescope costs less than half of a typical manual telescope of similar aperture is not surprising. It is much smaller (only 6.0 feet in height when pointing at the zenith), weighs much less, has little precision machining, and is strictly functional. When it comes to remotely controlled or automated astronomy, it does a much more efficient job than is possible with a conventional telescope.

Editors Note: After this paper was written and presented, but before this volume was sent to the printers, construction was started on three 0.75— meter telescopes.

INTENSIFIED CCDs FOR AUTOMATIC ACQUISITION AND CENTERING

Louis J. Boyd and Russell M. Genet

Fairborn Observatory

I. INTRODUCTION

The automatic photoelectric telescopes (APTs) developed by the Fairborn Observatory to date have all used the same approach to finding and centering stars, namely a square spiral search to find the stars, and an iterative four−position telescope movement to center the stars (Boyd, Genet, and Hall 1983). There would be some advantage to employing an intensified CCD camera for acquisition and centering because the efficiency of the observations could be significantly increased. Also, objects too faint or nebulous to center using the current method could be accurately offset from any reasonably bright star in the field.

Initial star acquisition requires some form of area sensor if a large number of telescope movements are to be avoided. Fortunately a number of low cost commercial (uncooled) CCD cameras exist. Because of the

---
New Generation Small Telescopes, ed. D. S. Hayes, R. M. Genet, & D. R. Genet.
© 1987 Fairborn Observatory.

limited sensitivity of these commercial, uncooled, fast-scan cameras, they are not suitable for operating near the faint limits of a photometer.

There are two approaches to increasing CCD camera sensitivity. One would be to add an intensifier in the optical path while the other would reduce the scan rate by increasing the integration time. Long integration times without cooling the CCD detector would yield only a limited improvement. Cooling the detector would be a major complication and expense and, as will be seen, is not required. However, as cooled CCDs get less expensive and bothersome, increasing the integration time over standard frame rates will become more attractive.

Alternatively, one could use a commercial, uncooled fast scan CCD camera with an intensifier. Such intensifiers are relatively inexpensive; they are currently much less expensive than special-purpose, cooled CCD cameras. Commercial intensifiers can typically provide a five-magnitude improvement over an unintensified CCD camera, and this will bring one into the range of magnitudes where the low photon-arrival rates would start to become a factor on reasonable length photometric exposures. In other words, a commercial, uncooled, but intensified CCD camera would be a good match for photometry on automatic photoelectric telescopes (APTs).

II. APPROACHES TO ACQUISITION

The algorithm used in acquisition would be affected by what one knows about the star field. For instance, if it were assumed that the star being sought would be the star closest to the center brighter than a given threshold, this would greatly simplify finding the star. Based on our experience with APTs to date, there should be sufficient telescope pointing accuracy to allow the object (or an offset object) to be acquired as the brightest object near the center of the field (or the offset position). It could therefore be assumed that pattern recognition would not be required to do a search. With this in mind, one would only need the object's coordinates and its effective magnitude in the pass band of the intensifier. If a brighter offset star were used, one would need this information for the offset star and the directions of the offset.

Assuming an intensified, uncooled CCD camera were used, and assuming that the star to be found was the brightest one above the threshold nearest the center (or the offset position), then one method to acquire the star would be to store the entire frame in RAM and start at the center address and work outwards in a square spiral. If there were 256 by 256 pixels with an 8-bit brightness resolution for each pixel, this would require 64K bytes of RAM. This could be implemented with two static RAM chips. Alternatively, one might store only one bit of intensity resolution, and check for intensity crossing of a threshold with a comparator. RAM requirements would then be only 64 K bits (i.e., 1/8th the storage). In either case, the time required for the computer to do the spiral search would be longer than the time to capture a frame, although much faster than doing a spiral search by mechanically moving the telescope.

An alternate approach to acquisition would be to store the addresses (locations) of all the pixels that cross a threshold. One might only store an address during the first rise that exceeds the threshold, not trying to use all the available information. Using the "first rise" would avoid downstream blooming.

A characteristic of intensifiers is sparkle due to sky background, cosmic rays, ions, *etc*. These sparkles could be overcome by taking four successive frames and noting those crossing the threshold. This could be repeated several times, sorted, and all those that did not have matches thrown away.

The scale of the image must be such that the pixel spacing yields adequate centering accuracy for the application at hand, and should be at least twice the centering accuracy desired so that one could be off one pixel and still have adequate centering. For example, if one arc−second resolution were needed, assuming that there were 300 pixels across the screen, then the telescope would require a pointing accuracy of at least five arc minutes.

Commercially available intensifiers have a current limiting feature to protect them against bright sources by adjusting the gain. However, this does not perform well on stellar images, because they have a high current density in a small area which could cause damage to the tube. For this reason it would be necessary to control the amount of light that reached the intensifier. Experiments we have conducted have shown that a range of three stellar magnitudes would be available between solid detection and unacceptable blooming. Three neutral density filters (or a continuously variable neutral density filter) should be adequate to cover the magnitude range of the stars normally observed.

Two approaches might be considered in selecting the neutral density filters. One could start with the densest filter and monitor the video level of the camera, switching to less dense filters until the appropriate one was found (or rotating the continuously variable neutral density filter until the right angle was found). Alternatively, one could calculate which filter should be used, based on the known magnitude of the expected object in the field. The first approach protects the camera better although it would be slightly slower than the latter approach. In any event, setup time would only be a few tenths of a second.

III. CENTERING

In centering, unlike searching, finding the centroid of the stellar image (as opposed to the leading edge) would be required. With centering, it would be important that the brightness of the object as seen by the system be held fairly uniform. One would not want a bright object to appear bloomed as this would affect the center location. The same approach to reading the addresses of where the images were found could be used for centering as was used in acquisition. With centering, however, all the addresses of images that exceeded the image threshold (but in a smaller area of the CCD) would need to be recorded. One could find the mean in both axes that exceeded the threshold and call this position the center, and if it were not in the center of the field, the telescope would be moved to the field center. The threshold itself would be set in software in all cases.

The duty cycle in recentering depends on the length of the observations, the allowable centering error, and how accurately the telescope tracks. Normal photometry would not require recentering in most cases. For more stringent centering requirements or unusually long integration times, one might want to recheck centering on occasion. This could be accomplished in under one second by flipping the mirror, making the CCD observation, calculating the correction, moving the telescope, and flipping the mirror back to the photometer or other instrument.

Accurate centering would require that the position of the diaphragm center relative to the center of the CCD frame be known, and it could not be assumed that they were coincident. A reasonably bright star would be used in an x—y scan pattern (made by moving the telescope in small increments) to establish the edges of the diaphragm and hence its center.

IV. OTHER CONSIDERATIONS

Many difficulties could be avoided by not attempting to acquire and center stars near the faint limit of the intensified CCD camera (this would not apply, of course, to offset objects, which could be any brightness). The software programs for acquisition and centering could be written in assembly language using integer arithmetic. Running the software would not require an auxiliary computer. In fact, the new search and center program should not require any more memory than the existing search and center program.

Our conclusion is that for our larger automatic photoelectric telescopes (APTs) an intensified CCD is desirable for acquisition and centering. Even on smaller APTs, the marked increase in efficiency may often warrant the additional expense of an intensified CCD.

There would be two useful side advantages of using an intensified CCD camera. One is automated focus, where the focus could be adjusted to minimize the number of pixels covered by a star image. (Although focusing could be done in other ways.) The other is having a video output available while on site for telescope alignment and monitoring system performance during acquisition and centering to diagnose any problems with the telescope drive system. Already mentioned was the added ability to accurately offset, and hence observe faint or nebulous objects.

ACKNOWLEDGEMENT

Boyd and Genet are grateful to the National Science Foundation for support through research grant AST 84-14594. David L. Crawford made several helpful suggestions which we are pleased to acknowledge.

REFERENCE

Boyd, L. J., Genet, R. M., and Hall, D. S. 1984. *I.A.P.P.P. Comm.* No. 15, 21.

CONTRIBUTIONS OF THE SLEEPING ASTRONOMER

Norman L. Markworth and John Mullikin

Department of Physics and Astronomy
Stephen F. Austin State University

I. INTRODUCTION

The progress we have made in computer control of telescopes at the Stephen F. Austin State University Observatory has caused us to consider the next logical step in the procedure. Will it be possible to ask the telescope/control system to locate objects automatically? The approach taken by the Fairborn Observatory has made the automated telescope possible, but is restricted to bright, isolated target objects near the program star. The 41-inch telescope of the SFASU Observatory would almost certainly cause this searching technique to fail, since it becomes more difficult to meet the criterion of "isolation" as the aperture of the telescope increases. Likewise, because the field of view of the 41-inch telescope through the photometer optics is only 4.5 arcmin, the chances of moving automatically to a suitable "bright object" to begin the search are

---

New Generation Small Telescopes, ed. D. S. Hayes, R. M. Genet, & D. R. Genet.
© 1987 Fairborn Observatory.

rather low. We can, however, reliably move to the near vicinity of any object in the sky. If we could digitize the video image and compare it to a previously stored master pattern of the region around the target object, the problem is then one of recognizing where in the master we are and making suitable adjustments to the telescope position to center the target object. The type of video processing required here is not at all like the normal approach. The major difference is the time allowed for the processing. We require a system that works in as close to realtime as possible. This means our process must take shortcuts and approximations that are generally frowned upon by most people doing video processing. What follows is the result of over a year of investigation into the problem of star pattern recognition.

Our system consists of an image digital sampling device which sends an M × N array of gray level values to the computer for data pre-processing and pattern recognition. Initially, the 8-inch viewfinder is centered on a target star of interest. This image is pre-processed and the reduced data are stored on diskette. Later when the telescope is moved to the neighborhood of this target star, the image of the 41-inch telescope is digitized and also pre-processed. Computer pattern recognition algorithms will then be used to compare the two images and determine the position of the telescope relative to the desired position. These data are then sent to the existing control programs to redirect the telescope to the proper position. This paper will concentrate on the pre-processing and pattern recognition algorithms under development.

II. IMAGE DATA PROCESSING

The digital image is completely specified by $G(i,j)$, where the value of the function is proportional to the average darkness, or gray level, of the input image within the cell located at row i and column j. The image is then stored in memory as an M × N array, where the contents of each memory location are the corresponding gray levels. Alternately, some sampling devices yield values proportional to an average brightness, $B(i,j)$. The memory locations or digitized image positions are referred to as picture elements or pixels. In this form the image is now ready for data processing.

The primary goal of pre-processing is to segment the data in such a manner that all the pertinent data are retained, while all the unnecessary data are discarded. In our case, the only pertinent data are the coordinates and relative brightness of each star. If, for example, there are 15 stars in the input image, then our goal is to reduce the entire image data set down to 15 vectors of the form $V(i,j,br)$ where i and j are the approximate coordinates of the center of the star and br, its approximate brightness. This will greatly streamline the pattern recognition process, since for a typical 256 × 256 array, the amount of data being manipulated will fall from (256)(256) or 65,536 bytes to a mere (15)(3) or 45 bytes.

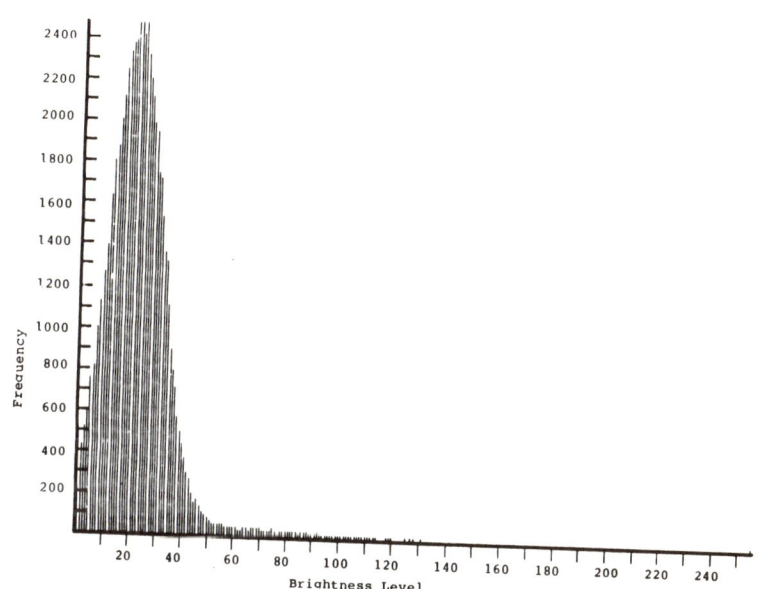

Figure 1. *A Histogram for typical star field. The background level is 20. The standard deviation of the noise is 10, with a mean of zero.*

The first step in this process is to determine what portion of the image is background light. An important tool in this determination is a plot of the frequency of occurrence of a given brightness level, which is called a histogram. Figure 1 shows the histogram for a typical star field. Ideally, all background pixels would have the same gray level and there would be a much larger peak in frequency at this value. Because of system noise, however, there is variation about this level. The noise-free image of a star can be modeled as a symmetric gaussian of the form:

$$G(r) = Ae^{-r},$$

where r is the distance from the center of the star, and A is the peak intensity as shown in Figure 2. The peak in the intensity does not correspond to one in the histogram, since the bright area in the center of the star is smaller, and its gray levels therefore less frequent, than the area of the dimmer outer pixels.

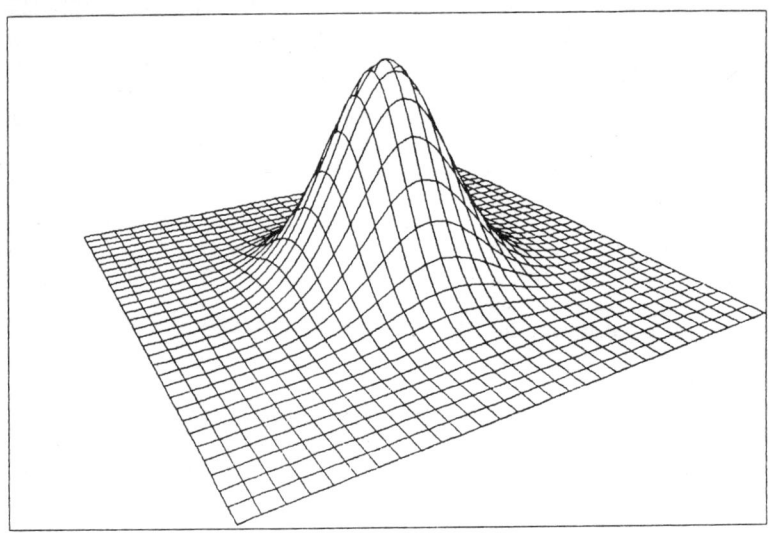

Figure 2. Gaussian model for a star intensity.

Thresholding the image is a very common technique used for image simplification. A brightness level is chosen and all pixels with levels below this value are made zero, while all of those above this threshold value are made one. A common case in image processing involves a single object and its uniform background. In these types of scenes, there are two peaks in the histogram corresponding to the foreground and

CONTRIBUTIONS OF THE SLEEPING ASTRONOMER

background. The threshold is usually chosen at the minimum between the two. For our model of a star image, though, this will not be the case, but because the average value of the background brightness level is the most frequent one, called the mode, then the threshold can be selected enough above this value to encompass most of the fluctuations due to system noise. Note, that no matter how the star images are modeled, as long as they are not all the same brightness while also taking up more area than the background, the mode will only be affected by the background level. Note, also, that the higher the threshold, the smaller the area corresponding to a star. There exists, then, the possibility of eliminating a dim star, although such stars are not usually important. Although the histogram plot does not need to be displayed, it must be constructed in order to determine the mode. The relative simplicity of the image will in fact allow us to use the mode obtained from sampling only one row (or column) to select an appropriate threshold.

Not all of the thresholded non−zero pixels can necessarily be classified as belonging to stars because some of the background may exceed the threshold level. Also it is possible that dust or other stray particles may be present, so another discriminating test must be used in order to further simplify the image. The shape of the "spot" may be considered, but its mere size is the simplest and probably the most useful feature discriminator. If after thresholding, the star images sometimes contain long projections away from the body of the star, then more pre−processing, such as low pass filtering or windowing, must be done prior to thresholding to localize the image. These projections, though, are likely to be dim, so that thresholding at higher levels above the background will probably eliminate the problem, if it exists. The situation must be evaluated experimentally and these additional routines will not be considered here.

The quickest way of finding the size of a star is to find the smallest box which will enclose it. If the stars are isolated "blobs," this should prove sufficient. An algorithm for boxing−in the spot is as follows:

1) Moving left to right, and top to bottom, if the pixel is on and it does not have at least one of the four neighbors as shown in Figure 3a, then continue scanning.

2) If the pixel has one of the neighbors, then move to the highest numbered neighbor. But only move to position 4 if there is a neighbor in position 3.

3) Repeat step two until a pixel without a neighbor is reached.

4) Subtract the starting row from the ending one to find one side of the box.

5) Repeat steps 2 through 4 for the other side of the box using the neighbors shown in Figure 3b.

Note that in some cases, such as that shown in Figure 4, the boxing process can break off prematurely, but the problem is easily handled by merging any boxes which have common sides. Note also that after a spot is found, the search must continue along other columns left and right of the detected spot.

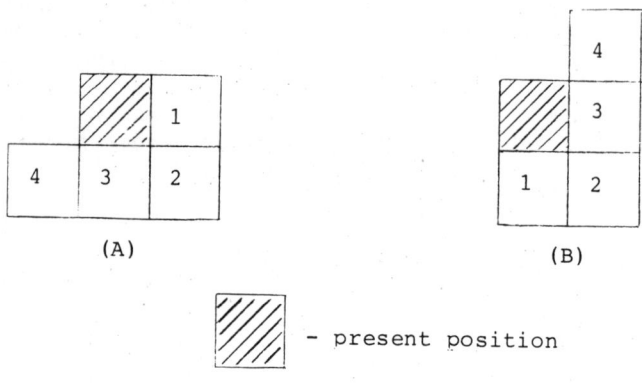

Figure 3. The numbered squares are candidates for the next pixel position.

Once the spot has been localized, a decision structure can then be implemented to classify the object. Total area of the box and the number of pixels inside the box may be used to determine membership. Also some qualifiers, such as the ratio of side lengths, will be useful in eliminating unwanted fingerlike structures.

The relative brightness of the star can also be roughly determined by its size, but recall that the size is dependent on the background light. Alternately, the brightness levels contained in the original image can be integrated over the area of the corresponding pixels, which have been localized, and divided by the total number of pixels to yield an average brightness per unit area, called the integrated optical density (IOD). The IOD is less sensitive to the background light, and improvements can be made by considering the more interior points and then subtracting the IOD of the background.

A center of brightness (X,Y), analogous to the center of mass, can be found for the object to determine the location used in the pattern recognition algorithm. The image should now be reduced to only star positions and brightnesses, as required.

III. PATTERN RECOGNITION ALGORITHM

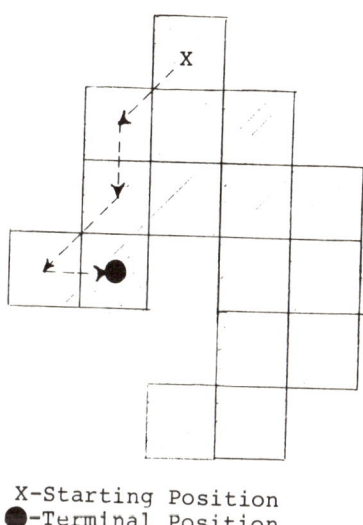

X-Starting Position
●-Terminal Position

Figure 4. *The boxing algorithm has terminated prematurely. When the outlining continues, the two boxes must be merged.*

Now that the data has been properly segmented, we can begin solving the problem of locating and identifying the star field. There are several requirements that a solution must meet. First, the process must be fast enough to work in realtime, although most of the work towards this goal will have been done in segmenting the image. Secondly, the process must be "robust," meaning that pre−processing failures or mistakes such as the appearance of false stars or the omission of known ones, will not necessarily cause system failure or misclassification. The process must also account for some errors in the positions of stars, and should not rely heavily on the background−dependent brightness data.

The method of solution developed at SFASU is actually rather simple, yet still meets all of the above criteria. Referring to Figure 5, each star in the larger, master field is numbered 1,2...i...S and assigned a set of vectors $s_{1,1}(x,y,br),...,s_{1,n}(x,y,br)$ corresponding to the position and brightness of its neighboring stars within a limited region. For the n_1 stars in this local region, there will be n_1-1 vectors associated with the i'th star. For the viewed region, the vectors $v_1(x,y,br),...,v_m(x,y,br)$ are likewise calculated from the centermost star to all others in the image, and if necessary, multiplied by a length scaling factor. The program then compares the set of vectors V to each set S_i to find the best correlation.

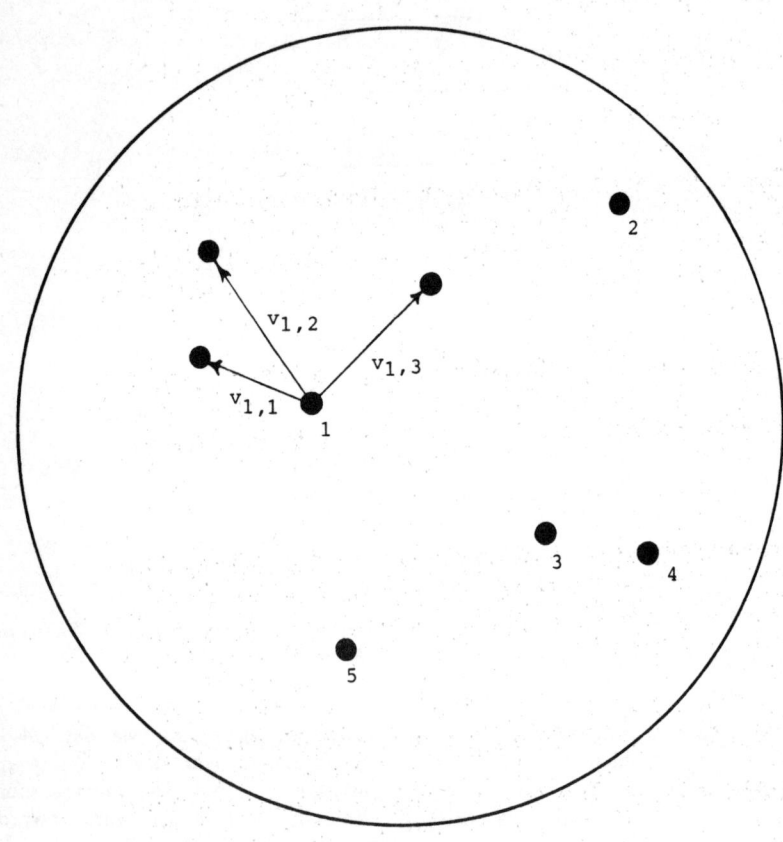

Figure 5. Vectors associated with star number 1.

CONTRIBUTIONS OF THE SLEEPING ASTRONOMER 49

The correlation can be defined in several ways, but to allow for a missing or inserted star, a scheme based on particular matching criteria between vectors is appropriate. Two vectors are considered to match if, first, their distances are the same within a given error, and, secondly, the difference in their IODs is not too great. The number of matches is then counted for the S comparisons of the set V to each set S_1. The star with which there is the greatest number of matches should be identified as the centermost star in the viewing region. From the coordinates of this star with respect to the center of the viewed field, and with respect to the programed star, the telescope can then be repositioned. In the event that two or more comparisons have the same number of matches, and a check for the number of stars in the regions cannot eliminate the choices, then the star with the minimum sum of differences in distances and IODs between vectors that matched should be selected. If only one star appears in the viewing region, a check can be made to see if a corresponding vacant region exists in the stored image. If not, then a star has probably been eliminated so the threshold must be readjusted.

Once the target star has been located and the telescope is positioned very near the object, the same pre-processing methods will be applied to a smaller central region of the image, and the coordinates of only the centermost star found. If the coordinates of the star are not close enough to the center, then the telescope can be adjusted. As long as this process is performed fast enough, the telescope will accurately track the object unaided. If the object for some reason is not the nearest star to the center, then the full routine can again be used.

The work done at SFASU is presently in its early development stages and not all of the procedures have been fully tested. Results and conclusions are therefore minimal. The next step in the project is to determine the actual signal characteristics. If a completely uniform light source is input to the sampling device, the standard deviation of the noise fluctuation can be determined from its histogram. By varying the input intensity, the relationship between the signal brightness and the system noise can be determined. Later, when a real image is sampled, the mode will be used as the estimate of the mean background, and from our tests, its noise fluctuation can also be estimated. An appropriate threshold can then be set, for example, corresponding to three standard deviations away from the estimated mean. At this point a decision can be made to use various filtering or image smoothing techniques. The likely error in position and brightness must then be established in order to set reasonable matching criteria for pattern recognition.

THE DETECTABILITY OF SUPERNOVAE AGAINST ELLIPTICAL GALACTIC DISKS

Eric C. Pearce

Digitized Astronomy, Department of Physics
New Mexico Institute of Mining and Technology

ABSTRACT: A 75 cm telescope has been automated with a Prime 300 mini–computer to search approximately 250 galaxies per hour for young supernovae. The high–speed star–location and comparison algorithms used in the Digitized Astronomy Supernova Search (DASS) system must be such that: (1) The SN is detectable over a reasonable luminosity–fraction of the galaxy, (2) a limiting absolute magnitude of less than 1/10 of a Type I–SN at maximum–light can be reached, and (3) a false detection rate of less than 1 per 1000 images can be realized. These three goals may be obtained by careful selection of four free parameters which define a "star" in the image processing software.

I. INTRODUCTION AND APPLICATION

The Digitized Astronomy Supernova Search (DASS) system utilizes a 75−cm automated Cassegrain telescope controlled remotely by a PRIME 300 minicomputer (Colgate, More, and Carlson 1975, Colgate and Thompson 1980, Colgate 1982). Software to orchestrate an efficient, orderly search is based on a multi−process design (Thompson 1980). One of these five processes performs the high−speed image processing necessary for the DASS system to maintain a search rate of 250 galaxies per hour with real−time image analysis. The details of the high−speed star−location and digital−image comparison system were presented at the Seventh Annual Fairborn/IAPPP Symposium (Pearce 1986).

In this paper, the more basic question of supernova (SN) detectability against a typical elliptical galactic disk with the current algorithms will be addressed. Specifically, the star−location algorithm must meet three design goals concerning the detectability. First, SN must be detectable over a large fraction of the host galaxy. Naturally, some percentage of the galaxy will be unobservable because of the brightness of the core, while still more will be lost outside the field of view. However, the remaining observable portion must be a reasonable fraction of the total galaxy (in terms of luminosity fraction, not area). Secondly, the algorithm must be able to detect a young supernova at 1/10 of its maximum luminosity. For a typical type I (M = −19.7 mag at maximum [Cadonau, Sandage and Tammann 1985]) this corresponds to a limiting absolute magnitude of −17.2 mag. Of course, the required limiting apparent magnitude will depend on the distance to the host galaxy. Finally, the false detection rate must be kept reasonably low without compromising the first two goals. A rate of 1 per 1000 images can be dealt with easily. Before these issues can be addressed, the methods and criteria used by the star location algorithms must be summarized. Details may be found in Pearce (1986). Essentially, the star−location algorithm, FIND, defines a "star" in the following way:

> A contiguous collection of at least MINSIZ pixels in excess of background by LOW times the standard deviation of the background, containing at least one pixel in excess of background by HIGH times the standard deviation.

The background is taken as the maximum of a smoothed histogram (the mode) of a regularly selected sample of pixels. The standard deviation of the background is calculated with a common grouped data method using the pixels in the lower half of the distribution about the mode.

After FIND tabulates all star−like objects in the field, the list of objects is compared to a reference list. Thus, galaxy fields are checked for new objects without actually subtracting a reference image. This approach has a strong advantage in the amount of data that must be dealt with since object−lists are much more compact than complete images and contain more robust information. The major disadvantage to this system is lack of detectability of SN in the core, as shall be discussed in Section III.

Clearly, the type of variation from the sea of pixels which FIND will detect as a star are confined by the three free parameters; HIGH, LOW, and MINSIZ. As we shall see, the values of these parameters determine the sensitivity of the algorithm to faint stars and the false detection rate. Along with these three parameters, the integration time, INTTIM, may also be varied. The correct selection of these four parameters (HIGH, LOW, MINSIZ, and INTTIM) are pivotal in allowing the high speed star location algorithm FIND meet the three design goals for the detection of young SN.

II. GALAXY AND STAR IMAGES

An intensified silicon target (SIT) vidicon is used in the DASS system. The SIT vidicon is preferable to a CCD because off its lower readout noise and for economic reasons. The readout electronics produces a (128×128) pixel format, where each pixel may have a value from 0 to 255 inclusive (8 bits, unsigned). The pixel value is a linear function of the number of incident photons. The 75−cm f/12 telescope optics produce a pixel−scale of approximately 2.6 arcsec/pixel.

Typically, stellar images cover a few−to−several pixels in this format. A slight defocusing of the readout beam causes stellar images to occupy more than a single pixel. Thus, the statistical reliability of detections is improved. The intensity of pixels adjacent to the central pixel of a stellar image is down by a factor 1/e on average.

The resultant signal produced by a star on the SIT vidicon, (henceforth called the vidicon signal) as calculated by the algorithm FIND in accordance to the definition given in section one, can be expressed as:

$$S^* = S_1 (10^{-0.4m}) t \qquad (1)$$

Here, S^* is the vidicon signal, or the intensity as reported by FIND. Additionally, m is the apparent stellar magnitude and t is the integration time. The constant S_1 defines the sensitivity of the image tube to stellar images and has units of inverse time. The most obvious method to establish S_1 is to observe several stars of known apparent magnitude for a known integration time and observe the resulting vidicon signal. When this is done for five different K and G stars with m = 9.3 to 10.5 mag, S_1 is found to be 8.7×10^7 S^{-1}.

Next, the behavior of the image tube to the background sky illumination must be considered. Typical values for the sky during the supernova search are about 20.7 mag per arc square second (at 470 nm). The background sky signal, S_{sky}, is:

$$S_{Sky} = S_1\theta^2(10^{-0.4\Sigma sky})t \qquad (2)$$

where Σsky is the background sky–magnitude per arc second and θ is the pixel size. With the current f/12 format of 2.59 pixels, this reduces to:

$$S_{Sky} = S_{Sky,o}\ t = (3.2s^{-1})t \qquad (3)$$

with some variation in the constant due to changing sky glow conditions.

Since FIND makes use of the standard deviation of the background as a discriminator when scanning for stars, the behavior of the standard deviation with increasing background must be understood. Clearly, if photon shot noise was the only source of noise, the two quantities would be related by:

$$\sigma_{sky} = (S_{sky})^{1/2} \qquad (4)$$

where σ is a constant due to the non–unity amplification of the analog to digital converters. However, there is another constant noise source caused by a variety of effects which behaves like a readout noise. Thus, the noise will be modeled by the expression:

$$\sigma_{sky} = (S_{sky} + \sigma)^{1/2} \qquad (5)$$

where $S_{sky} = 2.5$ and the constant noise term $\sigma^2_o = 6.25$.

To complete our model of the detection problem, a model–galaxy luminosity profile is needed. Since the principle goal of the DASS system is to locate young SN–I in elliptical (E) galaxies, an elliptical profile will be adopted. Several analytic functions have been suggested such as those by Hubble (1930) and Abell (1962). We chose the model proposed by de Vaucouleurs (1953) since it most closely fits observed profiles and is widely accepted by the community.

$$\Sigma(r) = [8.33\ (r/r_e)^{1/4} - 1] + \Sigma_e \qquad (6a)$$

$$S_{gal}(r) = S_1\theta^2(10^{-0.4\Sigma(r)})t \qquad (6b)$$

Here, r_e is the radius which encloses one–half the total luminosity and Σ_e is the surface–magnitude at that radius. Typically, r_e ranges from 2.5 to 30 kpc (Kormendy 1982). An average value of $r_e = 10$kpc and $\Sigma_e = 22.7$ will be used of for this discussion. (Kormendy [1982] has shown that Σ_e is a weak function of r_e).

III. THE SN–DETECTION PROBLEM

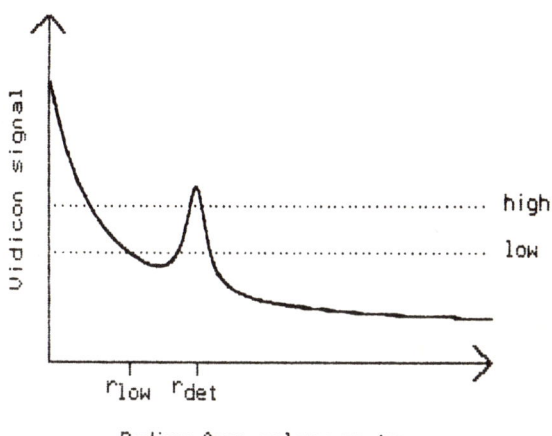

Figure 1. The luminosity profile of the model elliptical galaxy with an overlying supernova are shown graphically. The intensity levels HIGH and LOW used by FIND to discriminate pixels are indicated by dotted lines. Inside the radius r_{low} the galaxy core is detected by FIND as a "star", while SN can only be detected as discrete objects outside the larger radius r_{det}.

Now that the behavior of the image tube to stars, sky background, and galaxy background, is understood, and the specific problem of locating a young SN may be considered. This problem is shown schematically in Figure 1. The de Vaucouleurs–disk is shown in profile, with radius in pixels along the x–axis and intensity along the y–axis. A stellar object (perhaps a young SN or a foreground star) is also shown near the nucleus. Inside the radius r_{low}, the galaxy disk is bright enough to cause a per pixel illumination greater than LOW times the standard deviation of the background. Thus, the galaxy core is detected by FIND as a star (assuming that the central core is in excess by HIGH times the standard deviation). Clearly then, new stellar objects could not be detected as a discrete object inside this radius. Instead, it would cause a slight brightening of the already detected core. In fact, for an SN to be detected against this underlying disk, it must be outside a slightly larger radius r_{det}, so that at least one "dark" pixel is found between it and the LOW contour of the core. This problem obviously does not occur if images were first aligned and subtracted before scanning for new stellar objects. The added processing time and data storage required for image subtraction outweigh this minor loss of detectability in the inner core.

Secondly, because of the large angular extent of galaxies and the finite field of view of the telescope, some fraction of the galaxy will be unobservable unless we use multiple overlapping fields or a wider field instrument. Overlapping fields may be a reasonable alternative for nearby galaxies, but is not time-efficient for deeper galaxies. A wider-field instrument would compromise the core region, where star density, and hence SN density, is greater. Deeper than 50 Mpc, less than 16% of the galaxy luminosity is lost outside the field.

When considering the fraction of the galaxy which FIND can detect SN in, one must consider the fraction by luminosity rather than area or volume. Since SN-I are associated with well-evolved stars, the local SN rate may be assumed to be proportional to the local luminosity density. The intensity-form of equation (6b) may be integrated and normalized to find the "luminosity fraction" inside a radius r:

$$\mathcal{L}(r) = 1 - \exp[-a(r/re)^{1/4}] \sum_{n=0}^{7} \frac{[(r/re)^{1/4}a]^n}{n!} \quad (7)$$

$$a = 7.67$$

By subtraction, the "observed luminosity fraction" between two radii, r_{det} and r_{max} can easily be found.

Another aspect that must be considered is the limiting magnitude to which FIND can detect a stellar object. Clearly, FIND must reach an absolute magnitude of −17.2 mag in order to detect SN-I at one tenth of maximum luminosity. This corresponds to an apparent magnitude of 16.3 mag at a distance of 50 Mpc (the average distance to a galaxy in the current supernova search catalog). The limiting absolute magnitude of the search will be discussed in detail in the next section. However, the simpler question of finding the limiting vidicon signal for a star must be considered first.

The minimal vidicon signal required by FIND follows directly from the definition of "star" given in Section I. A detection must contain at least one "high" pixel (a pixel in excess of background by HIGH times the standard deviation of the background) and MINSIZ−1 "low" pixels. Thus, the minimum detectable signal is given by:

$$S^* = (h + [m-1]l)(S_{sky} + \frac{\sigma}{c})^{1/2} \quad (8)$$

where h, l, and m have replaced HIGH, LOW, and MINSIZ for brevity. This limit, however, is too optimistic since real stellar images do not distribute their light in such an ideal way for FIND. A more pessimistic assumption is that the light is evenly distributed over MINSIZ pixels. In this case, S^* is given by:

$$S^* = mh(S_{sky} + \sigma)^{1/2} \quad (9)$$

In practice, equation (9) more closely matches actual performance. Thus, the limiting magnitude reached will depend closely on the selection of HIGH and MINSIZ.

One final consideration must be the false detection rate. Since the goal of the DASS system is an automated search, false detections must be minimized. The false-detection rate can be estimated for a given value of MINSIZ by making two reasonable assumptions: (1) That false detections are caused by random noise in the background, and (2) That the probability of a random high pixel, P_{high}, and low pixel, P_{low}, are much less than unity. Since FIND also requires that a detection has pixels in two rows (to prevent sweep glitches from being detected), the possibilities are limited to MINSIZ > 1. Additionally, choices of MINSIZ > 3 are not considered since unsaturated stellar images do not cover such a large area. Thus, the false detection rates are, from combinatorial argument (including MINSIZ = 1 with the vertical extent condition relaxed):

$$\mathcal{R} = np_{high} \qquad \text{MINSIZ} = 1 \qquad (10a)$$

$$\mathcal{R} = 2np_{high}p_{low} \qquad \text{MINSIZ} = 2 \qquad (10b)$$

$$\mathcal{R} = 18np_{high}p^2_{low} \qquad \text{MINSIZ} = 3 \qquad (10c)$$

Where n is the size of the area considered. It is critical to note that P_{high} and P_{low} are the local probabilities of a high and low pixel respectively. This distinction is important because the background and standard deviation used by FIND to discriminate pixels are for the sky, with minimal galactic contribution. As FIND approaches r_{det}, the background and noise are both higher due to the underlying galactic disk. Thus, P_{high} and P_{low} must be adjusted accordingly. For example, the actual local deviation from the sky/galaxy background required to create a random high pixel in terms of the local standard deviation can be found from (5) as:

$$\frac{\Delta S}{\sigma_{local}} = \frac{h\sigma_{sky} - S_{gal}(r)}{(S(r) + \sigma_c^2)^{1/2}} \qquad (11)$$

where S(r) is the sky/galaxy signal from (2) and (6b). From the deviation in terms of local standard deviation, the local value of P_{high} may be easily found. Obviously, the same correction must be applied to P_{low}.

In order to calculate a per-image false-detection rate, one must integrate over the range of radii from r_{det} to the field edge with P_{high} and P_{low} as variables. However, since the local probability of a false detection is obviously highest at r_{det}, we can place an upper bound on the false-detection rate by applying this rate across the entire field. Due to its simplicity, this approach will be taken.

IV. THE SELECTION OF THE FIND PARAMETERS

In the previous sections, the effects of the four parameters, HIGH, LOW, MINSIZ, and the integration time, on the ability of the algorithm FIND to locate SN against an underlying elliptical galaxy have been discussed. In order to standardize the image–comparison process, the parameters HIGH, LOW, and MINSIZ should be the same for all galaxies observed; however, the integration time can be varied from galaxy to galaxy. Additionally, the selected parameters must work within the three design goals outlined earlier over the range of distances in the search catalog (from 10 Mpc, with most galaxies near 50 Mpc).

Clearly, one could examine the five–dimensional space (HIGH, LOW, MINSIZ, INTTIM, distance) and locate some "optimal" solution, but this approach is laborious and is unnecessary. Determining suitable values of the parameters one–by–one is a simpler approach. First, consider the parameter LOW. As discussed in Section III, LOW determines at what radius the galaxy core is detected as a star by FIND. Thus, inside a slightly larger radius, r_{det}, two pixels larger than r_{low}, new SN cannot be detected by FIND as discrete objects. Using the average search–catalog galaxy–distance of 50 Mpc, and allowing approximately 20% of the luminosity of the galaxy to be lost in this manner, r_{low} is found to be approximately 3 pixels (1.9 kpc). Using an integration time adequate to detect objects as faint as absolute magnitude -17.2 mag (5.0 seconds), equations (6b) and (5) show that LOW = 4.8 will result in a detected core of 3 pixels radius with the model E galaxy.

Next, the parameters HIGH and MINSIZ will be selected using the design goal of no more than one false detection per 1000 images. From equation (11), the deviation from the local mean in terms of the local standard deviation can be used with equations (10 a, b, c) to determine HIGH for a given MINSIZ and desired false detection rate. Starting from (6b) the vidicon signal due to the galaxy is 16.2 at the minimum detection radius r_{det} = 5 pixels. Then, from equation (11) a pixel at r_{det} must have a deviation of $1.76\sigma_{local}$. Thus p_{low} = 0.0392. Notice that p_{low} decreases when the background sky illumination is higher since the standard deviation of the sky increases more rapidly than local standard deviation at r_{det} in equation (11). Since actual star images typically occupy three or more pixels, MINSIZ = 3 is a logical choice. Also, from equation (10c), it allows a smaller value of HIGH to be used while achieving the same false detection rate. From the target false detection rate of 1/1000 images, equation (10c) predicts a P_{high} = 2.2×10^{-6}, or a local deviation of approximately 4.7 times the local standard deviation (Weast and Selby 1967). By rearranging equation (11), HIGH must at least be 8.8 to achieve the desired false detection rate.

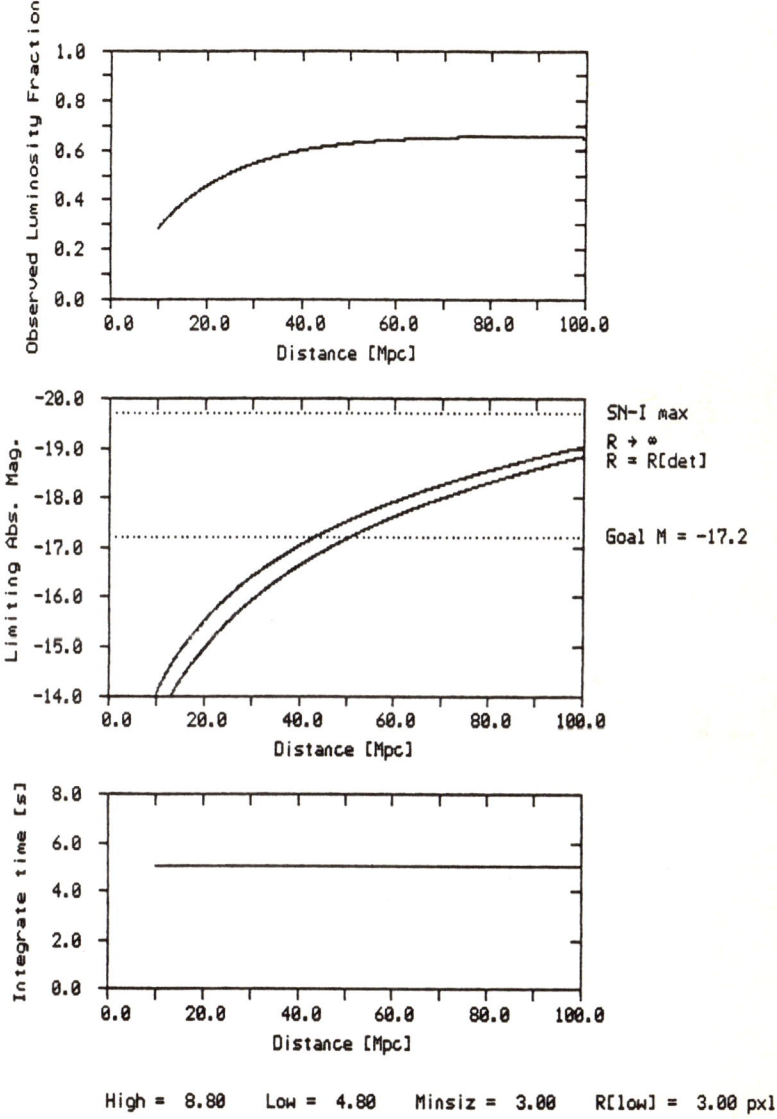

High = 8.80   Low = 4.80   Minsiz = 3.00   R[low] = 3.00 pxl

Figure 2. The resulting observed luminosity fraction and limiting magnitude for the selected FIND parameters using a constant 5.0 second integration time are shown above. The limiting absolute magnitude of detection is a function of location in the field. Note the rapid loss of observed luminosity fraction inside 40 Mpc.

All of the parameters have now been set for the model elliptical galaxy at 50 Mpc, however, the final design goal concerning limiting magnitude has not been checked. Using equations (9) and (1), the limiting apparent magnitude is 16.0 mag. Equation (9) does not account for the light from the underlying galaxy, though. Since this light adds to that of a star, the limiting magnitude is a function of underlying–galaxy surface luminosity. Thus, equation (9)c is valid only where the galactic contribution is negligible. Otherwise, equation (9) must be modified as follows:

$$S^* = mh(S_{sky} + \sigma^2)^{1/2} - m_{gal}(r) \qquad (12)$$

The limiting magnitude will be faintest, then, at r_{det}. At 50 Mpc this limiting apparent magnitude is 16.3 mag. Converting to absolute magnitudes, the limit ranges from -17.5 mag as $r \Rightarrow \infty$ to -17.2 mag at r_{det}. Although this limit is somewhat short of the design goals, the limit is adequate.

In the above discussion, the parameters HIGH, LOW, MINSIZ, and the integration time have all been selected for a model elliptical at 50 Mpc. Additionally, FIND has been shown to meet the three design goals with these parameters at this distance. Now, the final variable must be dealt with, the distance to the host galaxy. One possible approach would be to simply use the same 5.0–second integration time for every galaxy, regardless of distance. Figure 2 shows the effects of a constant 5.0–second integration time on the three design goals over the distance range 10 to 100 Mpc. In the top graph, the observed luminosity fraction drops off rapidly inside 40 Mpc while the limiting magnitude (center graph) increases superfluously. The rapid loss of observable galaxy at distances inside 40 Mpc is primarily due to the over–integrated core.

Clearly, a second approach would be to increase the integration time with distance. From a practical standpoint though, the distances to the galaxies in the search catalog are not known and cannot be easily measured by the DASS system. Yet, the model galaxy can be easily analyzed at any desired distance. Thus, a strategy to automatically adjust the final parameter, the integration time, must be developed which can be implemented without actually measuring or knowing the distance to each galaxy. If the integration time is adjusted so that the angular measure of r_{low} for all galaxies is less than or equal to some constant (say r_{low} for the average galaxy above, 3 pixels), the observed luminosity fraction decreases more slowly at the expense of the superfluous sensitivity. Furthermore, the integration time adjustment can be calculated from a single observation at some constant initial time in a straightforward way. Specifically, the galaxy is observed once with an integration time of two seconds, and the integration required to give a core size, r_{low}, is given by:

$$t_1 = ft_0 \left[\frac{l^2 S_{sky} + (l^{42} S^2_{sky} + 4[\overline{S_{gal}(r_{low})}]^2 l^2 \sigma_o^2)^{1/2}}{2[S_{gal}(r_{low})]^2} \right] \qquad (13)$$

where $\overline{S_{gal}(r_{low})}$ is the average vidicon signal in a ring of radius r_{low} about the center of the galaxy. The factor f is a constant somewhat less than unity to insure that the observed r_{low} with an integration of t_1 will be somewhat less than the desired constant value. This prevents repeated automatic attempts to adjust the integration time.

From an analytical standpoint, the integration time required for the model elliptical galaxy at a given distance R can be predicted with a formula similar to equation (13). The results of this adjustment are shown in Figure 3. The improvement in the observed luminosity fraction at R < 40 Mpc is clearly apparent. Beyond about 54 Mpc, the integration time reaches 6.0 seconds. Integration times longer than 6.0 seconds are both time-consuming and have additional sources of noise and are not used. The compromise in sensitivity is apparent by the discontinuity in the limiting magnitude curves. Nonetheless, sub-maximal SN−I can be detected out to distances beyond 100 Mpc.

V. CONCLUSION

The above discussion has addressed the ability of the algorithm FIND to locate young SN against model elliptical galaxies. It has been shown that the algorithm, with the proper values of the controlling parameters HIGH, LOW, MINSIZ, and the vidicon integration time, can locate pre-maxima SN over a reasonable fraction of the galaxy. Additionally, an upper bound on the false detection rate has been established and is manageable.

The variation of core size in the de Vaucouleurs model has not been discussed. Additionally, the effects of an exponential disk component to the elliptical galaxy and spiral galaxies have not been considered. Spiral galaxies have the additional complication of non-radial variations such as arms and HII regions. The author feels that a similar analysis with model spiral galaxies will generate similar results and calculations are currently underway to demonstrate this.

The high speed star location algorithm FIND, along with the rest of the DASS system allows galaxies to be searched for pre-maxima SN at a rate of 250 galaxies per hour while performing all image processing in real time. Assuming a mean supernova rate of approximately one per 25 years per galaxy, the DASS system should be capable of locating at least one SN per month when regular monitoring of a set of 1000 galaxies begins. A search program with the newly determined FIND parameters outlined in this paper is beginning in early 1987.

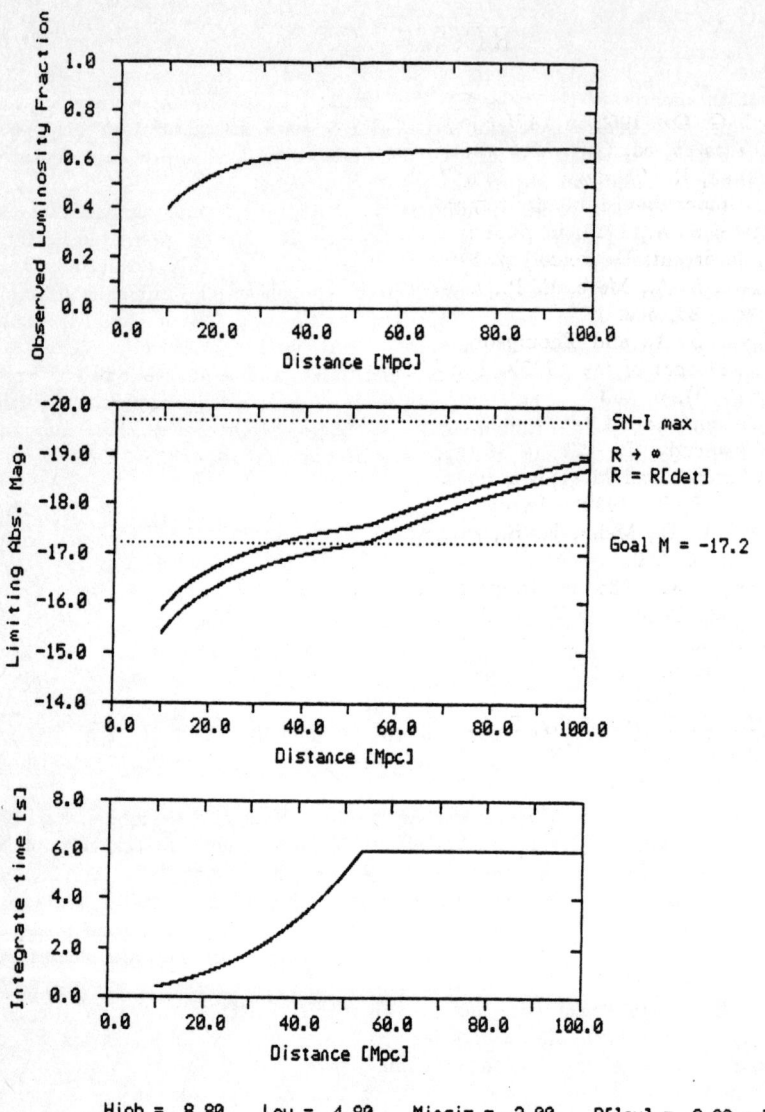

Figure 3. The resulting observed luminosity fraction and limiting magnitude are shown with the integration time adjustment described in the text. Note the improvement in observed fraction inside 40 Mpc at the expense of superfluous sensitivity as compared to Figure 1.

REFERENCES

Abell, G. O. 1962 in *IAU Symp. 15: Problems of Extra-Galactic Research*, ed. G.C. McVittie (IAU Symp. 15), p. 213.

Cadonau, R., Sandage, A., and Tammann, G. A. 1985 in *Supernovae as Distance Indicators*. (Springer–Verlag) p. 151.

Colgate, S. A., 1982, in *Supernovae*, ed. N. G. Ress and R. J. Stoneham (Dordrecht: D. Reidel) p. 319.

Colgate, S. A., More, E. P., and Carlson, R. 1975, *Pub. Astron. Soc. Pac.*, 87, 565.

Colgate, S. A., and Thompson, W. C. 1980, in *Optical and Infrared Telescopes of the 1990's, I*, ed. A. Hewitt and G. Burbidge (Tucson: KPNO), p. 480.

de Vaucouleurs, G. 1953, *Mon. Not. Roy. Astron. Soc.*, 113, 134.

Kormendy, J. 1982, in *Morphology and Dynamics of Galaxies*, ed. L. Martinet and M. Mayor (Geneva Observatory) p. 113.

Hubble, E. P. (1930), *Astrophys. J.*, 71, 231.

Pearce, E. C., Meier, K. B., and Colgate, S. A. 1985, *Bull. Amer. Astron. Soc.* 16, 909.

Pearce, E. C. 1986, in *Automatic Photoelectric Telescopes*, ed. D. S. Hall, R. M. Genet and B. L. Thurston (Mesa: Fairborn Press), p. 139.

Thompson, W. C. 1980, "Synchronization and Communication Considerations for a Digitally Controlled Telescope," M. S. thesis, New Mexico Tech.

Weast, R. C. and Selby, S. M. 1967, *Handbook of Tables for Mathematics* (CRC Press) p. 879.

AUTOMATED K−LINE PHOTOMETRY OF ACTIVE CHROMOSPHERE STARS

Sallie L. Baliunas

Harvard−Smithsonian Center for Astrophysics

Louis J. Boyd, Russell M. Genet, and Donald S. Hayes

Fairborn Observatory

I. INTRODUCTION

As suggested by Crawford (1987), an important feature of New Generation Small Telescopes (NGSTs) is their specialization for a particular task. If a particular task requires a dedicated instrument and the observing program is a long−term one, then the NGST (telescope and instrument) can be built for just this single task. As a result, for NGSTs the cost, both initial and recurring, can be quite low.

As an example of such a specialized, dedicated NGST, we will consider chromospheric Ca II K—line photometry of active chromosphere stars. We describe first the science that could be done with such a system, and how it would complement existing observational programs. Then we will consider the feasibility of filter photometry on a 0.75—meter telescope. Finally, we will describe a photometer that could make K—line observations using filters.

II. THE SCIENTIFIC OPPORTUNITY

Chromospheric and coronal radiation from a cool, solar—type star arises from a temperature inversion in the outer stellar atmosphere. In the Sun, the temperature above the visible surface drops slightly to a local minimum (about 6000 K), begins increasing again through the chromosphere (to about 10,000 K), steeply rises in the transition region (to several hundred thousand K) and finally levels out at two million degrees in the corona (cf. Linsky 1980; Avrett 1981).

The source of the temperature inversion and steepening is an input of nonradiative energy from both acoustic waves generated in the convective zone and magnetic subsurface waves (with fairly mysterious physical properties). Compounding the mysterious nature of the magnetism as a source of heating the outer stellar atmosphere is its variability. On the Sun, the most obvious magnetic variation is the (roughly) 11—year sunspot cycle, a somewhat periodic modulation in the average annual number of sunspots. The sunspots are intense (kilogauss), concentrated areas of magnetic activity, and their formation and variation remain inexplicable. A magnetic dynamo qualitatively describes the sunspot cycle by the interaction of an internal magnetic field with subsurface motions due to convection and differential rotation (Parker 1955).

Because the theory of the solar dynamo is complex and arcane, we are expanding the observation of magnetic variations to cool stars besides the Sun. Instead of the sunspot number we must be content to measure a disk—integrated, proxy indicator of the magnetic activity in the stars, whose surfaces remain spatially unresolved.

One powerful indication, by proxy, of stellar magnetic activity is the Ca II H (3967 A) and K (3934 A) emission cores. Formed in the chromosphere, these emission features are situated in the visible, and are accessible to ground—based telescopes (most other chromospheric and coronal emissions radiate in the ultraviolet and x—ray regimes). Historically, the Ca II H and K doublet have been the easily—available magnetic proxy monitored for time—series variations.

How powerfully H and K variations can reveal stellar magnetic activity is demonstrated, for example, with the Mt. Wilson Observatory time series. At the 100-inch telescope originally, and the 60-inch telescope currently, the relative HK fluxes in 99 lower-main-sequence stars detail starspot-cycle periods over years (Baliunas and Vaughan 1985) and rotation periods over weeks (Vaughan et al. 1981; Baliunas et al. 1983). Further, differential surface rotation can be inferred over several years' rotation time series (Baliunas et al. 1985). Rotation (and differential rotation) and cycle periods are important empirical constraints for magnetic dynamo theories.

Time-serial monitoring is the only way to document such magnetic variations as starspot cycles, and the timescales of magnetic variations hold the empirical key to magnetic dynamo theories.

It is important to emphasize that time series reveal physical details that are difficult or impossible to obtain by any other means. Most importantly in this case, telescope aperture cannot substitute for telescope time. A large telescope aperture can hinder the progress of a time series, because of interruptions for short-term studies of faint astronomical sources.

Our objective is to build a modest-aperture telescope that will be dedicated to chromospheric time-serial monitoring. The future of the Mt. Wilson HK project remains uncertain, so we are investigating other possibilities to continue the 20-year series of starspot-cycle variations on 99 lower-main-sequence stars. The aperture proposed here may not be adequate to observe the fainter G, K, and M stars in the Mt. Wilson survey. Our proposed design is, however, a step towards the continuation of the Mt. Wilson HK project. Realistically, we could monitor, for example, a magnitude-limited sample of evolved stars, or all the brighter stars in the HK survey, thereby freeing the Mt. Wilson facility for more of the fainter stars. Because the starspot cycles are years long, many more stars should be observed to explore statistically the long-term magnetic variations.

Another aspect of magnetic variations is introduced here. With the 0.75-meter telescope and K-line photometer, we propose additional photometric time series for each star: Hα and continuum variations. The strength of the correlation between the chromospheric and photospheric time series is an important astrophysical problem: Is the energy that is blocked by the cooler and visibly darker spots redistributed or stored somehow? (see Baliunas et al., 1987, in this Symposium). The only way to address this question is to obtain simultaneous photospheric, continuum and chromospheric K-time series.

III. FEASIBILITY

This analysis assumes that Ca II K-line photometry with respect to the continuum will be sufficient, i.e. that H- and K-line photometry are redundant and are not both needed. Our direct experience at Mt. Wilson with the HK project (Baliunas et al. 1981, 1983, 1985, 1987), plus theoretical understanding of the formation of the Ca II H and K doublet (Avrett 1981), show that the H and K emission variations contain similar information. This analysis also assumes, for simplicity, that all of the stars on the observing program are of 5th magnitude (V). In actuality, the stars would almost all be 5th magnitude and brighter, with only a few being fainter than this value.

a) Exoatmospheric photon flux

The analysis begins with the standard star, Vega. Hayes and Latham (1975) determine the exoatmospheric photon flux from Vega to be 948 photons/sec/cm^2/Angstrom at 5500 A. From this starting point, we adjust this value for spectral type (G2 instead of A0), wavelength (3934 A instead of 5500 A), depression of the chromospheric emission core at the bottom of the deep photospheric K−line absorption profile with respect to the apparent stellar continuum, and apparent magnitude (5.0 instead of 0.0).

For the Sun, Neckel and Labs (1981) determine the power flux versus wavelength. For a 20 A bandpass, they find it to be 376.2 micro−watts/cm^2 at 5500 A, and 153.1 micro−watts/cm^2 at 3933 A (approximately the K−line). Thus the power flux ratio of the Sun's continuum at 3933 vs. 5500 A is 0.41. To convert to photons requires an additional factor which is just the ratio of the wavelengths, *e.g.* 3933/5500, which is 0.71.

The K−line emission core is really just a small "pip" at the bottom of the deep, photospheric K−line absorption feature in the flux versus wavelength curve, and thus is well below the continuum value. Measurements made at Mt. Wilson are 20 A wide in the continuum at 3900 and 4000 A, and 1 A wide at both the H− and the K−line emission features. Integrations made on the K−line feature are 7.0 times longer than those on the continuum. The typical count ratio of the K−line feature to continuum for the Mt. Wilson solar−type program stars is 0.20. Thus the actual photon flux ratio of feature to continuum is 0.56.

Finally, we correct the photon flux from Vega (V=0.0 mag) to V=5.0 mag, for an additional reduction in the photon flux of a factor of 0.01. Thus, the expected exoatmospheric photon flux at the Ca II K−line from a 5th magnitude solar−type star is:

$$948 \times 0.41 \times 0.71 \times 0.56 \times 0.01 = 1.54 \text{ photons/sec/cm}^2/\text{A}.$$

b) Count Rate from the Detector

The detected photon flux will be smaller than the exoatmospheric photon flux because of atmospheric extinction. Hayes and Latham (1975) found the extinction at the wavelength of the K−line to be 0.40 mag at Palomar Observatory, or a factor of 0.69. Because the APT Service location on Mt. Hopkins is higher in elevation (at 7810 feet) than Mt. Wilson, this is probably a conservative estimate.

The collecting area of a 0.75–meter telescope is 4418 cm^2 minus some 507 cm^2 shadowed by the secondary mirror and its baffle, for a 3911 cm^2 effective area. Losses occur at each optical surface. Five optical surfaces are involved, with an assumed average loss of 5% each. Thus, a factor of 0.95^5, or 0.74, covers these losses. The Daystar K–line filter is assumed to have a bandwidth of 1.4 A, which introduces a factor of 1.4 compared to the per–Angstrom value. The transmission of the filter is 0.06 (Woods 1987). The filter is an Etalon with a mica spacer, blocking filters, and very accurate temperature control. The transmission of this type filter is never worse than 0.05, but rarely better than 0.07.

The detector is an EMI Gencom 9789 QA photomultiplier (PMT). This PMT has a bi–alkali photocathode whose sensitivity peaks in the blue and falls off rapidly in the red. While being a 2–inch diameter PMT, the 9789's photocathode is masked to only a 1–cm diameter to reduce the dark current. At an operating temperature of between 0 C and −10 C, the dark count rate will only be 1.0/sec (Young 1987). The use of a quartz window is not needed for ultraviolet transmission, but its use will reduce the potassium 40 decay–induced dark count. The photocathode quantum efficiency, as stated by the manufacturer, is 0.25 to 0.30, but the overall detector efficiency (accounting for photoelectrons that do not make it to the first dynode, *etc.*), is only 0.15 to 0.20 (Young 1987). The more conservative figure, 0.15, is used.

When combined together these factors give:

0.69 × 3911 × 0.74 × 1.4 × 0.06 × 0.15 = 25.16
counts.A.cm^2/photon.

The final count rate from a) and b) above is:

1.54 × 25.16 = 38.7 counts/second.

c) *The number of stars observed per night*

We desire a precision of 0.01 mag, which requires a total count above 10,000 to keep Poisson photon arrival noise to this level. It would take an integration time of 10,000/38.7 or 258.4 seconds to reach this precision. The Poisson noise from this measurement would be 100 counts. The sky background is assumed to be negligible (compared to a 5th magnitude star in a 30 arcsec diaphragm). At a dark count rate of 1.0/sec, the contribution of the dark current will be 258.4 counts per measurement, and the Poisson noise from the dark current will be 16 counts, sufficiently smaller than the 100–count star noise that it will not be a problem.

We assume the continuum filter to be 50 A wide, a factor of 35.7 wider than the emission–line filter. Unlike the faint emission core, the continuum is not depressed, so the 0.56 factor introduced above should be removed, or a factor of 1/0.56 = 1.78 introduced. Thus the count rate in the continuum is:

35.7 × 1.78 × 38.7 = 2459 counts/second.

The Poisson noise from the bright continuum should not be significant compared to the noise from the faint emission feature, for G−type stars. To achieve 50,000 counts will require an integration time of 20.3 sec.

Rather than having two long integrations it would be better to "chop" back and forth between the wide and narrow filters so that the effects of changes in atmospheric extinction, minor equipment drifts, *etc.*, will be minimized. We assume a complete cycle every 10 sec, and that 0.4 sec/cycle would be lost to changing mirrors, *etc.* The duty cycle for non−integration time will thus be:

0.4 × [(258.4 + 20.3)/10] = 11.1 sec.

In addition, time is required to slew to each star; we estimate it to be about 10 sec, at most, with DC servo motors. Time is also needed to find and center each star, 2.0 sec at most, with the intensified CCD and technique discussed by Boyd and Genet (1987). We recenter every minute or so, for a total of 5 times per star.

The total time taken per star is:

258.4 + 20.3 + 11.1 + 10.0 + 5 × 2.0 = 308.1 sec.

If a typical night were 8.0 hours long, and 1.0 hour was spent in calibrations of various sorts, then the number of stars observed per night would be:

7.0 × 60 × 60 /as 308.1 = 81.7 stars/night.

III. K−LINE PHOTOMETER DESIGN

The original design was first described by Baliunas *et al.* (1986). While the design described here resembles the earlier design in many ways, several significant changes have been made. The most important is the addition of an intensified CCD camera for faster and more precise acquisition and centering of the stars to be measured.

The photometer is as follows. Light enters through a quartz window that protects the photometer against dust and bugs. The light then encounters a number of possible elements on a ball slide positioned by a linear stepper. These include several mirror/neutral−density filter combinations that deflect the light to the left to the intensified CCD camera. There are also two fiber−fed calibration light sources. One is a broad−band incandescent lamp and the other is a very narrow−band calcium lamp. Also, the ball slide permits the selection of three sizes of diaphragms.

If an observing diaphragm is selected, the light proceeds to the first flip mirror. If the light is deflected to the right it proceeds through a collimating lens and then through each of two slots in two independent filter wheels. These filter wheels contain Johnson UBV, Strömgren $uvby$, and Hβ filters, and also have several neutral-density filters to allow observations of bright stars. The broad, continuum Ca K-line filter is also included as are open and closed positions. The light then reflects off of the fixed mirror, the second flip mirror, and is imaged on the photocathode of the photomultiplier by the Fabry lens. The detector is the EMI 9789 QA photomultiplier described above.

When both flip mirrors are rotated (they counter-rotate so the net change in angular momentum is zero, thereby reducing vibration), the light passes through the Daystar Ca K-line filter, the Fabry lens and onto the photomultiplier. The Daystar filter is precisely tiltable so that its central wavelength can be set to match the expected center of the incoming K-line from some particular star, having accounted for the differential Doppler shift between the earth and the star in question.

V. DISCUSSION

Our analysis suggests that the relative chromospheric emission strength in weak-emission-line solar-type stars is detectable with a 0.75-meter telescope and the K-line filter photometer described above. In fact, integration times for stars as faint as apparent magnitude V=5.0 are short enough for about 80 stars to be observed per night.

ACKNOWLEDGMENTS

This material is based upon work supported by the National Science Foundation, under Grant AST-8616545, and the Smithsonian Scholarly Studies Program.

REFERENCES

Avrett, E. H. 1981 in *Solar Phenomena in Stars and Stellar Systems*, ed. R. M. Bonnet and A. K. Dupree (Dordrecht: D. Reidel), p. 173.
Baliunas, S. L. 1983, *Astrophys. J.*, 275, 752.
Baliunas, S. L., Horne, J. H., Porter, A. H., Duncan, D. K., Frazer, J., et al. 1985, *Astrophys. J.*, 294, 310.
Baliunas, S. L., and Vaughan, A. H. 1985 *Ann. Rev. Astron. Astrophys.*, 23, 379.
Baliunas, S. L., Boyd, L. J., Genet, R. M., and Guinan, E. F. 1986 *IAPPP Comm.* No. 22, 32.
Baliunas, S. L., Donahue, R. A., Loeser, J. G., Guinan, E. F., Genet, R. M. and Boyd, L. J. 1987 in *New Generation Small Telescopes*, ed. D. S. Hayes, R. M. Genet and D. R. Genet, (Mesa: Fairborn Press), p. 97.
Boyd, L. J. and Genet, R. M. 1987 in *New Generation Small Telescopes*, ed. D. S. Hayes, R. M. Genet, and D. R. Genet, (Mesa: Fairborn Press), p. 35.
Crawford, D. L. 1987 in *New Generation Small Telescopes*, ed. D. S. Hayes, R. M. Genet and D. R. Genet, (Mesa: Fairborn Press), p. 7.
Hayes, D. S. and Latham, D. W. 1975. *Astrophys. J.*, 197, 593.
Linsky, J. L. 1980, *Ann. Rev. Astron. Astrophys.*, 18, 439.
Neckel, H. and Labs, D. 1981 *Solar Physics*, 74, 231.
Parker, E. N. 1955, *Astrophys. J.*, 122, 293.
Vaughan, A. H., Baliunas, S. L., Middelkoop, F., Hartmann, L., et al. 1981 *Astrophys. J.*, 250, 276.
Woods, D. 1987 Private communication with Genet.
Young, A. 1987 Private communication with Genet.

APPLICABILITY OF SPACE INSTRUMENTATION RELIABILITY CONCEPTS TO UNATTENDED AUTOMATIC TELESCOPES

G. Eichhorn, R.B. Piercey, J. McKisson, F. Giovane
and D. W. Ely

Space Astronomy Laboratory, University of Florida

I. INTRODUCTION

In the last few years, interest in unattended, automated ground-based telescopes has risen sharply. This sharp rise is due both to the increased need for comprehensive survey and monitoring data and to advances in technology which have substantially reduced the cost of reliable digital electronics. Compared to attended telescope systems, remotely operated systems have to be more reliable, since the opportunity for timely operator intervention does not exist. The amount of attention that has to be given to reliability aspects is determined largely by the degree to which the telescope is to be unattended.

Many of the requirements and operational constraints for these ground−based systems parallel those for space−based instrumentation. The need for long−term, fault−free operation of space instrumentation has resulted in the evolution of procedures and techniques which, when implemented, insure a high degree of reliability. When a high degree of reliability is required for ground−based systems, reliability concepts adapted from the space industry can provide useful guidelines.

In this paper, some relevant procedures and techniques that can improve system reliability are described. The list of topics included in the paper is not meant to be comprehensive, but should serve as a useful guide to anyone interested in system reliability. Since many of the concepts discussed are expensive and time consuming, only those that produce cost effective improvements in reliability should be implemented. Often, the most difficult project task is to decide which and to what extent the different concepts should be applied.

The Space Astronomy Laboratory has had extensive experience in both space−based and automated ground−based instrumentation, including instruments developed for Pioneer (Weinberg, et al. 1974), SkyLab (Weinberg and Hahn 1976; Giovane, Scheurman and Greenberg 1977), ISPM, Giotto (Levasseur−Regourd, et al. 1984) and the Space Shuttle (Weinberg, et al. 1980) as well as automated and semi−automated ground−based observatories for the South Pole (Giovane, et al. 1983) and atop Mt. Haleakala (Eichhorn and Giovane 1987; Giovane, et al. 1987).

II. RELIABILITY CONCEPTS

The procedures and techniques which evolved from the requirements of space instrumentation have been applied to achieve a high degree of reliability for these systems. When high reliability is required for ground−based systems, these concepts can guide project managers and engineers in all phases of the system implementation. Several of the concepts out of the wealth of procedures and techniques available are shown in Table I.

SPACE INSTRUMENTATION RELIABILITY CONCEPTS

As can be seen in the table, design activities constitute the crucial project phase in determining final system reliability. "Functional Tests", "Environmental Tests", and "Documentation" are included in the design phase since test plans and evaluation procedures must be drafted at an early stage in the project. Reliability—related activities in the later project phases primarily relate to tracking specifications and procedures laid out in the design phase.

TABLE I.

The Concept Applicability Matrix

	DESIGN	PROCURE	CONSTRUCT	TEST
MODULARITY	*			
TECHNOLOGY	*	*		
DERATING	*			
REDUNDANCY	*			
SPARING	*			
SOFTWARE HARDENING	*			
SPECIFICATION	*	*	*	*
INSPECTION	*	*	*	*
FUNCTIONAL TESTS	*	*	*	*
ENVIRONMENTAL TESTS	*	*	*	*
DOCUMENTATION	*	*	*	*

III. HARDWARE DESIGN

In this section, reliability concepts related to hardware design are discussed. These include concepts such as modularity, redundancy, sparing, derating and technology considerations. Although at times, the tasks associated with insuring system reliability may seem monumental, a systematic, step—wise approach can produce major improvements in system reliability without over—complicating a project. During the design process, development tools such as prototyping, testing and failure modes effects analysis (FMEA) can be used to evaluate proposed designs and help to systematize the approach to reliability.

a) Failure Modes Effects Analysis

A detailed FMEA can determine the subsystems or components most likely to contribute to system failures. Failure modes and their effects, ranging from catastrophic (which disable the entire system) to minor (which reduce performance) should be identified at an early stage. Once failure modes have been identified, design modifications can be implemented to eliminate or reduce the effects of these failures. Failure modes effects analysis should be continued throughout the entire design process in order to insure final system reliability.

b) Modularity

A modular architecture for both mechanical and electrical design will greatly simplify debugging, testing and validation of individual subsystems as well as the final system. Modularity is implemented by identifying the functional blocks of a system and by then defining physical modules which perform those functions. Modularity simplifies system specification by allowing a hierarchical approach. Precise requirements for modules and inter−modular interfaces can be developed more easily than for a distributed non−modular design.

Reliability is improved as a secondary effect of a modular design. Modules can be tested more thoroughly and more confidence in system function is gained with a modular approach. Furthermore, even at the later stages of design and construction, a modular system can be upgraded to include more reliable components if they become available. Modular design also allows the inclusion of commercially available boards. These are usually more reliable than custom−built boards because of the design effort and the testing built into the commercial boards. Modularity also can simplify the implementation of redundancy and can make the system easier and less expensive to maintain, repair and expand.

c) Redundancy and Sparing

Redundancy and sparing are two similar but distinctly different concepts for increasing system reliability by providing backups to support system operation in case of failures. A system or subsystem is redundant when two or more components are operating in parallel, accomplishing the same task but individually capable of performing the task if any one or more of the other redundant systems fail. Sparing is the concept wherein a secondary component is enabled to perform a task after the system has detected a failure in the primary component. The distinction between redundancy and sparing is whether the primary and secondary components (redundant) are active simultaneously or whether the secondary (spare) is enabled only after a failure is detected.

Redundancy and sparing are ubiquitous in the manned space program. The Space Shuttle uses five on−board guidance and control computers. Three are usually configured as redundant systems, all using the same data and running identical programs. The other two computers normally are running different programs but can quickly be switched over to perform the task of the first three in case of failure. These other two computers are configured as spares to the first three.

d) Derating

Derating involves specifying components so that their ratings exceed the expected stresses by a derating factor. Derating factors in space based systems range from 1.2 to 1.5 and even larger factors are applied to critical systems. Even though many parts are tested above their rated stress levels, further derating is a very cost effective way of improving reliability. Resistors are derated by power dissipation, power supplies by supplied current and temperature, and other components by some appropriate critical stress parameter. Derating increases reliability by reducing the stress on individual components and subsystems.

e) Technology Considerations

Space based systems often use only older technology because only after several years of use and evaluation and testing has its reliability been established. The automated telescope designer must make a tradeoff between the proven reliability of older technology and the newest technical advances. Especially for computer hardware, reliability history is extremely important to system design and component selection. While state-of-the-art components can be useful, they lack a history for reliability comparisons.

f) Complexity Versus Reliability

As complexity increases, so does the number of components which can fail. The additional hardware necessary to activate a spared CPU, for example, can also fail, perhaps leaving the system with no CPU operating. There is a point of diminished return as more ancillary circuitry is added to implement complex redundancy or sparing schemes. Ancillary circuits which are necessary must be painstakingly designed for simplicity and reliability.

IV. SOFTWARE DESIGN

One major design factor for software in space-based systems is the ability to tolerate memory failures and single event upsets in processor hardware. These failures are frequently encountered at high altitudes due to cosmic radiation. The telescope system designer can take advantage of the solutions that have been developed for space-based systems to improve reliability of ground-based systems. Design concepts applicable to software reliability are implemented in the software development phase and apply to system startup and runtime error checking.

a) Development Phase

The development of reliable software demands a strict adherance to established software engineering principles including a structured, modular program design. The requirements of the system have to be carefully analyzed and the functions of the necessary software modules defined. Only after all modules are defined, their interaction analyzed, and a final design generated, should code actually be written. Modules should correspond to functional units and be small enough to make this function apparent. This is extremely important in the validation process. Modules should interact by parameter passing in order to make the validation process manageable. Global variables that are accessed by several modules should be avoided wherever possible. In this way, most modules can be tested separately over the whole range of their input parameters. Such exhaustive validation is usually impossible in an integrated system.

Software documentation during the design and development phases is essential both for debugging and for failure analysis after the system is in service. Flowcharts, data flow diagrams, and source code listings are aids to the entire software system development and validation effort. A systematic trace of all software changes should be maintained, together with a record of revalidation.

b) System Startup Reliability Techniques

During system startup the software should check its own integrity and the status of the system hardware. In a stand-alone system, software boot-algorithms such as those implemented in the MRS (Maximally Redundant Software; Eichhorn and Piercey 1986) model help to make the boot procedure insensitive to memory failures. In MRS, the initial boot module is very short (typically ten instructions), and exists in several copies. The initial boot program calculates a checksum of several copies of the much longer major boot program. If it finds one with the correct checksum, control is transferred to the validated major boot program. Exception vectors restart a different copy of the initial boot program if an error occurs. The major boot module checks the available RAM memory and identifies bad memory blocks. A majority-voted copy of the main software is then downloaded into a validated area of RAM and control is transferred to the application code.

c) Runtime Error Checking

Once an automated system has been booted and is running, it is important to monitor the hardware and software status continuously to detect partial failures. Latent errors are particularly dangerous since a corrupted instruction or hardware malfunction might generate bad data without "bombing" the system. Such errors could, conceivably, cause a telescope mount to move in a wrong direction without indicating the erroneous motion in the data.

Runtime monitoring should check all hardware modules periodically, even if the module is not currently active. If a module does not generate the proper response, it should be flagged or disabled and its new status logged. If possible, spare hardware is enabled at this time to replace the failed module.

The runtime self-validation of software entails the detection of corrupted memory. Periodic checking of the code segments against a known checksum can detect most code changes. These errors can be corrected by downloading the affected code again into re-validated memory. Checking of the variable space is more difficult since checksums would have to be recalculated every time a variable is changed. Although this can be too time consuming to do for all variables, grouping less frequently accessed variables together can make checking feasible.

If damaged code is executed before it can be repaired, an exception interrupt usually (hopefully!) occurs. Depending on the type of exception, the error handling can involve repair of the code or rebooting the system. The exception handling should be designed so that the disturbance of the telescope operation is minimized. Code repair is much less disruptive than rebooting the system, since this usually involves repositioning motors to their default position. Only in cases where the location of the damaged code cannot be determined (e.g after a wrong branch) should the system be rebooted from scratch. In addition to self-validation, watchdog architectures in hardware can safeguard against infinite loops.

V. PROCUREMENT

Procurement should be included in even the briefest list of reliability concepts since the quality of components and subsystems often determines the overall system reliability. Procurement can be initiated as subsystem designs are completed and verified. The careful procurement of system components requires detailed specifications, attention to vendor selection, and a concern for quality control (QC).

Where component failures are the major contributor to system malfunction the designer should consider the use of MIL-Spec parts. Many common electronic components have military counterparts which operate more reliably and can be used in more extreme environments than their civilian counterparts. MIL-Spec components are usually available from the same vendor that provides civilian components. When possible, choose vendors which are approved for space-based and military applications. These vendors have consistently demonstrated their product reliability and can usually provide manufacturing specifications, test data and materials lists.

Once the specifications have been written and components have been purchased, additional steps are in order to ensure that the components meet the specification. For critical components, a 100% incoming inspection is called for. Subsystems that are procured as assemblies should be visually inspected as far as possible to find any obvious flaws. Functional tests should also be conducted to ensure the module meets the agreed upon specifications.

VI. CONSTRUCTION

After the system design is as reliable as is practical within the bounds of other project constraints, steps must be taken to ensure that this designed−in reliability is not compromised during construction. A number of NASA and military procedures governing the construction of electronics and mechanical systems have been developed to increase overall system reliability. These procedures should be obtained and reviewed, since some of them can be implemented by even a small laboratory. For instance, the use of NASA soldering procedures can increase the reliability of in−house constructed subassemblies. This is especially important since soldered connections are known to be particularly failure prone. Procedures concerning the assembly environment can also improve the transfer of reliability from the design to the final unit. The use of a clean facility has always been mandated for space instrumentation. A certain amount of attention to cleanliness during construction will also improve the reliability of ground−based instrumentation.

VII. TESTING

Intensive functional and environmental testing of an instrument is necessary to validate the design and construction of an unattended telescope. All tests should be conducted according to a written test plan that includes the test objectives, procedures, and preformatted logs. A full test plan is usually much easier to generate when the system has been designed with a high degree of modularity. Specialized software is often developed for the functional tests of individual subsystems.

a) Functional Tests

As testable modules are completed, and once again when the system is assembled, a full functional test should be conducted and the status of all subsystems noted. A test report should be filed for later comparison. A valid functional test consists of a complete mechanical, electrical and software check of the operation of the module or instrument. Functional tests should be conducted prior to, and after, each environmental test.

b) Environmental Tests

Space based instruments go through many repeated environmental tests. For a ground—based instrument such as an automated telescope, the environmental testing may not have to follow such a rigorous schedule of testing and retesting. Each test will, however, give the designer more confidence in his design or (at worst) will reveal a problem in the reliability of the hardware which can then be solved. The telescope designer will have to consider the environment that his instrument will be subjected to when it is deployed. The environmental tests that may be conducted include structural, vibration, thermal, and Electromagnetic Interference/Compatibility tests (EMI/EMC).

A structural test involves applying the design loads (*e.g.*, weights, dynamic motion) to the instrument and recording the responses of the system. The test may reveal over—stressed members, bad gear alignment or unbalanced loads. Structural tests on primary assemblies can be done before the instrument electronics and detectors are completed if mass models are substituted for the missing components.

Many latent problems associated with bad solder joints, loose or defective fasteners, weak wiring connections, *etc.*, will not be found during functional tests but may arise only after the instrument is in operation. Thermal and vibration tests are especially useful in identifying such latent faults.

For space—based instruments vibration tests provide assurance that the system will survive the launch environment. For ground—based telescopes, similar tests can be equally important in the detection of structural weaknesses or the latent problems mentioned above. It is not necessary to subject the system to the high vibration levels used for space instruments because ground—based environments are usually less severe than those for launch. However, actual test levels may need to be higher than expected environmental levels in order to reveal latent problems in the design or construction. Such tests are crucial in characterizing the behavior of the system and the results may suggest design modifications and/or require the rework of certain components.

Thermal tests consist of operating the instrument at the expected environmental temperature limits and verifying proper system behavior. It is often prudent to test beyond these limits in order to identify latent problems. Thermal tests may reveal mechanical or electrical problems and may demonstrate the effectiveness of the heating or cooling systems. Incompatibilities at interfaces between differing materials may also be revealed. This can be important to ground—based telescopes as well as to space based instruments since the temperature environment for both can be extremely harsh.

An EMI/EMC test can be important in insuring that telescope electrical systems do not interfere with data collection, telescope control or with each other. The results of this test may indicate a need for better grounding, shielding or filtering. Problems with EMC are usually extremely difficult to anticipate completely, and their effects are often intermittent or well hidden. An EMC test plan should be carefully developed and the procedures should be executed with the unit in its true final configuration.

VIII. ECONOMIC CONSIDERATIONS

The cost of a reliable telescope system can be substantial in terms of both components and manpower. Typically in planning a space project, the cost of a system to be flown can be estimated by multiplying by a **hundred** the cost of what it would take to build a functionally equivalent ground system. This factor is due in part to the cost of implementing procedures which improve the reliability of the system. In the development of ground-based systems, steps can be taken to keep costs within reasonable bounds while still utilizing the experience of the space industry. The clear specification of reasonable goals and their proper documentation are necessary first steps. The need to fully understand the nature of the system being designed, and its potential failure points is essential in the planning operation. By paying "upfront" for vigorous planning, savings in construction and procurement can be realized. Careful use of specifications in procurements can also result in substantial savings of both money and time. The training of construction personnel and a defined process of quality assurance on construction is generally considered cost effective.

In the long term a reliable automatic system can provide a large reduction in the cost of operation and maintenance. Only a reliable system can avoid the loss of critical data when rare observation opportunities occur. Three questions that have to be answered by any scientist who is contemplating an automatic system are: How important are his observations and time? Is he willing to spend the money to assure a reliable system? and Can he afford not to?

IX. CONCLUSION

The requirements for reliability of an unattended automated telescope can be loosely compared to those of a space-based system. There are a great number of procedures and techniques developed for the space industry which can be applied to any system, including the ground-based telescope, to improve its reliability. It is the system designers task to evaluate his requirements and to determine the appropriateness of any particular high reliability concept to the design, procurement, construction, or testing efforts. At the start, the tasks required of a system designer may seem monumental, but in fact, many procedures can be applied, to an extent, by even a small laboratory.

System reliability is best introduced by applying appropriate concepts early in the design stages. Failure Modes Analysis is used in an iterative design cycle prior to writing vendor specifications. Considerations during procurement and construction help ensure that the designed—in reliability is not compromised by a flawed component or poor assembly practice. Throughout the project, testing is used to further prove designs and assure reliability of the final assembly. The level of effort required to fully implement all of these concepts is high and the system designer must judge to what extent they can be implemented within financial constraints. Any of the concepts presented above can be implemented individually to a greater or lesser extent. The systematic application of the concepts in a carefully planned design effort will greatly enhance the overall reliability.

The degree to which a system is to be unattended determines, in large part, the extent to which reliability concepts should be considered in the design. The less often service can be provided to an instrument the closer to a space—based system it becomes in terms of reliability requirements. A related aspect of the concepts discussed here is that some of them not only improve the reliability of the instrument, but also its maintainability. Maintainability hinges on the modularity concepts in both hardware and software design and depends heavily on vigorous documentation. The better your documentation of the whole system from design to construction, the easier the maintenance will be later on. The maintainability aspect may be the pivotal argument for applying some of the reliability design concepts in the design stage.

There is no need to space qualify a ground—based instrument; however a certain level of space qualification concepts and procedures is necessary to obtain a reliable instrument or subsystem.

REFERENCES

Eichhorn, G. and Piercey, R. B. 1986 *IEEE Trans. Nuclear Sci.*, Vol. NS 33, No. 4, 1100.

Eichhorn, G. and Giovane, F. 1987, in *New Generation Small Telescopes*, D. S. Hayes, R. M. Genet and D. R. Genet, eds., (Mesa: Fairborn Press), p. 317.

Giovane, F., Schuerman, D. W., and Greenberg, J. M. 1977, *Applied Optics*, 16, 993.

Giovane, F., Wood, F. B., Oliver, J. P., and Chen, K. Y. 1983, in *Microcomputers In Astronomy*, ed. R. M. Genet, (Fairborn: Fairborn Obs.), p. 86.

Giovane, F., Ely, D., Weisenberger, A., Eichhorn, G., Rilum, J. and Soderberg, B. 1987, in *New Generation Small Telescopes*, D. S. Hayes, R. M. Genet and D. R. Genet, eds., (Mesa: Fairborn Press), p. 307.

Levasseur-Regourd, A. C., Bertaux, J. L., Dumont, R., Festou, M., Giese, R. H., Giovane, F., Lamy, P., Llebaria, A. and Weinberg, J. L. 1984 *Adv. Space Res.* 4, 9, 287. Weinberg, J. L., Hanner, M. S., Beeson, D. E., DeShields II, L. M., and Green, B. A. 1974 *J. Geophys. Res.*, 79, 3665-3670.

Weinberg, J. L. and Hahn, R. C. 1976 in *Proc. AIAA/AGU Conference on Scientific Experiments of Skylab, 1974: Progress in Astronautics and Aeronautics*, Volume 48, ed. M. I. Kent, E. Stuhlinger, and S. Wu, (New York: Amer. Inst. Aeronautics and Astronautics), p. 223.

Weinberg, J. L., Hahn, R. C., Giovane, F., and Schuerman, D. W. 1980 in: *IAU Symposium 90: Solid Particles in the Solar System*, ed. I. Halliday and B. A. McIntosh (Dordrecht: D. Reidel), p. 25.

THE APT SERVICE: A BRIEF PROGRESS REPORT

Russell M. Genet and Louis J. Boyd

Fairborn Observatory

The beginnings of the Automatic Photoelectric Telescope Service were reported in the proceedings of last year's symposium (Genet, Boyd, and Baliunas 1986). The APT Service is now well into its second year of operation. Located at the Smithsonian Institution's Fred L. Whipple Observatory, the facilities, shown in Figure 1, are at the 7800-foot level on Mt. Hopkins. The facilities and site support, such as the 4-wheel drive vehicle shown on the left, are provided by the Smithsonian Institution. Since the last symposium, most of the equipment formerly used with the laser ranger has been removed, making way for the increasing amount of automatic photoelectric telescope (APT) equipment, test equipment, etc. The Baker-Nunn camera is being disassembled to provide the space for the four 0.75-meter APTs that will be installed in the south end of the roll-off roof area.

The first APT operational on Mt. Hopkins was the Fairborn-10 system. This APT first saw operation in Fairborn, Ohio, and it was moved to Mt. Hopkins where it began regular operations on 1 January 1986. The bulk of the observational program on this telescope has been that of Sallie L. Baliunas of the Harvard-Smithsonian Center for

Astrophysics. She has described elsewhere in these proceedings some of the results already obtained from this APT (Baliunas *et al.* 1987).

Figure 1. The APT Service facilities. The building on the left contains the office, shop, computer room, bath, three bedrooms, and storage. The observatory area on the right currently houses three APTs. Four 0.75-meter APTs and two smaller APTs will be joining the existing systems for a total of nine APTs in the existing roll-off roof area. The two buildings are connected with a covered stairway.

The most recent APT to achieve operation on Mt. Hopkins is the Vanderbilt–16. Shown in Figure 3 along with Douglas S. Hall, this system is devoted entirely to observations of RS CVn binary stars. This telescope belongs to Dyer Observatory at Vanderbilt University (funded by the National Science Foundation), and it provides a compliment to the Dyer Observatory facilities in Nashville.

Figure 2. The Vanderbilt−16 APT and its main user, Douglas S. Hall. The photometer on this system utilizes a GaAs photomultiplier (thermoelectrically cooled) which allows UBVRI observations to be made with a single detector. Two 10−position filter wheels allow UBVRI, uvby, Hα, and Hβ observations to be made with the same photometer. This photograph was taken by Dan Brocious of the Fred L. Whipple Observatory staff.

With three APTs operating at the same time in close proximity to each other, the sound produced is a unique symphony that only a robot (or data hungry astronomer) could really cherish. The three operational APTs are shown together in Figure 3 in a photograph taken in May 1987, just before the proceedings went to press. Left to right they are the Fairborn−10, the Phoenix−10, and the Vanderbilt−16.

Figure 3. The three currently operational APTs at the APT Service. They occupy about half of the northern third of the roll—off—roof area. The Phoenix—10 (center) was moved to Mt. Hopkins after a freak thunderstorm destroyed its housing in Phoenix. As part of the move, the electronics for its photometer were overhauled and a completely new control system was added to bring it up in performance and reliability with the other APTs.

In front of the Vanderbilt—16 can be seen a portion of the roof control system. The aluminum box attached to the top of the stairway rail contains the warning horn which sounds prior to roof movement (warning humans to get out of the way), an auto/manual switch that allows humans to take control away from the site control computer, and buttons to manually open and close the two roofs (main roll—off and north tilt).

THE APT SERVICE: A PROGRESS REPORT   89

A true roof story cannot be resisted. In the early days of the APT Service when the site control software was not fully debugged, one of the daytime maintenance personnel noticed that the roof was opening and closing repeatedly. Rushing over to the APT Service, he took manual control, shut the roof, and called Lou Boyd. Computers are very efficient, but they sometimes do weird things!

Figure 4. *Participants in the 1986 APT Service Workshop, held for nine days in early August. Left to right are Russ Genet, Don Hayes, Rusty Genet, Doug Hall, Ken Kissell, and Judith Lee. The photograph was taken by Don Brocious of the F. W. Whipple Observatory staff.*

The Phoenix–10 APT is currently making observations for a variety of institutions including Armagh Observatory (Mart de Groot), the University of Arizona (Raymond E. White), Franklin and Marshall College (Harold L. Nations and Michael A. Seeds), Limber Observatory (David McDavid), the Lines Observatory (Helen and Richard Lines), Maria

Mitchell Observatory (E. P. Belserene), the University of Montreal (Anthony Moffat), the University of Toronto (J. Donald Fernie and John R. Percy), Virginia Military Institute (David L. DuPuy), Wesleyan University (William Herbst), and Williams College (Jay M. Paschoff).

As is well known, the summer weather in Arizona is frequented by thunderstorms, with the mountain peaks getting more than their fair share of lightning. During the "monsoon season" the winds come from the southeast, and the air is moist. In the lowlands it is very hot and humid — a nasty combination — while on the mountaintops it is moist and cool — a nice combination for humans but hopeless for photometry. It is during the height of the monsoon season that we hold the APT Service Workshop on Mt. Hopkins. The workshop provides an opportunity for APT Service staff members and astronomers from other institutions who have APTs at the service to work together on software and data analysis and to discuss equipment, etc. The 1986 APT Service Workshop concentrated on issues related to quality control.

The main computer room is shown in Figure 5. There are currently four computers, four terminals, and a modem. One computer and the modem is for site control, while the other three are for controlling the three APTs. The hallway on the right leads up to the telescopes. Off to the left, just out of the picture, is the main power unit. It contains a large set of batteries and a power inverter that allows operation to continue for many hours without power.

The modem on the left of the picture allows the site control computer to call in status reports to the Phoenix area some 160 miles away, and to make emergency calls. The normal status calls are made each morning at 0730 (local time) after the roof has been closed and all the APTs "put to bed" for the day. These calls are the *Morning Report*, and information is provided on the times the roof opened and closed, weather sensor readings every hour, the number of groups successfully observed by each APT, the number of groups each APT attempted to observe but could not, and, for each color band on each APT, the number of observations of stars that had an internal consistency of 0.02 mag or better, and the number less than 0.02 mag.

If a human should intervene at the site, this is also included in the morning report. We noted one morning that a human had overridden the site control computer and opened the roof. The mountain staff was surprised that we knew about this from the Phoenix area. Who had told us?

If the site control computer thinks an emergency situation exists, it will call us in the Phoenix area. If it thinks the weather has turned bad and it can't get the roof closed, for instance, it will give us a call. This has happened a few times, and a quick call to someone on the mountain usually takes care of the problem.

Figure 5. The site control and APT control computers. Normally the screens on the terminals are turned down as there are no humans around to watch them.

The site control computer obtains its information about the weather from the cluster of weather sensors. These are located about four feet above the roof of the building which contains the computers, office, *etc.* The site control computer uses these sensors rather than the APTs themselves to make all determinations about the weather, including whether or not it is cloudy.

The cloud sensor consists of two aluminum disks, one looking at the sky, and the other looking at the ground. A Peltier element senses the temperature difference between the two plates. When it is clear, the sky is much colder than the ground (in the IR where this sensor is most sensitive), while when it is cloudy the temperatures are nearly equal. The cloud sensor can be surprisingly fast acting. We have been in the upper part of the observatory when a small, lower—altitude cloud was sucked over the top of the ridge. Within a second or two the warning horn would sound and the roof would start shutting.

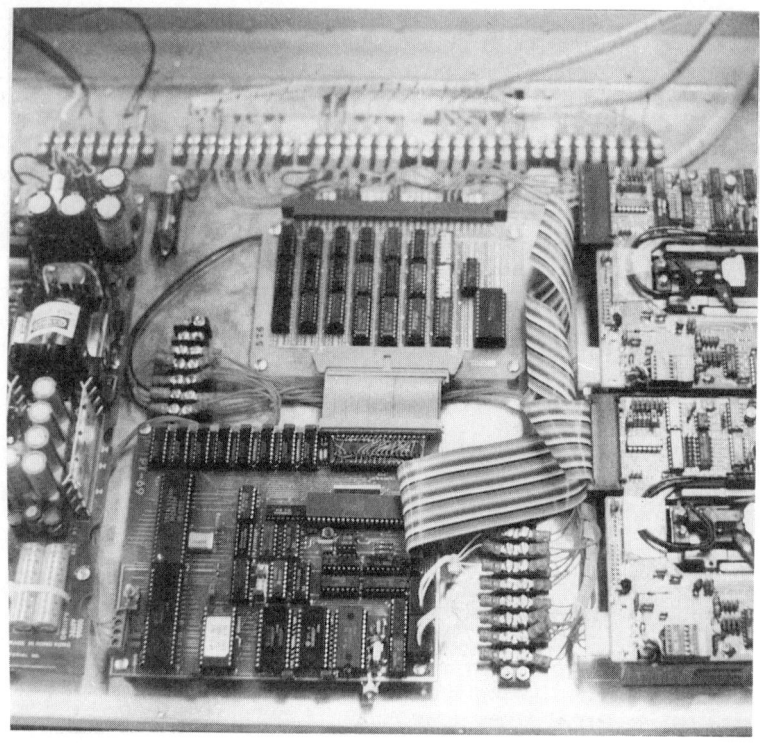

Figure 6. An APT control system. Except for the stepper drivers and power supply, this is the complete APT control system. All the items shown in the photo are commercially available except for the custom logic board (center rear).

Our emphasis in designing hardware continues to be to keep it simple. Shown in Figure 6 is one of the APT control systems. It consists of a power supply (on the left), a single—board computer (center front), a small custom logic board (center rear), and two floppy disk drives (on the right). The front floppy is used to store the control program and star list, while the back one stores the observational results.

Currently, the APT Service is concentrating on keeping the existing equipment in proper operational shape, monitoring the quality of the observational results, and working on improvements to the existing systems.

ACKNOWLEDGEMENTS

We are pleased to acknowledge the site, facilities, and site support provided by the Smithsonian Institution, a true partner in providing the APT Service. We are also pleased to acknowledge the support for the purchase, installation, and operation of the Vanderbilt−16 APT by the National Science Foundation grant AST 84−14594.

REFERENCES

Baliunas, S. L., Boyd, L. J., and Genet, R. M. 1987 in *New Generation Small Telescopes*, ed. D. S. Hayes, R. M. Genet and D. R. Genet, (Mesa: Fairborn Press), p. 97.

Genet, R. M., Boyd, L. J., and Baliunas, S. L. 1986 in *Automatic Photoelectric Telescopes*, ed. D. S. Hall, R. M. Genet, and B. L. Thurston, (Mesa: Fairborn Press), p. 15.

Robert Stebbins, a sociologist who studies amateurs in fields ranging from sports to science.

Douglas Hall, co-founder of the IAPPP. Doug has chaired or co-chaired eight annual IAPPP symposia in a row.

The Bulldog Mountains as viewed from the symposium grounds. In the foreground, George McCook, Joel Eaton, and Sallie Baliunas are listening to Saul Adelman. In the distant left, Michael Seeds is talking to Harold Nations.

Michael Seeds (beard) makes a point as Harold Nations (left) and Don Hayes (with camera) listen.

BROADBAND PHOTOMETRY OF BRIGHT STARS: THE FIRST YEAR OF APTS AT THE F. L. WHIPPLE OBSERVATORY

Sallie L. Baliunas, Robert A. Donahue, Jennifer G. Loeser

Harvard—Smithsonian Center for Astrophysics

Edward F. Guinan

Department of Astronomy and Astrophysics, Villanova University

Russell M. Genet and Louis J. Boyd

Fairborn Observatory

New Generation Small Telescopes, ed. D. S. Hayes, R. M. Genet, & D. R. Genet.
© 1987 Fairborn Observatory.

ABSTRACT. We present time series of broadband V, R, and I photometry obtained with the Fairborn Observatory 10-inch Automatic Photoelectric Telescope in its first year of operation at the F. L. Whipple Observatory. Time serial measurements of photometric variations of bright stars contain insight into the physical processes of stellar pulsation, winds, and mass loss in stars of a wide range of spectral type across the H R diagram. Additionally, photometric variations in cool stars have signaled the presence of large starspots. Photometric time series of spotted stars have revealed stellar axial rotation. Continued photometry of the spotted stars will likely reveal starspot cycles.

I. INTRODUCTION

Bright stars are scientifically no less interesting than fainter astronomical sources. In fact, their brightness is a virtue because they can be studied with enormous detail, or "scientific resolution" (Crawford 1987). With telescopes of large aperture (for example, several meters), faint objects have often been pursued, justifiably; many classes of objects, such as quasars or pulsars can only be studied in the visible with a large telescope, whereas bright stars often can be examined with modest apertures.

Technology and ambition have uncovered a new facet to our physical insight into bright stars. One important property of a star is its behavior with time – not changes on an evolutionary timescale, but variations during a stable episode of a star's life. One example is the average 11-year sunspot cycle. Although the Sun will remain "quietly" a lower main sequence star for billions of years, it is, nevertheless, variable. The cause of the sunspot variations is not understood in quantitative detail. We believe a magnetic dynamo, through the subsurface motions of convection and differential rotation, twist and refresh poloidal and toroidal magnetic fields (Parker 1955, 1979). The theoretical models of the solar dynamo are complex and unproven. Yet the empirical fact of the sunspot cycle taunts us: sunspot numbers vary with an average 11-year period. The sunspot period presumably depends upon the motions and conditions of the solar interior, as well as the physics of the magnetic dynamo. Therefore, studies of starspot cycles, a type of stellar variability on a timescale shorter than gross evolutionary effects, may lead to a better understanding of the solar interior and the magnetic dynamo.

Stellar variability is not only a macroscopic property of a star but also a tool to unlock the physics of the variability process, through time serial monitoring. Because time is an unaffordable luxury on large-aperture telescopes, modest-aperture telescopes have been at the forefront of monitoring efforts. Further, "automatic" telescopes have alleviated the boredom and inefficiency that often accompany long-term monitoring projects.

Thus, the Automatic Photoelectric Telescope (APT) is making a grand debut into the realm of astrophysics. This paper describes some of our initial scientific results from photometric monitoring of bright stars during the first full year of operation at the F. L. Whipple Observatory.

II. INSTRUMENTATION

The Fairborn Observatory and the Smithsonian Institution cooperate in a venture called the Automatic Photoelectric Telescope Service atop Mt. Hopkins, the location of the Smithsonian's F. L. Whipple Observatory. The site, maintained by the Smithsonian, and the equipment operated by Fairborn, have been described previously (Baliunas et al. 1985; Boyd et al. 1986a,b). Our projects are undertaken on the 10−inch (0.25m) telescope designated the Fairborn−10 and equipped with V, R, and I broadband filters for Johnson photometry with an Optec SSP−3a solid−state photometer. While the Optec photometer eventually has proven to be a reliable, accurate, and inexpensive device, its problems had to be tackled (Genet and Boyd 1987). Although our photometry from 1986 has a precision of only 1−2%, larger variations can be confidently investigated. The Optec can now perform at a level of precision of several milli−magnitudes.

III. DATA AND ANALYSIS

We have two broad scientific programs: (1) pulsation, winds and mass loss, and (2) starspots and activity cycles. The former category includes hot stars such as Be stars and P Cygni, the yellow supergiants 89 Herculis and ρ Cassiopeiae, and the red giants and supergiants classed as semi−regular or irregular variables. In the latter program, we are monitoring a sample of evolved stars (mostly giants and subgiants) for starspots and starspot cycles.

We give a few examples of time series obtained for these programs with the APT in 1986. A sample of program stars are listed in Table I, along with their spectral types. The data are differential broadband V, R, and I magnitudes. The amplitude (in magnitudes) of the variation observed in a particular filter is listed, along with the number of data points and the precision of the differential magnitudes of the comparison stars. An estimate of the precision is given for each time series in Table I by the sample standard deviation, σ:

$$\sigma = \sqrt{\frac{\sum (X - \bar{X})^2}{N - 1}}$$

where Xbar is the mean of the (check−comparison) time series, X is the value at each data point and N is the total number of observations. The amplitudes of variation in the target stars are well in excess of the precisions of their comparison stars.

We searched for periodicities in some of our target stars. Power spectra of the time series were computed according to Scargle (1982) and Horne and Baliunas (1986). This method treats unevenly sampled data without the bias introduced by other techniques when gaps are present in the time series. In the Scargle (1982) technique, the power spectrum indicates the weighted sinusoid that best fits the time series. Additionally, the significance of the height of the peak in the power spectrum can be calculated. The significance of a peak is the false alarm probability, defined as the probability that a peak at least a certain height in the power spectrum would be caused by Gaussian noise with the same variance as the data (see Horne and Baliunas 1986). Specific examples are discussed below.

IV. RESULTS

a) Pulsation, Winds, and Mass Loss

1) Semi−Regular Variables

Pulsation likely causes photometric variability in the red giants and supergiants. The red giants listed in Table I are classified as irregular and semi−regular variables whose amplitudes of variation are only several tenths of a magnitude. While many Mira variables, red supergiants with large light amplitudes (over 1−2 magnitudes), have been monitored by the AAVSO for over half a century, the small variations of our program stars require dedicated photoelectric photometry. The AAVSO observations span a luxurious interval, and demonstrate the power of long time series in studies of pulsation, and underscore the importance of a similar long−term data base for the smaller−amplitude red variables. For example, in a study of the AAVSO visual magnitudes of the semi−regular variable, V Canum Venaticorum (M4−M6IIIe), Loeser et al. (1986) discovered that one primary pulsation period (192^d) could be fit in phase throughout the entire time series (1926−1975). Second pulsation periods, between 175 and 195 days, appear frequently in the time series. A transitory, second pulsation interferes with the primary pulsation and produces a light curve that appears semi-regular when examined in intervals of several years. With the small-amplitude red giants, we hope to investigate the longevity of periodicities in our future photoelectric photometry.

Table 1 – *Selected Program Stars*

Name	Spectral Type	Filter	# Points	Range (Peak-to-peak)	σ (Check-Comp)
ψ^1 Ori	B1 Vpe	V	30	~0.19	0.015
		R	30	~0.27	0.012
P Cyg	B2 pe	V	62	~0.28	0.007
ω Ori	B3 IIIe	V	27	~0.18	0.012
		R	27	~0.30	0.013
89 Her	F2 Ibe	V	140	0.17	0.016
ρ Cas	F8 Iap	V	53	>0.27	0.013
δ CrB	G3.5III-IV	V	63	0.09	0.017
TV Psc	M3 IIIv	V	15	0.32	0.016
BC CMi	M4 III	V	15	0.37	0.017
UX Lyn	M4 III	V	36	0.37	0.024
RZ Ari	M6 III	V	22	~0.20	0.013
RX Lep	M6 IIIe	V	12	>0.27	0.018
RS Cnc	M6 IIIase	V	11	>0.12	0.011
Z Psc	N7.7	V	9	>0.30	0.008

The pulsations may result from shock waves that propagate through the tenuous outer atmosphere of the red variables. Radial velocity variations accompany the pulsations. In addition, emission lines, for example, hydrogen Balmer, CaII, or MgII, can strengthen dramatically in response to the energy dissipated from the shock waves (Glasby 1969; Willson 1976; Willson and Hill 1979; Brugel *et al.* 1986; Querci and Querci 1986). The shock waves can heat and extend the outer atmospheres, and thereby lead to mass loss, with rates near 10^{-06} solar masses per year. Knowledge about the characteristics of pulsations in visible light can lead to better understanding of the mass loss mechanisms.

Three examples of variations observed in V-band light with the Fairborn-10 APT from semi-regular variables are shown. The differential V-band magnitude of the variable (in the sense of variable star – comparison star) and the comparison (check star – comparison star) are plotted as a function of time (Julian Date). The star TV Piscium (M3IIIv) has varied by 0.3 mag in V (Figure 1) while the standard deviation of the differential magnitude of its comparison star, σ(V), is small, 0.016 mag. The previously-reported pulsation period of 50^d (Kukarkin *et al.* 1969) agrees with data so far. The semi-regular variable UX Lyncis (M4 III) has varied by almost 0.4 mag in V light while the standard deviation of its comparison star is only 0.024 mag (Figure 2). Note the different shapes and slightly different periods of the two pulses observed in UX Lyn. The carbon star Z Piscium (N7.7) varied by 0.3 mag in V; its comparison star's σ(V) is 0.008 mag (Figure

3). A period of 144^d (Kukarkin et al. 1969) could fit this short time series.

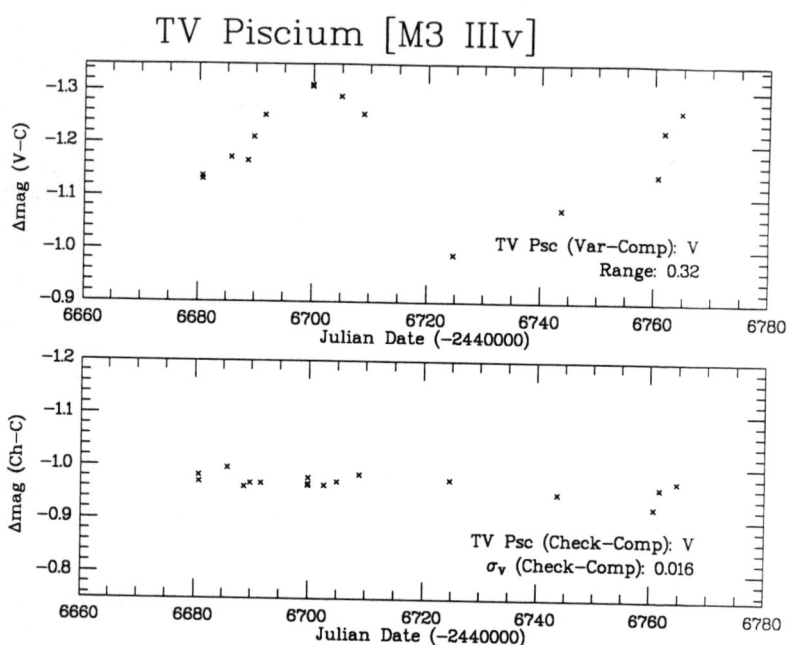

Figure 1. The differential magnitudes in V−band light plotted as a function of Julian Date in 1986. The semi−regular (pulsating) variable TV Psc (relative to a comparison star) varied by about 0.3 mag and the comparison star (relative to a check star) remained constant. The sample standard deviation of the comparison star magnitudes is low: $\sigma(V) = 0.016$ mag.

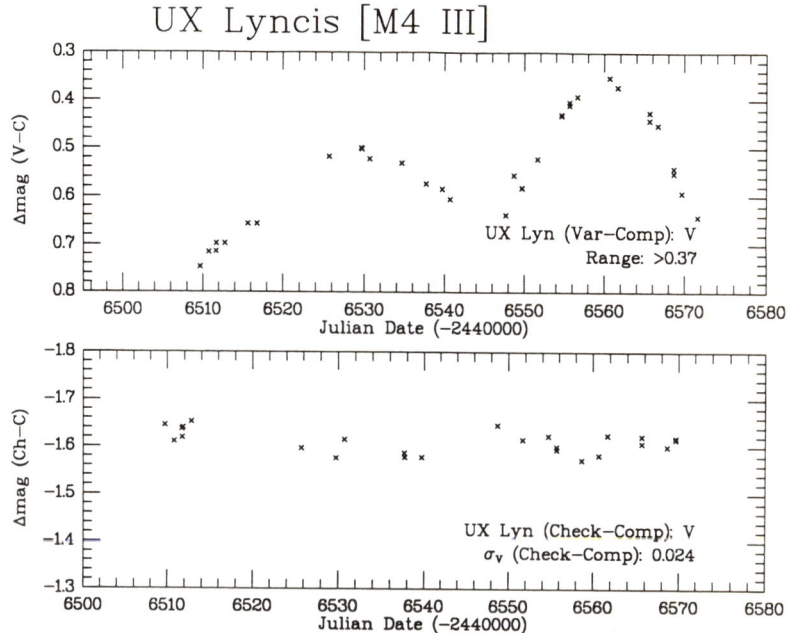

Figure 2. Differential magnitudes in V–light of the semi–regular variable UX Lyn (upper panel) and its comparison star (lower panel). The pulsations in UX Lyn varied in amplitude and period.

2). Yellow Supergiants

Pulsation is also imputed as the cause of photometric variability in the yellow supergiants 89 (V441) Herculis (F2Ibe) and ρ Cassiopeiae (F8Iap). A noticeable pulsation of amplitude 0.2 mag is evident in the V–band light of 89 Her; σ(V) is 0.016 mag in its comparison star (Figure 4). The lower panel of Figure 4 shows the power spectrum (Horne and Baliunas 1986) whose highly–significant peak corresponds to a period of 64.5 days. Although usually variable with pulses of about 58–70 days (see for example, Worley 1956; Percy and Welch 1979; Percy et al. 1979,

1981; Fernie 1986a), 89 Her occasionally quiets and exhibits only random fluttering (Fernie 1981). Radial velocity measurements by Arellano Ferro (1984) indicate that it may be a binary with a 285^d period. 89 Her is a member of the UU Herculis class of stars, F–type supergiants with high galactic latitudes that are, for the most part, indistinguishable from their low–latitude counterparts. The remarkable eclipsing binary system ϵ Aurigae contains a supergiant primary component similar in spectral type to 89 Her. The ϵ Aurigae supergiant also pulsates, with a period on the order of $120-160^d$ (Arellano Ferro 1985).

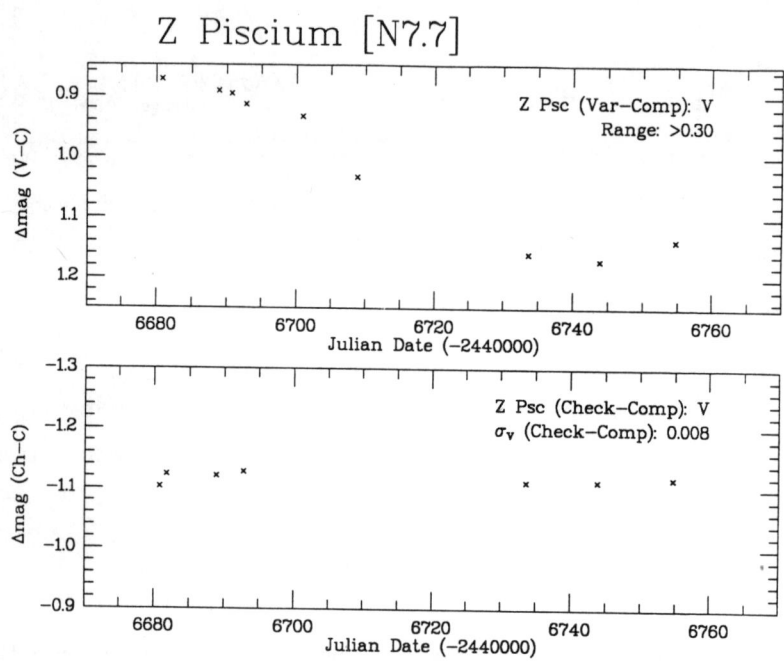

Figure 3. Differential magnitudes in V−light of the semi−regular variable Z Psc (upper panel) range 0.3 mag, while the comparison remained constant (lower panel).

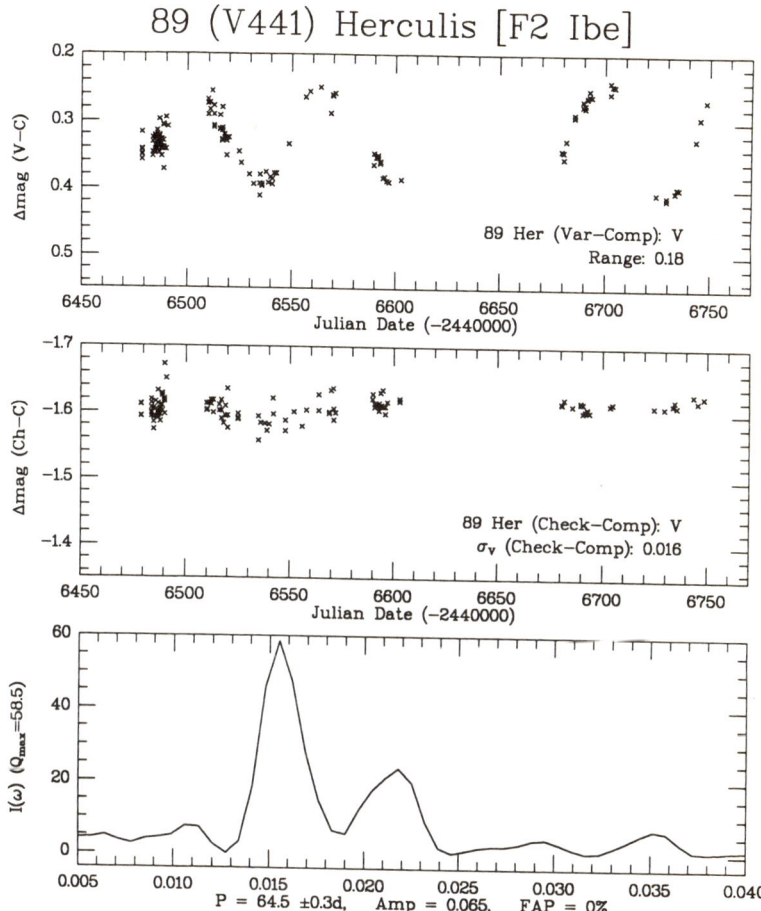

Figure 4. The yellow supergiant 89 Her varies almost 0.2 mag in V light (upper panel). Its comparison star was constant (middle panel). The power spectrum of the time series has a highly significant peak at a period of 64.5^d (lower panel). The power (normalized by the variance of the data) is plotted as a function of frequency (cycles per day). The period, P, is listed along with the semi-amplitude of the best-fitting sinusoid, Amp, and the false alarm probability, FAP. A value of FAP = 0% indicates that there is essentially no chance that the peak in the power spectrum could be simulated by Gaussian noise with the same variance as the data.

The behavior of the "hypergiant" ρ Cassiopeiae is characterized by slow light variations, punctuated by occasional deep minima and dramatic changes in apparent spectral class. The presence of these deep minima would suggest ρ Cas is an R Coronae Borealis type variable, but its spectrum does not resemble those of other R CrB stars (Glasby 1969). Nevertheless, the observed changes in the brightness and color of ρ Cas over time are intriguing. For example, although ρ Cas is currently classified in spectral type as F8 Iap, it resembled a K-star prior to 1930 and an M-star in 1946-47 after it dimmed by two magnitudes (Querci and Querci 1986).

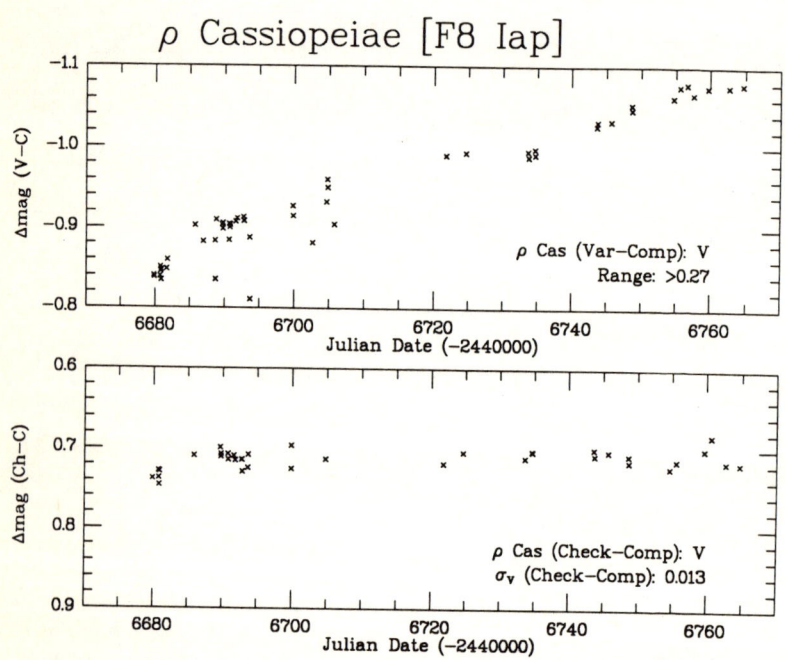

Figure 5. The yellow supergiant ρ Cas brightened by 0.3 mag in V-light (upper panel) in 1986. The light of its comparison star is essentially constant (lower panel).

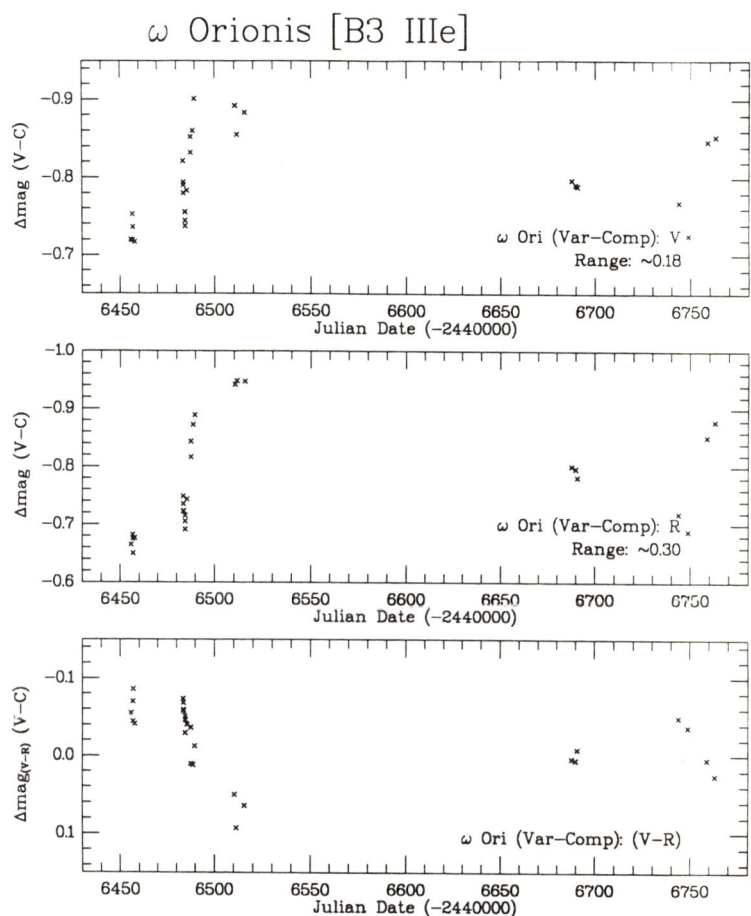

Figure 6. The Be star ω Ori brightened quickly in V (upper panel) and R (middle panel). The (V−R) color index (bottom panel) numerically increased (or, became redder) when the star brightened.

Dramatic fading in brightness in variables such as R CrB and ρ Cas herald abrupt changes in the spectrum accompanied by evidence for atmospheric flows and perhaps clouds of condensing graphite particles. In 1986 the V−magnitude brightened by almost 0.3 mag, while the comparison star has remained constant with σ(V) = 0.013 mag (Figure 5).

Rapid light diminution in ρ Cas would signal the need for intensive spectroscopic and photometric studies. Fourier analysis of photoelectric data through 1981 by Arellano Ferro (1985) suggests a period of 483 days, a period too long to identify in our short time series.

3) Be Stars

The Be stars have strong and variable hydrogen Balmer emission lines. Both the emission lines and the visible light vary dramatically and (apparently) sporadically, presumably as a result of episodic mass loss (Slettebak 1982; Adelman and Peters 1986; Peters 1986). The presence of the emission lines is often correlated with brightening visual light and changing energy distributions in the color indices. Non-radial pulsations may help to initiate the mass-losing episodes (Peters 1986; McDavid 1987). The broadband photometry could be investigated for evidence of short period non-radial pulsations and the character of the abrupt changes signaling mass loss.

The photometric time series for the Be star ω Orionis (B3IIIe; Hayes and Guinan 1984) dramatically and rapidly brightened in V (0.2 mag) and R (0.3 mag) light early in 1986 (Figure 6, upper panels). The bright state of the star was accompanied by a redder energy distribution in the $(V-R)$ color index (bottom panel, Figure 6).

4) P Cygni

The hot star P Cygni (B2pe) has an extremely extended atmosphere and a prodigious mass loss rate; P cygni may be the progenitor of a Wolf-Rayet star (Lamers et al. 1983). Its Hα line profile is the archetypal "P Cygni" profile that signals a dense wind and copious mass loss. The strong Hα emission and the blue-shifted absorption profile (Figure 7) is the result of an extended, dense region with outflow velocities up to 300 km/s and mass loss rates near 10^{-05} solar masses per year (de Jager 1980; White and Becker 1982).

In order to study mass loss in P Cygni, we are monitoring the Hα line profile as well as the broadband V, R, and I magnitudes that detail photospheric behavior. The high-resolution (0.015A) Hα spectra were obtained at the 61-inch reflector at the Smithsonian's Oak Ridge Observatory (Harvard, MA) with an echelle spectrograph and intensified Reticon detector. The two Hα spectra in Figure 7 demonstrate the large changes in the Hα emission (over a factor of two) that can occur in a short period of time (5 days). Rapid variations are also observed in variable linear polarization of the visible continuum (Hayes 1985).

The change in Hα emission strength can be translated into a change in the mass loss rate. For a linear velocity flow, the Hα flux will be proportional to the mass loss rate (Scoville et al. 1983); a factor of two decrease in Hα emission within 5 days corresponds to a factor of two decrease in the mass loss rate during this time.

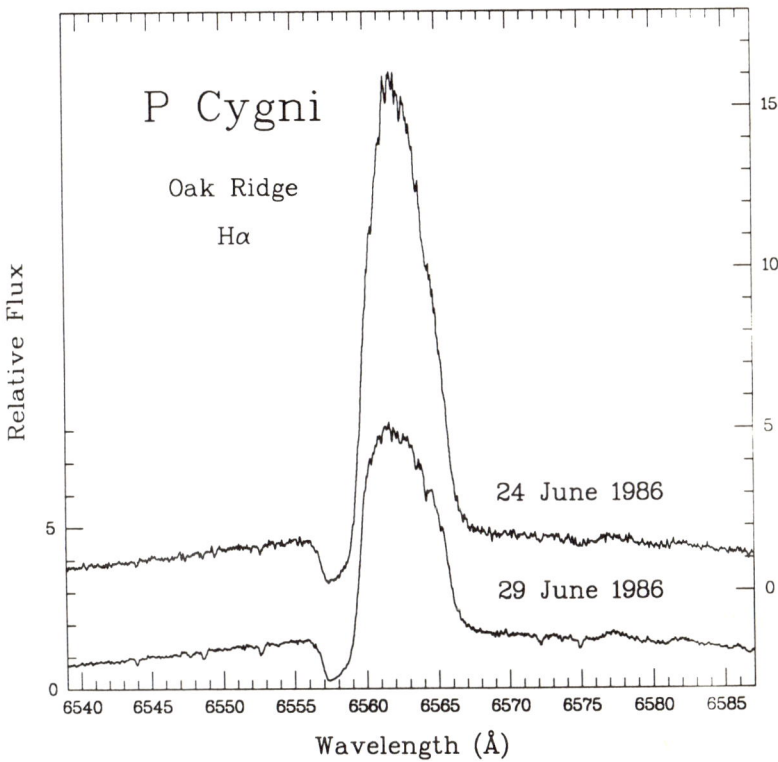

Figure 7. Two Hα profiles of P Cygni show an increase in the Hα emission strength of a factor of two in just 5 days. The two profiles are vertically offset; the left flux scale refers to the lower profile and the right to the upper profile. The blue-shifted Hα absorption component associated with the strong, outflowing stellar wind is centered near 6557A.

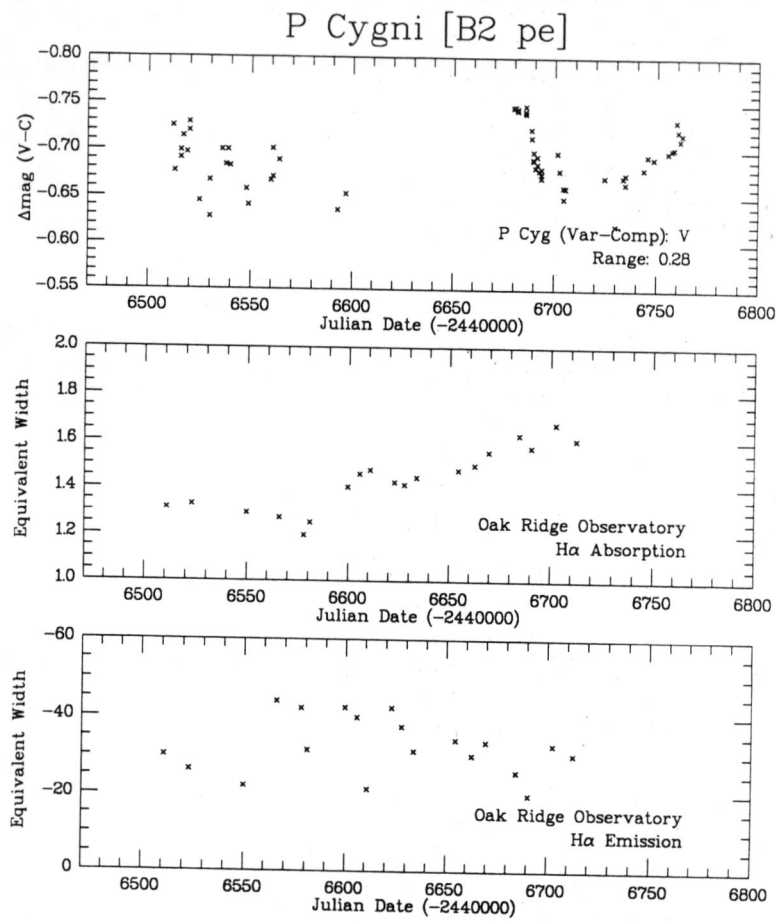

Figure 8. The differential V−magnitudes of P Cygni (upper panel), which detail photospheric changes, can be compared to changes in the Hα spectra, which describe the variable wind and extended atmosphere. The equivalent width of the blue−shifted absorption component (middle panel) increased by about 20% in 1986. The relative strength of the Hα emission component (plotted as a negative equivalent width) often varied dramatically (up to a factor of two) over several days.

The equivalent width of the absorption and emission components of the Hα spectra are plotted as a function of time in Figure 8. The V−band differential magnitudes are also plotted in the top panel. The R− and I−band magnitudes vary similarly to those in V: the colors remain farly constant. The variations in the Hα spectra (wind) and photometry (photosphere) so far appear to be erratic and unrelated.

b) Starspots

In the introduction, we described our motivation to understand the sunspot cycle by investigating the starspot cycles on other stars. Part of our APT program searches evolved stars, predominantly luminosity class IV and III, for evidence of starspots in broadband light. Analogous to the programs on other APTs that monitor the active chromosphere RS CVn variables (Hall *et al.* 1986a,b; Seeds and Nations 1986; Nations and Seeds 1986), we are sampling 200 subgiants and giants for photometric starspot activity. Since the giants in our sample are not as active as the RS CVn variables, we would expect much smaller variations. In this respect, the improved performance of the Optec, Inc. photometer will be critical to our success.

Time series monitoring of stars can reveal axial rotation because the light is modulated by an inhomogeneous stellar surface rotating through our field of view. The photometric light fades when the visibly darker and cooler spotted areas appear on the stellar hemisphere facing us. The spots must be relatively long−lived (longer than the rotation period), otherwise the signal will reveal the growth and decay of the spots. Actually, both these kinds of behavior are present, and several long time series must be investigated to identify rotation accurately. In addition, time series collected over years can reveal starspot cycles. Furthermore, rotation periods monitored over the course of starspot cycles may reveal differential surface rotation: spots at different latitudes would mark different rotation periods.

At the 60 and 100 inch telescopes at the Mt. Wilson Observatory, chromospheric variations in about 100 lower main sequence stars have been monitored for over 20 years. The relative strength of the CaII H (3967A) and K (3934A) emission lines are sensitive indicators of the presence of chromospheric activity such as active regions and plages, which can be associated with photometrically darker starspots. At Mt. Wilson, stellar rotation, differential rotation, and activity cycles have been documented (Wilson 1978; Baliunas and Vaughan 1985). The evolved stars being measured photometrically with the APT have also been added to the chromospheric monitoring program at Mt. Wilson. We are investigating the timescales of chromospheric and photospheric (V, R, and I photometry) activity. Are the chromospheric active areas and photospheric starspots physically associated and therefore always varying in concert? When the darker starspots appear, does the photospheric energy that is blocked reappear at other wavelengths or is it temporarily stored (say, in the convection zone), to be released later? The question of energy redistribution on the Sun is intensely debated (Newkirk 1983).

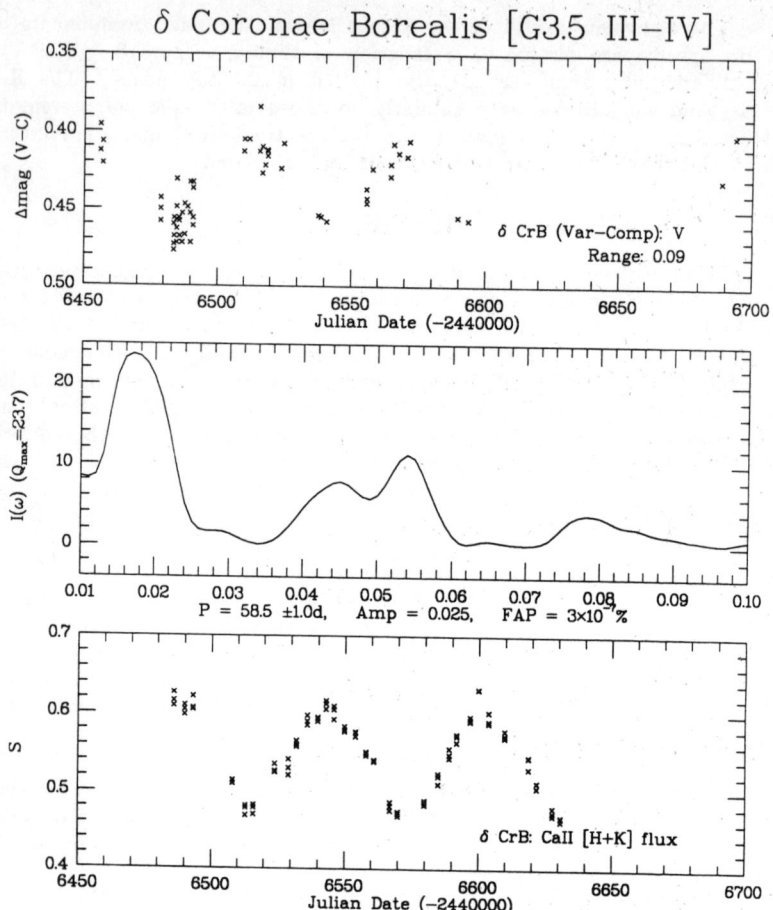

Figure 9. The photometric variation in V light (upper panel) in δ CrB marks stellar rotation when starspots are distributed inhomogeneously on the stellar surface. The power spectrum (middle panel) suggests a rotation period, P, of 58^d. Formally, the false alarm probability, FAP, suggests the peak has a vanishingly small probability, $3.0\times10^{-07}\%$, that it could be caused by Gaussian noise with the same variance as the data. The relative CaII H and K chromospheric emission strength, S, is also shown (bottom panel). The same 58^d rotation period is present in the time series, but the two curves vary in opposition (see text).

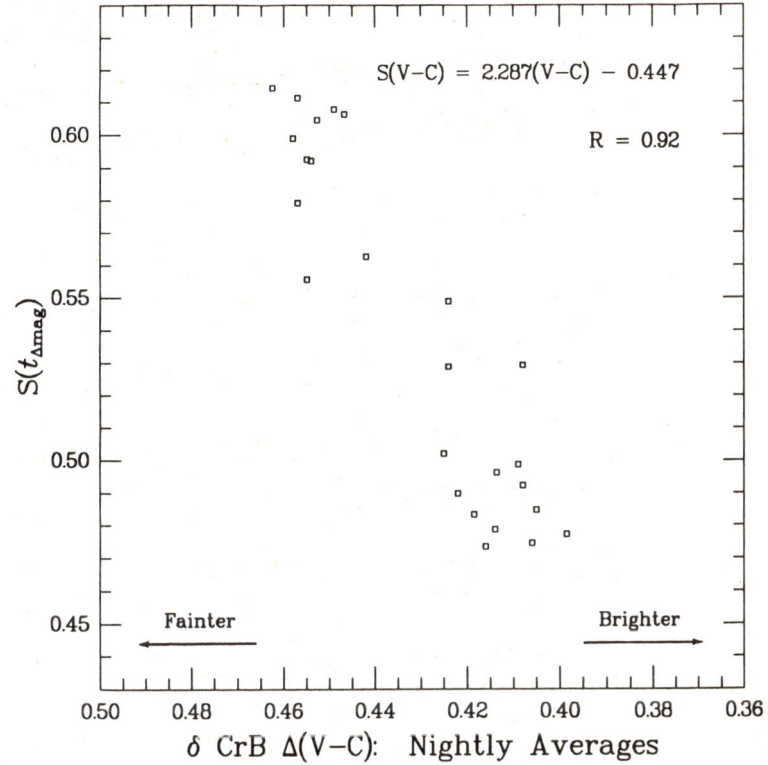

Figure 10. The S–values are linearly related to (and anti–correlated with) the differential magnitudes in V–band light in δ CrB (see Figure 9). The values of S were interpolated to the times of V–light measurements. The linear function is given, along with the linear (anti–) correlation coefficient, $r = 0.92$, suggesting an extremely good fit.

Dramatic spottedness on an evolved star, δ Coronae Borealis (G3.5 III–IV), is shown in Figure 9. In another long–term APT program, Fernie (1986b) reported V–light variations that portray rotation modulation in δ CrB. The top panel shows V–band modulation produced by starspots during stellar rotation. The amplitude is small, 0.1 mag, but significant compared to the standard deviation of the comparison star, $\sigma(V) = 0.017$ mag. The power spectrum of the V–band data (middle panel) reveals an extremely likely 58–day rotation period. The star was also observed at Mt. Wilson during 1986. Those data, the relative chromospheric CaII emission strength, S (bottom panel), reveals the same rotation period as the photospheric data. The two light curves, however, are anti—phased so the photospherically dark spots must be associated with the chromospherically bright active areas. The S—values are plotted as a function of V—band light in Figure 10. The S—values were interpolated to the nights when V—band data were present. The linear (anti—) correlation coefficient is extremely high (r = 0.92), and suggests the (chromospheric) active areas and (photospheric) starspots are contemporary and cospatial.

V. SUMMARY

Even small—aperture telescopes such as the Fairborn—10 APT are progressing our knowledge about stellar pulsation, mass loss, starspots, rotation and activity cycles. Furthermore, the complete automation achieved by the APTs provides efficient, inexpensive, and reliable time series. The time series document variability, which is an important attribute of stars. Time series provide physical insight that is difficult or impossible to obtain by any means.

Remember: "The trend is your friend!"

ACKNOWLEDGEMENTS

We deeply appreciate the efforts of James Frazer, Tony Misch, and Laura Woodard who kindly obtained the Mt. Wilson data for δ CrB. We thank R. McCrosky and the Oak Ridge Observatory staff for the Hα spectra. We appreciate discussions on P Cygni with John Hillier. This material is based upon work supported by the National Science Foundation under Grant AST—8616545, and the Smithsonian Scholarly Studies Program. We also thank S. Criswell, D. Latham, and I. Shapiro for their support and encouragement.

REFERENCES

Adelman, S. J. and Peters, G. J. 1986, *I.A.P.P.P.Comm.*, 25, 71.
Arellano Ferro, A. 1984, *Publ. Astron. Soc. Pacific*, 86, 641.
Arellano Ferro, A. 1985, *Mon. Not. Roy. Astron. Soc.*, 216, 571.
Baliunas, S. L., and Vaughan, A. H. 1985, *Ann. Rev. Astron. Astrophys.*, 23, 379.
Baliunas, S. L., Boyd, L. J., Genet, R. M., Hall, D. S., Criswell, S. 1985, *I.A.P.P.P.Comm.*, 22, 47.
Boyd, L. J., Genet, R. M., and Baliunas, S. L. 1986, *I.A.P.P.P.Comm.*, 25, 15.
Boyd, L. J., Genet, R. M., and Hall, D. S. 1986, *Publ. Astron. Soc. Pacific*, 98, 618.
Brugel, E. W., Willson, L. A., and Cadmus, R. 1986, in *New Insights in Astrophysics: 8 Years of Research with IUE*, ed. E. Rolfe, ESA−SP 263, p. 213.
Crawford, D. L. 1987, in *New Generation Small Telescopes*, D. S. Hayes, R. M. Genet and D. R. Genet, eds., (Mesa: Fairborn Press), p. 7.
de Jager, C. 1980, *The Brightest Stars*, (Dordrecht: Reidel).
Fernie, J. D. 1986a, *Astrophys. J.*, 306, 642.
Fernie, J. D. 1986b, preprint.
Fernie, J. D. 1981, *Astrophys. J.*, 243, 576.
Genet, R. M., and Boyd L. J. 1987, in *New Generation Small Telescopes*, D. S. Hayes, R. M. Genet and D. R. Genet, eds., (Mesa: Fairborn Press), p. 85.
Glasby, J. S. 1969, *Variable Stars*, (Cambridge: Harvard University Press), p. 161.
Hall, D. S., Kirkpatrick, J. D., Seufert, E. R., and Henry, G. W., 1986, *I.A.P.P.P.Comm.*, 25, 43.
Hayes, D. P. 1985, *Astrophys. J.*, 289, 726.
Hayes, D. P., and Guinan, E.F. 1984, *Astrophys. J.*, 279, 721.
Horne, J. H., and Baliunas, S. L. 1986, *Astrophys. J.*, 302, 757.
Kukarkin, B. V, *et al.* 1969, *General Catalogue of Variable Stars 3ed.* (Astronomical Council and Sternberg Astron. Institute, Moskva)
Lamers, H. G. J. M., de Groot, M., and Cassatella, A. 1983, *Astron. Astrophys.*, 123, L8.
Loeser, J. G., Baliunas, S. L., Guinan, E. F., Mattei, J.A., and Wacker, S. 1986 in *Cool Stars, Stellar Systems, and the Sun*, ed. M Zeilik and D. M. Gibson (New York: Springer−Verlag), p. 460.
McDavid, D. 1987, in *New Generation Small Telescopes*, D. S. Hayes, R. M. Genet and D. R. Genet, eds., (Mesa: Fairborn Press), p. 237.
Nations, H. L. and Seeds, M. A. 1986, *I.A.P.P.P.Comm.*, 25, 56.
Newkirk, G. Jr. 1983, *Ann. Rev. Astron. Astrophys.*, 21, 429.
Parker, E.N. 1955, *Astrophys. J.*, 122, 293.
Parker, E.N. 1979, *Cosmical Magnetic Fields: Their Origin and Their Activity*, (Oxford: Claredon).
Percy, J.R., Baskerville, I., and Tervorow, D. 1979, *Publ. Astron. Soc. Pacific*, 91, 368.

Percy, J. R., and Welch, D. 1981, *Publ. Astron. Soc. Pacific*, 93, 367.
Peters, G. J., 1986, *Astrophys. J. Lett.*, 301, L61.
Querci, F., and Querci, M. 1986, *I.A.P.P.P.Comm.*, 25, 156.
Seeds, M. A. and Nations, H. L. 1986, *I.A.P.P.P.Comm.*, 25, 50.
Scargle, J. D. 1982, *Astrophys. J.*, 263, 835.
Scoville, N., Kleinmann, S. G., Hall, D. N. B., and Ridgway, S. T. 1983, *Astrophys. J.*, 275, 201.
Slettebak, A. 1982, *Astrophys. J.Suppl.*, 50, 55.
White, R. L., and Becker, R. H. 1982, *Astrophys. J.*, 262, 657.
Willson, L. A. 1976, *Astrophys. J.*, 205, 172.
Willson, L. A., and Hill. S. J. 1979, *Astrophys. J.*, 228, 854.
Wilson, O. C. 1978, *Astrophys. J.*, 226, 379.
Worley, C. E. 1956, *Publ. Astron. Soc. Pacific*, 68, 62.

PLANNING FOR A CONSORTIUM APT—A PROGRESS REPORT

Robert J. Dukes, Jr.

Physics Department, The College of Charleston

I. INTRODUCTION

At last year's symposium we described plans for an automatic telescope operated by the Automatic Photoelectric Telescope (APT) Service for a consortium of four schools (Dukes and Adelman 1986). In this paper we report on the progress of these plans and our efforts to obtain funding for this venture. The College of Charleston is acting as the proposer, financial manager, and coordinator for the project. The other institutions involved are The Citadel, The University of Nevada, Las Vegas, and Villanova University. Participants from the various institutions and the APT Service are listed in Table I. In our paper last year we provided a description of the background of our institutions, a description of our needs, why these needs are difficult to meet by currently available astronomical research facilities, and the reasons why we feel that these needs would be met by an APT. In this paper we provide

---

New Generation Small Telescopes, ed. D. S. Hayes, R. M. Genet, & D. R. Genet.
© 1987 Fairborn Observatory.

details on our plans for utilizing the APT as well as our attempts to obtain funding for the project.

TABLE I

INSTITUTION	PARTICIPANT
The College of Charleston	Robert J. Dukes, Jr.
	William R. Kubinec
	Terry R. Richardson
The Citadel	Saul J. Adelman
Villinova University	George McCook
	Edward Guinan
The University of Nevada, Las Vegas	Edwin Grayzeck
	Diane Pyper—Smith
The APT Service	Louis Boyd
	Russell Genet
	Donald Hayes

II. THE SCIENTIFIC RESEARCH

A list of the projects which consortium members plan to carry out with the APT is given in Table II. This table includes the astronomers involved in the project as well as an initial time allocation. The time requirement for each project is given in terms of equivalent 8 hour nights where we have assumed that there will be 200 such usable nights per year. Based on Kitt Peak records, this is a reasonable expectation for the Tucson area. If the expected number of nights does not materialize, there will be adjustments made. A brief description of each of the projects is given on the next page.

Table II

PROJECT	PARTICIPANTS	NIGHTS
Solar Type Stars	McCook and Guinan	50
Multi–Mode Variables	Dukes and Kubinec	50
Chemically Peculiar Stars	Smith and Adelman	50
Long Period Cepheids	Grayzeck	15
Be Stars	Adelman	13
Photometric Properties of the Ursa Majoris System	Adelman	7
Determination of Photometric Standards	Consortium members with the APT Service	15

Up to this time APTs have only been used for differential photometry. In order to reduce observations to a standard system it is necessary to observe standard system together with the program stars. As part of this project Donald Hayes of the APT Service will devise algorithms for using an APT to obtain all–sky photometry which may be reduced to a standard system. These algorithms will be used in a number of the projects described below. Adelman and Peters (1986) described the ways in which observations of standard stars with an APT can be used to both define new standard systems as well as to improve existing ones. Since all–sky photometry will be performed with different photometric systems which each have their own standards, initial measurements of each standard in all systems must be made. Such measurements will allow derivation of standard quality magnitudes for any missing values in the systems being used. Once this is done a unified set of standards will be established and observing time will be saved.

McCook and Guinan plan to work on photometry of solar type stars. The variation in the Sun's magnetic field and activity cycle remain one off the most challenging problems in astrophysics. Studies of spotted and active regions on other stars of varying temperatures, rotation rates and ages will elucidate the mechanism involved in maintaining and generating stellar magnetic activity. These observations should provide additional constraints on dynamo theories and lead to a better understanding of magnetically driven activity for our sun and for stars in general.

The proposed program will investigate the periodicities and light variations, at selected wavelengths, for a larger sample of single, solar–like stars. The extent and temperatures of the starspot regions and their relation to plage regions will be found through modeling. Whenever possible, differential rotation rates will be estimated from the evolution of the light curve with time. This will require extensive photoelectric observations over a long time interval (> 2 yrs.).

Dukes plans to monitor multi-mode variables. Details of these plans were presented at last year's symposium. This is a particularly interesting problem for an APT. These stars have several simultaneous excited pulsational modes. They have the potential for providing a check between theoretical calculations and observation since the presence of simultaneous excited modes imposes severe restraints on the properties of the atmosphere and envelope. Pulsational modes of single mode variables match observations fairly well, but, unfortunately, there are relatively few multi-mode variables whose properties have been analyzed in sufficient detail for an adequate comparison with theory.

Program stars will be selected from short period variables (both δ Scuti and β Cephei types) for which solutions have been obtained. For these stars the same solution will be sought independently by observing them over one to two seasons with an APT. Another set of program stars for which no adequate observations exist will be selected and observed over three seasons with the last season as a check on earlier seasons.

Pyper and Adelman (1986) reported plans for using the Consortium APT for observing chemically peculiar stars of the upper main sequence at last year's symposium. These stars, which constitute a substantial fraction of B and A type stars, are divided into two groups based on the presence of detectable surface magnetic fields: the magnetic and non-magnetic CP stars. In order of decreasing temperature, the magnetic types are: the helium-rich $(He-s)$, the helium-weak $(He-w)$ and silicon (Si) stars, and the strontium-chromium-europium (SrCrEu) stars, while the non-magnetic types are: the PGa He-w, the mercury-manganese (HgMn), the hot metallic-lined (Am) and the Am stars.

The basic thrust of the planned observations is to study the periodic small amplitude light variations of these stars. This requires high precision results obtained over a long period of time and thus is very appropriate for an APT.

Grayzeck plans to monitor long period Cepheids. These stars are important as distance indicators as they are luminous, easily identifiable, and obey a well determined Period-Luminosity relation. Two problems still plague the use of these variables as distance probes; determining the "true color" of the star, and applying appropriate reddening corrections to those cluster stars used to calibrate the P-L relation. The variation of color over the cycle must be averaged in some physically meaningful way to find the temperature and luminosity that corresponds to a "normal" supergiant. Madore (1977) pointed out that color/color plots, e.g., U-B/B-V, often show loops that may indicate companion stars for approximately 20% of the sample. The combination of these stellar colors affects not only the individual distances, but, for cluster cepheids, may also cause errors in the P-L calibration. In the red part of the spectrum, e.g., VRI colors, the loops are narrow but still cause uncertainty in choosing the meaningful "true color" (De yoreo and Karp, 1979).

The APT will be used to provide continual monitoring of 23 bright northern Cepheids with periods of >= 10 days in the broad band UBVRI system of Cousins. The measured color/color loops will indicate the presence of possible companions and provide the necessary corrections to estimate the "true color." The proposed sample of bright Cepheids covers the period range of most of these objects (10 days−27 days). They are ideal candidates for differential photometry measurements with a sample frequency of one or two times nightly. The comparison stars for each variable will be measured by all sky photometry. Although the Cepheid project accounts for about two weeks of quality observing, it requires a complete winter observing season.

Adelman and Peters (1986) reported plans last year to use the APT for a variety of projects including monitoring Be stars, determining parameters or Ursa Majoris Stream stars, and to improve several standard photometric systems.

Photometric observations of a selected set of Be stars will be obtained to catch these objects in the earliest stages of activity, and to investigate their photometric behavior. In addition to securing data with the standard Stromgren filters, Hα observations will be used to assess the state of activity of the star and, as well, to detect the onset of an emission phase. The initial observational plan is to observe several currently active Be stars every few nights throughout their observing seasons. Additional objects will then be monitored less intensively, at least once a year, to pursue the second objective. In these observations there will be cooperation with other observers. Those comparison stars already found to be non−variable by others will be used. With the help of other observers, important high resolution spectroscopy and polarimetry will be obtained once activity at Hα is detected in a program star. Since time is an important factor in obtaining the observations of a star in its earliest stages of activity, an algorithm will eventually be devised to detect the onset of such a stage. This might involve storing the nightly values of the Hα index (line−continuum) and comparing on line with these values. If statistically significant changes are found, then it should be possible to begin a more intensive monitoring of the star.

The study of stellar groups can yield information about the stars themselves, the interstellar medium, and the current status and history of the groups. The planned study concerns the Ursa Major Stream, but the techniques are applicable to other stellar groups such as those defined by Eggen (1985). This study, which will be supplemented by parallax data from other observers, has the goal of constructing an accurate Hertzsprung−Russell diagram of the stream. This will then be compared with theoretical isochrones in order to better determine the Stream age.

As there is not, at present, an Hα system of comparable quality to β photometry, the APT will be used to create one. The standard star candidates are from existing Hα systems and from the Stromgren system. Homogeneous observations of these stars in β will be obtained. This will permit the transformations of the older systems to the new one as well as the establishment of a tight connection to the β system. This is an extension of previous work by Guinan (1975). In the first two years of operation, we plan to make on the order of 20 measures in Hα and 10 measures in β for approximately 200 stars.

III. LOGISTICS

During operation the Project Director will coordinate scheduling and communicate the observing programs to the APT service. Scheduling will be accomplished by agreed upon telescope control algorithms with appropriate star lists. Since different projects are included, several different algorithms are necessary. With the flexibility of an automatic telescope, constraints involved in scheduling by nights or even blocks of time will be removed. Rather than allocating a certain number of nights per school or observing project, a list of projects in order of priority for each night will be prepared. Each project will have an algorithm and list of stars associated with it.

In devising the algorithms consideration will be given to the types of measurement protocols possible for photometry. The first type involves observing standard stars to tie into some established photometric system. This will require a list of program stars as well as a list of standard stars and an algorithm for selecting each. Appropriate algorithms will intersperse observations of program stars with those of standard stars. These algorithms will select the program stars from a list of possible program stars (including criteria such as the current air mass of each star and its position in the sky). Standard stars will be selected that bracket program stars in color, magnitude, and air mass. Several photometric systems will be available: Cousins UBVRI, Strömgren uvby, β, and H α. Since the standards for these systems are not always the same, the algorithms will choose standards appropriate to the system being used. For stars to be observed differentially, an algorithm which includes information on the required frequencies of observation will be used. For very short period stars (< 1 day) we will observe a given star as long as it is above some specified air mass. For short period stars (1 day $< P <$ 10 days) the algorithm will observe the star several times when it is above some specified air mass. Finally, for long period stars ($P > 10$ days) the algorithm will specify one observation per star per night.

A master algorithm will govern which algorithms will operate during a given night. For each night there will be a prime algorithm as well as one or more secondary algorithms which will be executed in the event all of the observations of the prime algorithm are satisfied. For example, if the prime algorithm involves observation of one short period variable star continuously as long as it is above a specified air mass, the master algorithm would select a secondary algorithm to run in that portion of the night when the conditions for the prime algorithm are not satisfied (i.e. the prime program star has set or has not risen).

Once obtained, data will be recorded on floppy disks and sent to the Project Director by APT Service personnel. This data will in turn be sent to each of the member institutions. Here the data will be reduced, analyzed, and archivved. The Project Director will also inspect the data to insure that each program is receiving its planned share of observing time and make adjustments to the schedule to insure that periods of bad weather affect all programs equally.

I would like to express my appreciation to the members of the Consortium listed in Table I who have contributed to our proposal to the National Science Foundation. Many of their ideas have been incorporated into this paper.

REFERENCES

Adelman, S. J., and Peters, G. J. 1986, in *Automatic Photoelectric Telescopes* ed. D. S. Hall, R. M. Genet, and B. L. Thurston (Mesa: Fairborn Press), p. 71.
De Yoreo, J. J., and Karp, A. H. 1979, *Astrophys. J.*, 232, 205.
Dukes, R. J., Jr. and Adelman, S. J. 1986, in *Automatic Photoelectric Telescopes* (Mesa: Fairborn Press), p. 65.
Eggen, O. J. 1985, *Pub. Astron. Soc. Pacific*, 97, 807.
Guinan, E. F., Baliunas, S. L., and Ciccone, M. A. 1975, *Pub. Astron. Soc. Pacific*, 87, 969.
Madore, B. 1977, *Pub. Astron. Soc. Pacific*, 90, 315.
Pyper, D. M., and Adelman, S. J. 1986, in *Automatic Photoelectric Telescopes* (Mesa: Fairborn Press), p. 76.

Astronomers milling among Saguarro cacti.

DELTA SCUTI STARS AS APT SUBJECTS

Robert J. Dukes

Physics Department, College of Charleston

The δ Scuti stars are a class of variable stars located near or on the main sequence in the region of the late A− and early F− stars. They are characterized by short periods (< 0.3 days) and small amplitudes (< 0.1 mag usually; sometimes up to 0.7 mag).

Historically, these stars have been known by a variety of names as well as having been categorized into several distinct types. The various designations included ultrashort period Cepheids, RRs stars, dwarf Cepheids, Al Velorum stars, and δ Scuti stars. Over the last few years, however, it has been realized that the only truly distinguishing characteristic between the types was a relatively large pulsational amplitude for the so−called dwarf Cepheids. Thus it is now common to refer to all of these types of variable stars as δ Scuti stars. Generalizations seldom work, and this is true here. Four members of the former dwarf Cepheid class do exhibit other distinguishing characteristics, including shorter periods, lower mass, lower luminosity, and higher space velocity than the "standard" δ Scuti stars which have solar composition. These stars, which are apparently Population II objects, are GD 428, SX Phe, CY Aqu, and DY Peg (Wolff 1983).

New Generation Small Telescopes, ed. D. S. Hayes, R. M. Genet, & D. R. Genet.
© 1987 Fairborn Observatory.

Many of the δ Scuti stars exhibit unstable light curves. One of the greatest unresolved questions concerning these stars is the cause of this instability. The basic suggestions which have been made are: (1) The stars are only quasi−periodic due to non−linear effects in the atmosphere such as an interaction between pulsation and convection (LeContel, et al. 1974) and (2) The stars are truly multi−periodic and difficulties in determining the period structure are due to inadequate observational material (Fitch 1976). The second explanation is currently the more accepted one. It accounts for the observed light curve instability by the interaction of two or more modes of pulsation. To date it has been difficult to resolve this question because of the large quantity of observational data required to properly analyze the variation of any one δ Scuti star. Fitch has suggested that a proper analysis would require data acquisition on at least 10 closely spaced nights in a single season while Breger (1980) recommends a total of 30 nights. Even in those few cases where sufficient data exist, the observations have generally been used for a single solution of the period structure. Seldom are these solutions repeated for independent sets of observational data by the same worker using the same techniques.

There have been several lists of δ Scuti stars published (Baglin et al. 1973; Breger 1979). However, none of these are complete, and new δ Scuti stars are discovered regularly. Observationally it is very difficult to recognize stars as members of the δ Scuti class since the variational amplitudes are generally so small. In fact, one of the most common ways of discovering a δ Scuti star is to find that a comparison star used in a program of differential photometry is variable. The current work is based on a database of δ Scuti stars prepared over the last several years at the College of Charleston.

We now examine the distribution of δ Scuti stars as a function of apparent magnitude and amplitude of variation. Figure 1 shows the number of δ Scuti stars in each apparent magnitude class from first through seventh magnitude. Interestingly, the majority of those known are naked eye stars. Assuming that the actual number of δ Scuti stars increases with increasing apparent magnitude in a manner similar to the increase observed in total star numbers we infer that there are many δ Scuti stars just below the limit of naked eye visibility which remain to be discovered. We shall return to this point below. Figure 2 shows the number of δ Scuti stars as a function of amplitude of variation in steps of 0.01 mag. Three points should be noted concerning this plot. First, the amplitude data is very uncertain for the majority of the stars; as we discuss below, the light curves of these stars show large cycle−to−cycle variations. Second, most reported amplitudes have been determined from a limited set of observations and, hence, are probably not reliable. Third, the vast majority of δ Scuti stars have very low amplitudes of variation. This means that they have primarily been found by accident. Again, the data set is not complete.

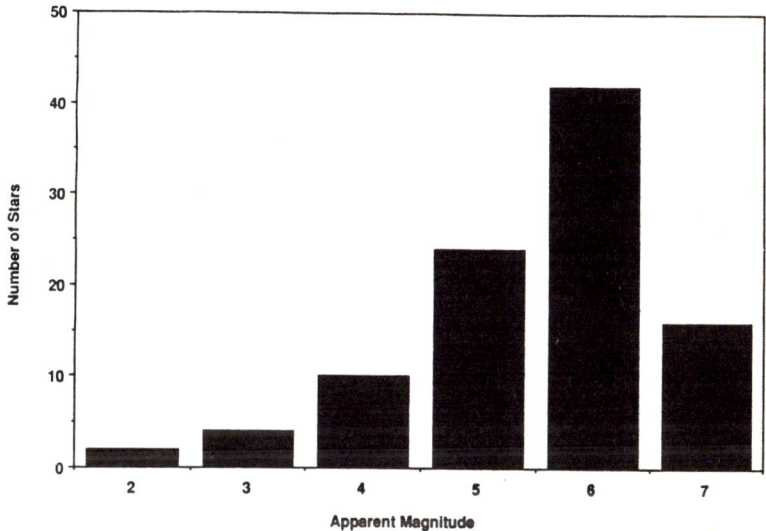

Figure 1. Plot of number of δ Scuti stars as a function of apparent magnitude. Magnitude scale is divided into bins of width one magnitude centered on the labeled magnitude.

We now consider further the data in Figure 1. Table 1 compares the number of δ Scuti stars in each magnitude interval with the total number of stars in that interval. The latter data is taken from the summary in the introduction to *Sky Catalog 2000.0* (Hirshfeld and Sinnot 1982). Column 1 gives the magnitude interval. The first interval includes all stars brighter than 0.5 mag, while the remaining intervals include all stars within 0.5 mag of the midpoint which is listed. Column 2 gives the number of known δ Scuti stars and Column 3 gives the total number of stars. If we assume that all δ Scuti stars brighter than third magnitude have been discovered then we can use these numbers to estimate the total number of δ Scuti stars in each magnitude interval.

We take the data in Columns 2 and 3 and normalize these so that there is only one star brighter than third magnitude in each category. The results of this normalization are given in Columns 4 and 5 respectively and plotted in Figure 3. It is obvious from Figure 3 that δ Scuti stars fainter than approximately fourth magnitude are greatly underabundant relative to all stars. We quantify this by recognizing from the data in columns two and three that about two percent of all stars brighter than third magnitude are δ Scuti stars. We predict then that in any magnitude interval two percent of all stars should be δ Scuti stars. This predicted number is given in Column 6. Column 7 gives the difference between the predicted and observed numbers. These are our estimates of the number of δ Scuti stars remaining to be discovered in each magnitude interval.

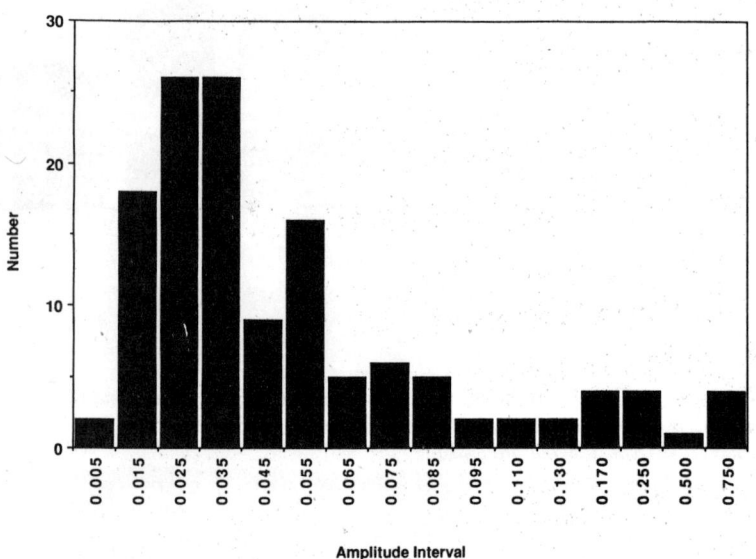

Figure 2. Plot of number of δ Scuti stars as a function of amplitude. Amplitude scale is divided into bins of width 0.01 mag centered on the labeled magnitude.

One of the major reasons for the lack of an adequate understanding of the δ Scuti mechanism is, as mentioned above, the large quantity of observational data required. Commonly, a worker will observe a δ Scuti star for several nights over the course of a season and report a determination of the strongest of two strongest apparent periodicities. In subsequent seasons the star may be observed by other workers and the reported periodicities modified. It is rare for the necessary quantity of observational material to be gathered in a single season and rarer still for this data to be gathered in each of several consecutive seasons. One concludes that known members of this class must receive more extensive observations and that more members must be discovered.

As we suggested at the last symposium, (Dukes and Adelman 1986) a successful observing program for a δ Scuti star is not easy to conduct. First, it requires frequent observations of the star. In order to construct a good light curve at least twenty observations per period are needed. Since periods as short as 43 min have been reported, this could require observations spaced by only two min. We note that some δ Scuti stars have periods reported as long as six hours. Obviously, for these stars this requirement is much easier to satisfy. For the shorter period stars this requirement restricts us to observing in a single color. Second, the small amplitude variation of these stars (in most instances less than 0.01 mag) requires good quality control which necessitates frequent observations of a comparison star, a check star, and from most observing sites, the sky. Third, the requirement that is most often violated is that many nights of observation be obtained. As mentioned above it has been suggested that at least 30 nights per season are required in order to permit an adequate frequency analysis. Despite this proviso, a large number of papers have been published based on ten or fewer nights which purport to present a reliable or adequate frequency analysis of a star. Fourth, even if thirty nights of observations and an adequate frequency solution are obtained in one season, this solution may not be valid for predicting the future behavior of a star. In order to determine if the period structure is stable with time, it is necessary to repeat the observational process during a second season in order to refine the periods determined and to enable us to maintain season−to−season cycle counts. Finally, the behavior of the star during a third season should be predicted and observations obtained to test the prediction. Thus, we see that at least three extensive seasons of observations are required on one star in order to present a convincing case that the variation is truly periodic.

In light of the preceding discussion, a number of advantages of APTs become apparent for working on δ Scuti stars. Of course, APTs permit a relatively rapid observing sequence although probably not better than that which can be obtained by a skilled photometrist. More importantly, APTs do not get tired and don't require breaks for food, drink, or any other purpose. Finally, APTs don't get bored with the "assembly line" nature of the described observing program.

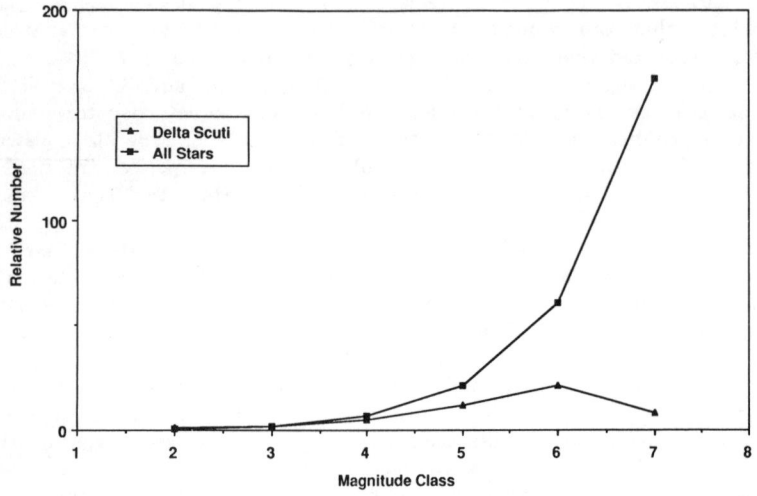

Figure 3. Plot of the number of δ Scuti stars and number of all stars as a function of apparent magnitude. Magnitude scale is divided into bins of width one magnitude centered on the labeled magnitude. Numbers have been normalized for each category so that there is one star brighter than third magnitude.

Summarizing, there are a number of points which should be considered in selecting a δ Scuti star as a program star for observation with an APT. The ideal star will be relatively bright (but not too bright), have a large amplitude and relatively long period of variation, be multi−periodic, and be well situated in the sky. Ideally, it should be one which has not been previously observed, but ours is not a program of discovery. The requirements regarding amplitude and period are relatively important and are related. Stars of low amplitude require numerous, very precise observations in order to adequately define the variation. If the period is short, the frequency of observations necessary is very difficult to meet, even with an APT. Therefore, it is better to observe the larger amplitude stars. (Note that no δ Scuti star has a really large amplitude compared with the more familiar types of variables).

TABLE I

m (1)	# δ Scuti (2)	# All Stars (3)	Normalized (4)	(5)	Predicted (6)	Remaining (7)
0	0	9		0.096		
1	0	13		0.14		
2	2	71	1	0.76	2	0
3	4	192	2	2.1	4	0
4	10	625	5	6.7	13	3
5	24	1963	12	21	42	18
6	42	5606	21	60	120	78
7	16	15565	8	170	335	324

    It is also advantageous to observe a star that is visible for as much of a given night as possible and for as much of the year as possible. For northern observers, this means observing a star north of the equator. There has never been an observing program as ambitious as the one outlined above carried out for a δ Scuti star, although a number of such stars have been fairly extensively observed. By judiciously selecting a program star and observing it systematically as often as possible with an APT, it should be possible to provide a definitive answer to the question: "Are δ Scuti stars strictly periodic?"

    I would like to acknowledge the assistance of my students Alan Johnson, Timothy Smith, and Michael Francisco with the preparation of the δ Scuti database mentioned above and my colleague William Kubinec for many helpful discussions concerning approaches to take in observing these stars. Part of this work was supported by a College Grant from the College of Charleston Faculty Research and Development Committee.

REFERENCES

Baglin, A., Breger, M., Chevalier, C., Hauck, B., le Contel, J. M., Sareyan, J. P., and Valtier, J. C. 1973, *Astron. Astrophys.*, 23, 221.

Breger, M. 1979, *Pub. Astron. Soc. Pacific*, 91, 5.

Breger, M. 1980, *Space Sci. Rev.*, 27, 361.

Dukes, R. J., and Adelman, S. J. 1986, in *Automatic Photoelectric Telescopes*, ed. D. S. Hall, R. M. Genet and B. Thurston (Mesa: Fairborn Press), p. 65.

Fitch, W. S. 1976, in *Multiple Periodic Variable Stars*, ed. W. S. Fitch (Dordrecht: D. Reidel), p. 167.

Hirshfeld, A. and Sinnot, R. W. 1982, *Sky Catalogue 2000.0 Volume 1* (Cambridge: Cambridge University Press), p. ix.

LeContel, J. M., Valtier, J. C., Sareyan, J. P., Baglin, A., and Zribi, G. 1974, *Astron. Astrophys. Suppl.*, 15, 115.

Wolff, S. C. 1983, *The A–Stars: Problems and Perspectives*, NASA SP–463 (Washington: NASA and Paris: CNRS).

THE CASE FOR PHOTOMETRY OF NEARBY STARS

Arthur R. Upgren

Van Vleck Observatory, Wesleyan University

Absolute photometry of reasonably good quality has seldom been extended to complete or even representative unbiased samples of stars. Far from being routine or makework projects, such systematic collection of data can often be used to provide new insights into the stars, given an adequate experimental design. The nearest solar neighbors represent a case in point and the case is made here for photometry of well−defined groups of stars.

For years, the most complete and representative survey of faint low−mass dwarf stars has been the survey made by Vyssotsky (1963) and his colleagues at the McCormick Observatory. The concept underlying the formation of this survey is that proper−motion surveys naturally incorporate a bias toward high−velocity stars and are therefore not kinematically representative of the galactic disk population in general. Since its completion, it and an extension made to cover the deep southern sky (Upgren et al. 1972) have proved very useful in many kinematical analyses. The principal (unavoidable) constraint of these objective−prism surveys is that they are limited in apparent magnitude and the limit introduces a small selection effect of its own. The only other major sources for either a complete or a representative sample of faint dwarfs are the distance—limited catalogues of nearby stars by Gliese (1969, with a supplement by Gliese and Jahreiss 1979) and Woolley et al. (1970). But these must necessarily be compilations from many sources since distance is not a directly observable parameter. They incorporate whatever gaps exist in the literature in the basic stellar parameters such as magnitude and color.

Now a new objective—prism survey of dwarf K and M stars has been completed and published by Stephenson (1986a,b) using plates taken with the Case Schmidt telescope. Although constrained to galactic latitudes of more than ten degrees on either side of the plane, it covers the remaining sky north of Dec. $-25°$ using objective—prism material far superior in quality to that available to Vyssotsky. Stephenson's two lists include some 4000 stars as opposed to 900 in the five McCormick lists. This suggests that Stephenson was able to detect and distinguish nearby K and M dwarfs to a fainter limiting magnitude, or that his stars cover a wider range of spectral classes.

Since the McCormick and Case investigations cover about the same area of sky (apart from the galactic plane) they provide, in theory, the means for the determination of the degree of incompleteness of each one separately as well as the number of stars missed by both. In practice, the magnitudes and spectral classes are not easy to compare since they are understandably not of high quality. Vyssotsky's magnitudes have been calibrated against the photoelectric V magnitude; this and the distribution in magnitude for his stars with and without photometry have recently been published (Upgren and Lu 1986) and show a small systematic deviation and a large individual scatter. The spectral types which he assigned have been calibrated against MK spectral classes (Upgren et al. 1972) and are systematically later than the MK types with a large scatter. The means and dispersions of the McCormick types are also known in terms of the $B-V$ and $R-I$ color indices as shown in Table I where the $R-I$ colors are on the Kron system. The last four columns give the mean color index and standard error for an individual star for each type and color. It is noteworthy that the M0 stars classed by Vyssotsky as peculiar are somewhat bluer than the normal M0 stars in each color.

Calibration of the Case data is currently underway at the Van Vleck Observatory and also at the Astronomisches Rechen—Institut. For this purpose, Upgren and Weis have obtained photoelectric observations of a large number of Stephenson's stars and will combine these data with photometry from other sources which happen to be available. Nevertheless, without photometric data for all stars in both lists, the comparison between the Vyssotsky and Stephenson lists is difficult at best. This is mostly due to the lack of equivalence of the magnitudes and spectral types for the majority of the stars which do not now have photometry.

TABLE I

Colors of McCormick Stars

McCormick Spectral Class	Total	Number of stars with B−V	R−I	<B−V>	s.e.	<R−I>	s.e.
K5	31	11	12	1.00	0.09	0.42	0.05
K8	269	151	157	1.13	0.11	0.45	0.10
M0	380	142	119	1.36	0.12	0.65	0.16
M0p	42	38	36	1.31	0.10	0.59	0.16
M2	95	82	82	1.47	0.08	0.84	0.16
M5	8	8	8	1.64	0.15	1.12	0.15

A first preliminary comparison of the two lists has been made. If we consider the ranges in apparent magnitude and spectral class within which both surveys are most likely to be complete; that is, the stars least likely to have been missed by either survey, we can estimate the degree of completeness of both of them separately and together. Vyssotsky gave classifications of either K8, M0 or M2 to almost all of his stars and the distribution in magnitudes of his stars shows a pronounced peak at 10.5 mag. indicating that his survey is not complete for stars fainter than this. We shall consider the region defined by m <= 10.5 mag. and sp >= M0 as the optimum region for his survey. The equivalent region for the Stephenson stars is roughly set by m <= 10.5 mag. and sp >= K7. In this region we find that Vyssotsky found 120 of the 190 Stephenson stars or 63% and Stephenson found 230 of the 302 Vyssotsky stars or 76%. Based only upon these percentages, a conclusion could be made that of all stars which fulfill these conditions in magnitude and spectral class, 48% were found by both investigators, 28% by Stephenson alone, 15% by Vyssotsky alone and 9% by neither. No attempt need be made to set an upper limit to spectral class because almost no dwarf stars much later than M0 are as bright as 10.5 mag. But if earlier spectral types or fainter stars are considered, the completeness of either source is likely to diminish rapidly.

These figures are very disturbing if they can be believed. They would suggest that stellar densities in the solar neighborhood should be increased by some 60% over densities derived from the Vyssotsky data for the dK−M stars. Such an increase in the stellar luminosity function in the range of absolute visual magnitude between about +7 to +10 mag. over that of Wielen (1974) based on Gliese's catalog of nearby stars, is scarcely credible. There are, however, several reasons for regarding the results with suspicion. First and most obvious is the probable non−

equivalence of the spectral classifications of the two surveys, no matter how they may be defined. This is evident from the difference in the numbers of stars from each survey quoted above. Second is the fact that the magnitude cutoff abruptly at 10.5 mag. introduces a systematic error, especially given the large scatter in the magnitudes of the individual stars and the systematic error in the mean magnitudes. These errors are known for the Vyssotsky sample; the difference between his magnitudes and photoelectric V magnitudes is given by $(m - V) = 0.63 - 0.07V$ with a dispersion for an individual star of 0.4 mag (Upgren and Lu 1986). The equivalent figures for the Stephenson sample are not yet available, but from a preliminary examination it appears that the individual scatter is of about the same size and that the systematic deviations are larger.

A possible third reason for the large percentages given above may derive from the nature of the observations used. The Vyssotsky survey was made from searches of plates centered in the blue region of the spectrum where criteria are readily available for luminosity classification in the K and M spectral regions and giants and dwarfs can be easily distinguished. The principal criterion is the appearance of the so-called Lindblad depression, a suppression of the continuum between the calcium line at $\lambda 4227$ and the G band (Vyssotsky 1963). The relative intensity of this feature compared to the cyanogen band immediately to its shortwave side is quite reliable, even at very low dispersion. The material available for the Case survey is of superior quality but is of 103a−F emulsion extending across the visual portion of the spectrum. Whereas the dwarf nature of virtually all of the stars in the Vyssotsky lists have stood the test of time, some preliminary indication has been found that some of the stars in the latter survey are actually luminous evolved stars. Some yet unknown part of the discrepancy between the two sources may be due to unrecognized giants in the sample.

Although a more careful and detailed analysis of the data may well improve the confidence that may be placed in the results, the very poor magnitudes and spectral types still limit their validity. All of the problems arising from the use of the two studies to obtain a valid estimate of the number of yet unrecognized nearby $K-M$ dwarfs could be eliminated or at least greatly reduced if photometry were available for the stars. The V magnitudes and the $B-V$ and $R-I$ colors could lead to an order−of−magnitude reduction in the uncertainties involved in the definition of the regions of the apparent magnitude−spectral class plane common to both lists. Furthermore, broad−band photometry in $BVRI$ colors can be used to separate the evolved stars of the late spectral classes from those on the main sequence.

A second closely related problem involving the stars in the Vyssotsky and Gliese catalogs has been discussed by Upgren and Lu (1986). That paper describes a program now underway which will obtain photometry in $BVRI$ colors of all of the remaining stars in the Vyssotsky lists as well as the stars in the Gliese catalog which are either brighter than $M_v = +9$ mag. or closer than parallax = +0."080 arcsec. The reasons for these limits are discussed in the paper and also in more detail by Gliese, Jahreiss and Upgren (1986). The limits are established by the fact that the brighter stars are statistically complete to the limiting distance of the catalog of 22 pc, whereas stars fainter than this magnitude limit are complete only to the distance af approximately 12.5 pc, as shown by several statistical tests.

The advantage in photometry of all of the stars lies in the accurate placement of each of them on the main sequence (or above it for any dwarfs that turn out to be giants or subgiants). The photoelectric color indices have the double advantage of avoiding the confusion between the several spectral classification schemes used for K and M dwarfs, and the reduction of the error in the abscissa of the HR or color–magnitude diagram to well below the bin size into which the stars are divided in the analysis. This advantage will be useful for comparing the magnitude–limited McCormick stars with the distance–limited Gliese stars in order to evaluate and remove the systematic errors inherent in each. It will also help to evaluate the degree to which subsets of either group (such as the stars with parallaxes and proper motions) are representative of the entire group.

Any observatory which has the capacity to obtain absolute photometry as well as differential photometry with reasonable precision, has a valuable contribution to make to the distributions and the physical and kinematical properties of stars of many kinds. This paper is an attempt to illustrate some of those contributions.

ACKNOWLEDGMENT

This research is supported by Research Grants AST−8318649 and AST−8610424 from the National Science Foundation.

REFERENCES

Gliese, W. 1969, *Veroff. Astron. Inst. Heidelberg*, No. 22.
Gliese, W. and Jahreiss, H. 1979, *Astron. Astrophys. Suppl.* 38l, 423.
Gliese, W., Jahreiss, H. and Upgren, A. R. 1986, *The Galaxy and the Solar System*, ed. R. Smoluchowski, J. N. Bahcall and M. S. Matthews, (Tucson: Univ. of Arizona Press), p. 13.
Stephenson, C. B. 1986a, *Astron. J.* 91, 144.
Stephenson, C. B. 1986b, *Astron. J.* 92, 139.
Upgren, A. R. , Grossenbacher, R., Penhallow, W. S., MacConnell, D. J. and Frye, R. L. 1972, *Astron. J.* 77, 486.
Upgren, A. R. and Lu, P. K. 1986, *Astron. J.* 92, 903.
Vyssotsky, A. N. 1963, *Stars and Stellar Systems*, Vol. III, ed. K.Aa. Strand, (Chicago: Univ. of Chicago Press), p. 192.
Wielen, R . 1974 *Highlights of Astronomy 3*, 395.
Woolley, R., Epps, E. A. Penston, M. J. and Pocock, S. B. 1970, *Royal Obs. Ann.* No. 5.

SECTION 2a—INTRODUCTION: AUTOMATED SPECTROPHOTOMETRY

Donald S. Hayes

Fairborn Observatory

Low—resolution spectrophotometry is a branch of photometry in which a relative or absolute energy distribution is measured with a resolution (1.0–100A) much higher than the bandpasses of typical filter photometry (150–1000A), but much lower than is typically used in spectroscopy (<1.0A). At present, spectrophotometry is most commonly done using some type of array (intensified or "bare" reticon; CCD) or scanning detector (image—dissector scanner) on a grating spectrograph; in the recent past, devices which used a photomultiplier as a detector and which scanned the spectrum by rotating the grating were used. In the latter case, some regions of the spectrum were skipped to save time; it is most common now to produce a continuous spectrum from which the less useful portions are eliminated in the analysis phase. The wavelength coverage is generally a subregion of the range from 3200A to about 11000A. As opposed to spectroscopy, spectrophotometry is done without a slit, in order to preserve information on the relative flux as a function of wavelength, and to gain efficiency. The use of an aperture, instead of a slit, results in a loss of resolution because of "seeing", image "blowup" and image motion due to tracking errors.

An advantage of spectrophotometry with respect to filter photometry is the ability to measure strong lines as well as slopes of the continuum; it is also possible to design "photometric systems" to be applied to classes of special objects which are not well measured by the standard filter systems. At the same time, it is possible to synthesize conventional filter photometry from the energy distributions. An advantage of spectrophotometry over spectroscopy is the higher sensitivity due to the lower resolution and the lack of loss of light at a slit (assuming the same detector is used). The higher sensitivity makes possible the measurement of more stars in a given time or fainter stars in the same time. Another advantage is the preservation of photometric information, which allows determining effective temperatures, surface gravities, and metallicities from the slopes and gross features of the spectrum.

Spectrophotometry is generally applied to specialized projects, so there has been little true standardization in the field. Most of the spectrophotometry found in the literature can be characterized as having a resolution of 10−50A, a wavelength coverage of a subregion of the wavelength range 3500−8000A, and involving a survey of a number of objects in some more−or−less well−defined class. The energy distributions are calibrated by reference to the absolute calibration of Vega by Hayes and Latham (1975) *via* secondary standards published by Taylor (1984), which supersedes those published by Breger (1976), and fainter secondary standards by Oke and Gunn (1983). A particular disadvantage of the calibration and brighter secondary standards discussed above is that they are given at a limited list of bandpasses which are intended to sample the continuum of early−type stars; a list of wavelengths derived from that published by Oke (1964) is used. In this age of continuous spectra derived with array detectors, a calibration of Vega and energy distributions of secondary standards which are continuous is badly needed. The continuous calibration of Vega has been addressed by Hayes (1985); the continuous calibration of the secondary standards has been addressed by Hayes, Barnes and Jacoby (1987) at the Kitt Peak National Observatory; a list of 45 secondary standard stars ranging from Vega, at the bright end, to 12th mag, at the faint end, has been observed. Most of Taylor's secondary standards, all of the secondary standards by Oke and Gunn, plus a number of other early−type stars are included.

Although the use of array detectors has improved the efficiency of spectrophotometric observations, it is still true that acquiring the additional information provided over conventional photometry takes more time. A consequence of this is that larger telescopes must be used for observations of stars of a given brightness than for filter photometry. In addition, spectrophotometry, like photometry, often requires surveys of many stars, or requires monitoring a variable star over an extended period of time. As a result of these characteristics, spectrophotometry tends to be very routine and requires extended observing runs. The increasing demands upon the time of the investigators, plus the increasing difficulty of getting extended observing runs at the national observatories, means that an alternate mode of observing would be welcomed. Such an alternate mode of observing is provided by fully automatic telescopes. Examples of such telescopes are provided by the automatic photoelectric telescopes (APTs) developed by the Fairborn Observatory; they are made available to all observers by the Automatic Photoelectric Telescope Service (APT Service; Boyd, Genet and Hall 1985 a, b; Boyd, Genet and Baliunas 1986).

So far, the APTs have only done differential filter photometry of variable stars, but they are operated in a mode which allows extending their program to include all-sky photometry without great effort (Hayes and Crawford 1987; Hayes 1986). It can be expected that full all-sky operation will be achieved in 1987. A 0.75-m telescope is currently under development by the Fairborn Observatory; additional telescopes in this aperture range can be expected to be built in the future.

The session on "Automatic Spectrophotometry" presented at this Symposium was organized to explore the idea of constructing an Automatic Spectrophotometric Telescope (ASPT), to be operated by the APT Service for a consortium of astronomers at several institutions. This consortium is, as yet, only an informal idea being discussed by a few of us; it eventually would include other interested investigators. Our intention is to produce a proposal whose purpose is to attract participants and funding for the project.

A few people were invited to this Symposium to present preliminary ideas about both the equipment and the scientific projects which could be tackled by an ASPT. Papers which describe the ideas contained in the oral presentations are given in the section which follows. Included are papers which describe some scientific objectives and their implications in the design of the equipment (Crawford; Adelman; Philip), some approaches to the design of a spectrophotometer for use on a small automatic or semi-automatic telescope (Hayes; White), and an algorithm for use in the reduction of spectrophotometric data (Hayes and Schmidtke).

REFERENCES

Boyd, L. J., Genet, R. M. and Hall, D.S. 1985a, *IAPPP Comm*, No 19, 1.
Boyd, L. J., Genet, R. M. and Hall, D.S. 1985b, *IAPPP Comm*, No 12, 59.
Breger, M. 1976 *Astrophys. J. Suppl.* 32, 1.
Genet, R. M., Boyd, L. J., and Baliunas, S. L. 1986 in: *Automatic Photoelectric Telescopes*, ed. D. S. Hall, R. M. Genet and B. L. Thurston (Mesa: Fairborn Press), p. 15.
Hayes, D. S. 1985 in: *IAU Symp. No. 111: Calibration of Fundamental Stellar Quantities*, ed. D.S. Hayes, L.E. Pasinetti and A.G. Davis Philip (Dordrecht: Reidel), p. 225.
Hayes, D. S 1986, *Precision Photometry of Bright Stars*, proposal submitted to the National Science Foundation, June, 1986.
Hayes, D. S. and Latham, D. W. 1975, *Astrophys. J.* 197, 593.
Hayes, D.S. and Crawford, D.L. 1986 in: *Automatic Photoelectric Telescopes*, ed. D. S. Hall, R. M. Genet and B. L. Thurston (Mesa: Fairborn Press), p. 87.
Hayes, D. S., Barnes, J. V. and Jacoby, G. 1987, in prep.
Oke, J. B. 1984, *Astrophys. J.* 140, 689.
Oke, J. B. and Gunn, J. 1983 *Astrophys. J.* 266, 713.
Taylor, B. J. 1984 *Astrophys. J. Suppl.* 54, 259.

Russell Genet and Frank Bradshaw Wood discuss the possibilities of larger automatic telescopes at the south pole.

SYNTHETIC PHOTOMETRY: MODELING A PHOTOMETRIC SYSTEM

David L. Crawford

Kitt Peak National Observatory
National Optical Astronomy Observatories[*]

ABSTRACT: This paper looks at the question of synthetic photometry from the position of modeling a photometric system, given the bandpasses that define the system and some source of stellar "fluxes." A number of insights can be gained into the nature of photometry this way, and to the characteristics of photometric systems. Further extensions of the approach will use more realistic bandpasses for the systems, more realistic stellar fluxes, and other terms in the integration equation to allow for atmospheric extinction, interstellar reddening, and so forth. The current study is a simple approach, designed to illustrate the concepts involved, and to look at some of the simple questions.

---

[*] Operated by the Association of Universities for Research in Astronomy, Inc. under contract with the National Science Foundation.

New Generation Small Telescopes, ed. D. S. Hayes, R. M. Genet, & D. R. Genet.
© 1987 Fairborn Observatory.

I. INTRODUCTION

There have been a number of previous studies of synthetic photometry, and even a conference on the subject. Let me reference only the latter and one other paper, as these two give many additional references to the topic, and well illustrate many of the concepts involved: *Theoretical uvby H−beta Indices*, by J. B. Lester, R. O. Gray, and R. L. Kurucz (1986); and the IAU Joint Commission Meeting on *Synthetic Photometry*, edited by Roland Buser (1986).

In this paper, I take a very simple-minded approach, to illustrate the issues involved. Even such a simple approach gives quite a bit of insight into the nature of synthetic photometry, and it offers some good understanding of photometric systems and the modeling thereof.

Such synthetic photometry is nearly identical to that which is and will be used to produce comparisons between spectrophotometric data (especially that at moderate to low resolution) and current photometric systems. We can think of an $I(\lambda)$ machine, which produces flux measurements for every wavelength, with a moderate spectral resolution. An example would be a "monochromatic" magnitude for each 10A wavelength bin from 3000A to 6000A. We want then to compare such measures with what we would observe with standard filters systems, such as the *UBV* or *uvby* system.

It matters little whether such fluxes are observed values, or values produced from theoretical models, or by some other technique. We integrate (convolute) the values along with the system bandpasses (and any other limiting factors) to estimate what we would observe with a photometric system. The system can be an existing one, or one we are investigating for possible use.

II. THE PROCEDURE

The flux that we are measuring at the telescope with our photometer can be thought of as the result of the following integration:

$$I(\lambda) = \int C_i(\lambda) * F_j(\lambda) \, d\lambda$$

if we use only those two terms for our initial investigation. Here C_i represents the different stellar flux distributions and F_j represents the photometric system bandpasses.

Other terms should be used, of course, for any detailed work such as one to represent atmospheric extinction, one for the detector sensitivity, *etc.* We will add more such terms in our upcoming papers on synthetic photometry, but these two are sufficient for a first look at the issue.

We may think of **Wide−band Photometry** as that in which the F_j term is of the order of 500A to 1000A wide, of **Intermediate−band**

Photometry when the term is about 200A to 300A wide, and **Narrow-band Photometry** when the term is approximately 10A to 50A wide.

We will limit ourselves, in this first approach, to the wavelength region from 3000A to 6000A, as that is sufficient to look at both the *UBV* and the *uvby* systems. We will use 20 points of 50A resolution each to represent our system bandpasses. All integrations will cover 1000A with 50A resolution no matter how wide the filter. The stellar fluxes used represent 60 points, with 50A resolution covering the range 3000A to 6000A. They have been taken from a paper by Straizvys and Sviderskiene (1972). Several of the energy distributions are shown in Figures 1 and 2, as examples. It is very useful to visually inspect the energy distributions when looking at output from integrations such as done here.

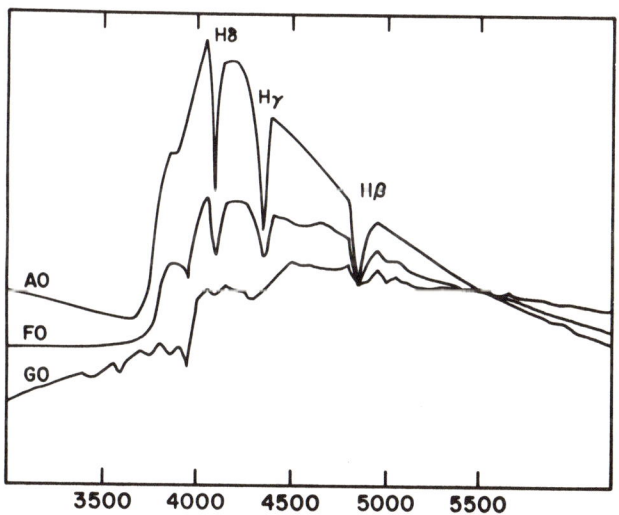

Figure 1. Energy distributions for the early type stars.

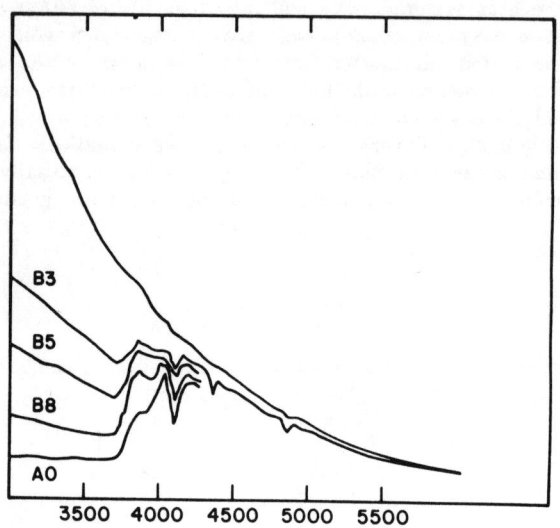

Figure 2. Energy distributions for the later type stars.

We used a number of simple "filter" bandpasses to represent the photometric systems. Naturally, a more detailed investigation should use more realistic filter curves. The ones used here are adequate to illustrate the techniques. The "filters" are:

1. Square band pass   1000A wide.
2. Square band pass   500A wide.
3. Square band pass   300A wide.
4. Square band pass   200A wide.
5. Square band pass   100A wide.
6. Monochromatic   "0"A wide.
7. Bell shaped   800A FWHM. ("Full width, half maximum")
8. Bell shaped   200A FWHM.
9. Gaussian shape   300A FWHM.

Two narrow-band systems were looked at, one comparing a filter of type #6 with two filters of type #6 50A on each side of the central one, and one comparing a filter of type #6 with one of type #5 centered at the same wavelength.

For comparison with *UBV* photometry, we generally used filters of type #7, but also looked at ones of type #1 and #2. For band−width comparisons, we used filters of types #1, 2, 3, 4, 5, and 6. For comparison with *uvby* photometry, we used filters of type #3 for *u*, and of type #4 for *v*, *b*, and *y*. We also compared using filters of type #8, and checked for band−width effects with filters of type #5 and #6. Occasional comparisons were made using filter # 9 as well.

Given these distributions for C_i and F_j, it is easy to do an integration. A simple computer program was written in BASIC to do the data handling, the integration, and cycling to cover the various filters and stellar distributions. The program then calculates the "observed" magnitude, as:

mag for star i and filter j = -2.5 * log [$I_{i,j}(\lambda)$].

Tables of the V magnitude and *(B−V)* and *(U−B)* were output, as well as coding for star and filter. Similar outputs were done for the uvby photometry, and for H β indices.

We also calculated effective wavelengths for each "filter." Here a lambda term was added to an integration, and that integration divided by one discussed above. This way we could see shifts in effective wavelength for any star and any bandwidth.

Naturally, even with this simple−minded approach, there is lots of output. We will discuss some of it in the next section, chosen to illustrate some of the concepts. It is a great way to learn some of the concepts of photometry, and to understand photometric systems. An approach like this will be used as the core of on upcoming book on Astronomical Photometry, by Crawford and Hayes.

III. SOME RESULTS

Most of the filters where used to calculate "UBV" indices, even the narrow filters, as such data helps one understand band−width effects on observed indices. In addition to the normal central wavelengths of the filters, each was run with some decentering, so as to investigate the effects of "non−standard" filters. The same was done for the *uvby* indices. (I put the system in quotes, for many of the outputs are for "filters" that are not at all like the standard ones, nor are the central wavelengths.)

Figure 3 shows the results of using the 800A−wide bell−shaped filter, centered at 3500A, 4500A, and 5500A, the nominal centers of the *UBV* filters. The *(U−B)* vs. *(B−V)* plot looks quite like the observational results, of course. Table I shows the results of transforming a few of the "non-standard" filters to the system defined by the "filters" used to produce the data shown in Figure 3.

TABLE I: "UBV" TRANSFORMATION RESULTS.

Filter set	Scale term	Standard error* of transformation	
For an 800A wide bell-shaped filter:			
4450/5500	0.928	5.0	(Quite good, of course)
4400/5500	0.874	4.9	(Both < 1.0, as expected)
4500/5400	1.101	3.1	(Shifts in V less critical)
3500/4400	1.054	31	(Much larger than in $B-V$)
3500/4450	1.026	18	(Better, but worse than $B-V$)
3600/4500	1.118	40	(U shift worse than B shift)
For a 200A wide bell-shaped filter.			
3500/4400	1.084	16	(Larger coeff, smaller scatter)
3600/4500	1.100	35	(U shift much worse)
4400/5500	0.864	24	(! Due to H γ effect)
4600/5500	1.164	4.5	

Careful study of such synthetic photometry (modeling) can help one understand photometric systems much better, and help see why one sometimes gets funny results with non-standard filters.

Figure 4 shows plots of the $uvby$ indices, calculated using a 300A wide square-top filter for u, and 200A wide square-top filters for v, b, and y. Figure 5 shows the $(u-b)$ vs. $(b-y)$ plot for the same filters, and Figure 6 shows the $(v-b)$ and $(u-v)$ plots vs. $(b-y)$. In each of these, I used a central wavelength of 4200A for v so as to minimize the adverse effects of H δ on the indices.

---

* The std errors are in units of 0.001 mag.

SYNTHETIC PHOTOMETRY 151

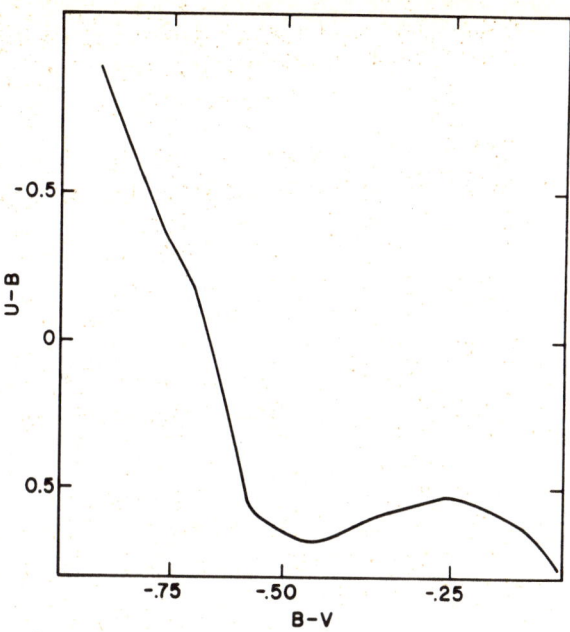

Figure 3. A $(B-V)$ vs. $(U-B)$ diagram from the computer output.

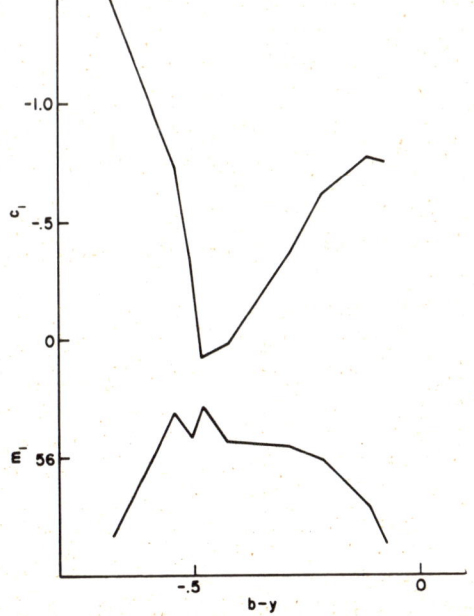

Figure 4. The c_1 and m_1 vs. $(b-y)$ diagrams.

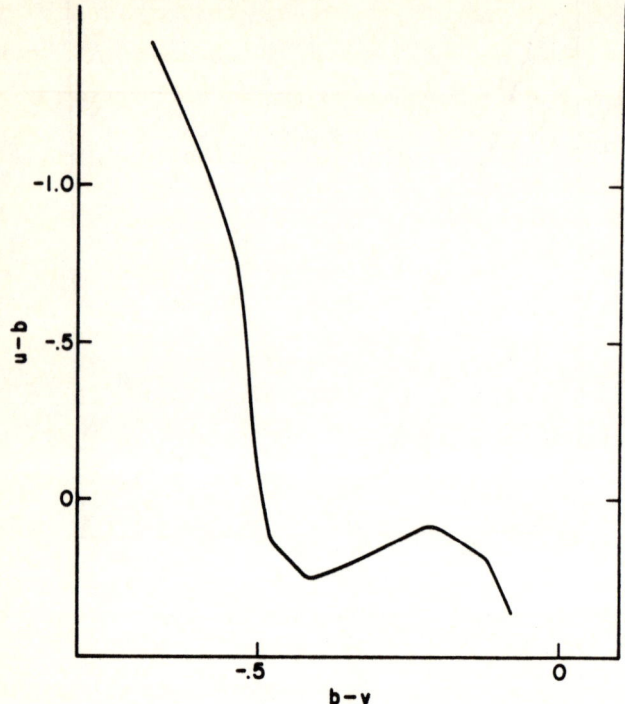

Figure 5. The $(u-b)$ vs. $(b-y)$ diagram.

In Figure 7, I show the synthetic Hδ narrow band indices. The top plot uses monochromatic bands at 4050A and at 4150A to compare with the intensity at 4100A, while the bottom plot uses the sum of the intensities at 4050, 4100, and 4150A to compare to the intensity at 4100A. These are two different observational ways to measure the intensity of the Hδ line. One would use a similar approach to measure Hγ, Hβ, or Hα, and such has in fact been done by observers.

Finally, in Figure 8 I, show the effect of bandwidth difference on the effective wavelength of the U filter bandpass of the *UBV* system. The plot shows the shift in effective wavelength, as a function of spectral type, for square bandpass filters of width 1000A, 500A, and 100A. For the latter, the shifts are less than 2A, while for the former the shifts are as large as 70A.

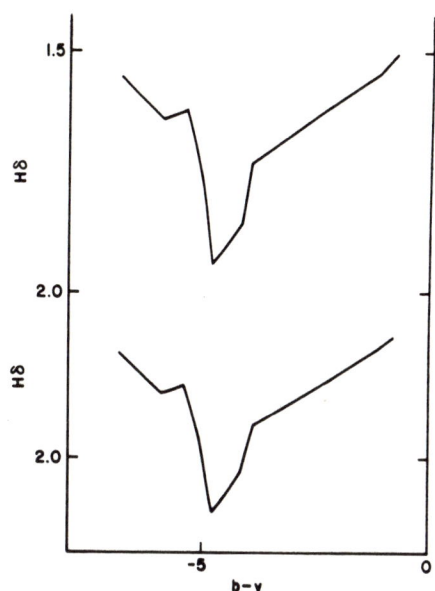

Figure 6. Two different H δ indices vs. the (b−y) index.

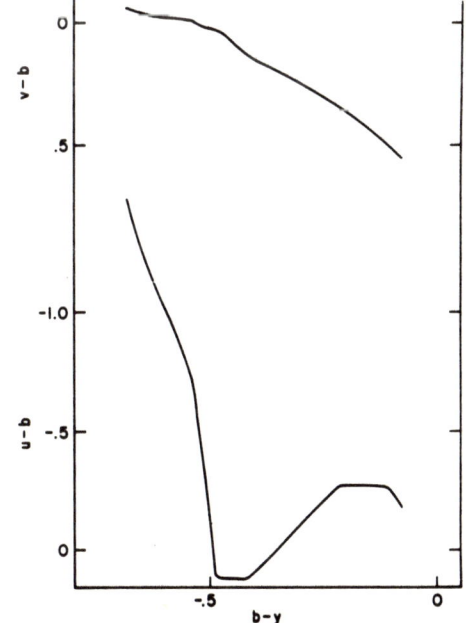

Figure 7. Band width effects on the U magnitude of the UBV set.

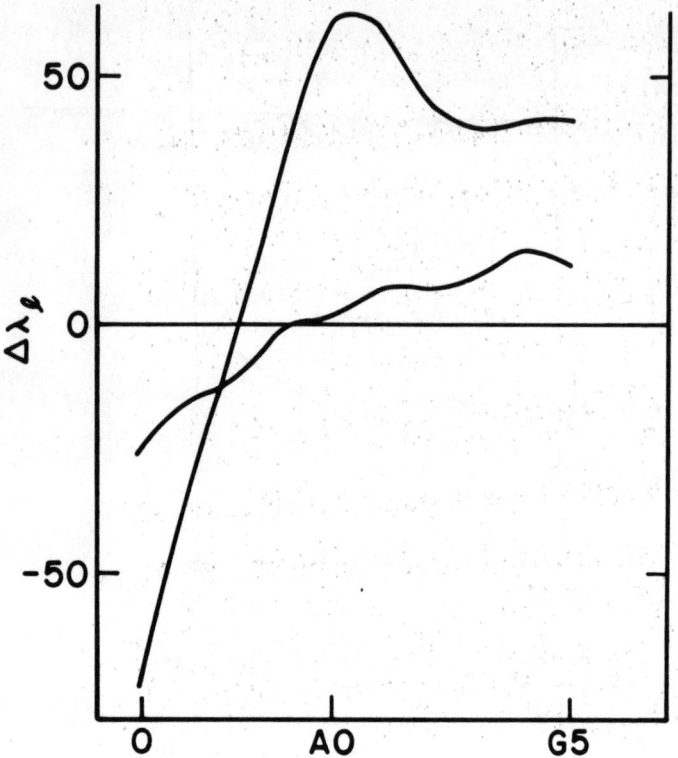

Figure 8. The effect of bandwidth difference on the effective wavelength of the U-filter bandpass of the UBV system.

I have barely scratched the surface of what can be seen and done with synthetic photometry, even using very simple assumptions, to look at characteristics of photometric systems. I urge everyone interested in doing photometry to spend some time doing similar inspections.

IV. SUMMARY

The techniques of synthetic photometry, even in a very simple-minded approach, are very helpful in illustrating and understanding some of the basics of photometry, and of understanding the characteristics of photometric systems. When done in more full-blown glory, a deep insight can be gained into the nature of any photometric system, into the how photometry can and should be done, and of how many mistakes have arisen in the past by lack of such insight. In addition, such techniques can be used to great advantage in investigating and designing future new photometric systems, or variations on existing systems. Synthetic photometry, of course, is how one will handle the output from $I(\lambda)$ machines (moderate resolution, accurate spectrometers of the present and future), to summarize and use the data for astronomy and astrophysics. It is also the way to compare output from theoretical models with observations.

We plan, in the near future, to elaborate the concept of this paper, with at least the following improvements (note that even the simple approach of this paper can offer great insight into what is going on with photometric systems !):

1. Better stellar fluxes, and at higher spectral resolution.
2. Such fluxes for other type stars, including:
   a. Metallic line stars.
   b. Weak line stars.
   c. Reddened O− and B−type stars.
   d. Double stars.
   e. Supergiants.
   f. Emission line stars.
   g. Others of special interest.
3. Additional filter curves, including asymmetric ones.
4. Terms for at least:
   a. Interstellar extinction.
   b. Atmospheric extinction.
   c. The telescope.
   d. The detector.
5. Tests of effects of radial velocity shifts on Narrow Band Photometry.
6. An R band; others as well.
7. Better resolution on the input, so as to look at Narrow Band Photometry.
8. Comparison to actual measurements, in detail.

The overall goals are:

1. Understanding filter and photometric systems.
2. Understanding the process of photometry.
3. Use with $I(\lambda)$ machines.
4. Possible tuning up of present systems and development of any needed new systems.
5. All this means: better astronomy, with more understanding and less pain.

REFERENCES

Buser, R. (ed) 1986 IAU Joint Meeting on Synthetic Photometry, in *Highlights of Astronomy* Vol. 7, ed. J.-P. Swings (Dordrecht: D. Reidel), pp 799−843.

Lester, J. B., Gray, R. O., and Kurucz, R. L. 1986 *Astrophys. J. Suppl.* 61, 509.

Straizvys, V. and Sviderskienė, Z. *1972 Bull. Vilnius Astron. Obs.*, No. 35, 3.

SOME THOUGHTS ABOUT THE NEXT GENERATION OF SPECTROPHOTOMETRIC INSTRUMENTS

Saul J. Adelman

Department of Physics, The Citadel

ABSTRACT. Some observational and astrophysical considerations are discussed which impact the design and operation of an Automated Spectrophotometric Telescope.

I. INTRODUCTION

Spectrophotometric instruments measure the absolute or relative energy distributions of astronomical objects with resolutions of a few to 100A. The bandpasses are thus intermediate between those of filter photometry and high dispersion spectroscopy. To measure the relative flux as a function of wavelength requires a slitless mode of operation. In the optical region such observations have been fundamental to determinations of stellar effective temperatures and surface gravities.

Within the last decade, Kitt Peak National Observatory, Mt. Wilson Observatory, and Palomar Observatory have retired their Cassegrain single− or dual−channel photomultiplier rotating−grating scanning spectrophotometers. Similar instruments are still in operation at other observatories and are performing useful scientific measurements. Multichannel spectrophotometers using some type of array detector have been placed into operation at several observatories. But these instruments have not been as accurate and/or have had restricted wavelength ranges. However, it is now possible to construct multichannel spectrophotometric instruments, which can observe the entire optical window, by employing high quantum efficiency detectors with the accuracy of photomultiplier tubes. Observing programs using these next generation instruments, while obtaining better accuracies and precisions, should also obtain considerably more data per night while avoiding some of the failings of the rotating− grating scanners, in particular the inability to adequately derive the extinction in the photometric sense without spending all night in making the necessary measurements.

II. OBSERVATIONAL CONSIDERATIONS

The advent and success of the Automated Photoelectric Telescopes (Boyd, Genet, and Baliunas 1986) suggests that the time has come for the application of automated telescopes and of automated instrumentation to a wider field of astronomical observations from the Earth's surface. The major advantages of an automated telescope−instrument combination are a lower cost of operation, more efficient data taking, and very high degree of consistency in how the data are obtained. The extension of APT type operations to the near infrared, the use of CCDs to replace photomultipliers as detectors on APTs, an automated spectrographic telescope (ASGT) operating at classification or higher dispersion, and an automated spectrophotometric telescope (ASPT) are all possibilities to be considered. Of these I believe that the ASPT is likely to have the greatest scientific impact in the next decade (see Hayes 1987 for a preliminary design of such an instrument).

Typically, grating scanners were mounted at the Cassegrain foci of 1− or 2−m class telescopes. These instruments incorporated heavy moderate dispersion spectrographs. If more compact instruments are designed with modern electronic detectors instead of photomultipliers, then a considerable reduction in instrumental mass can be expected. This would permit the use of less massive telescopes. But still, spectrophotometric instruments will be somewhat heavier than simple photometers. Although they could be mounted on smaller telescopes directly or coupled via fiber optic cables, the desirability of collecting as many photons as possible and thus the ability to observe as many objects as possible supports the use of 1–m class or larger telescopes. As the observations are going to be made near the center of the field of view, the optical quality of the images has to be high only for this region.

Based on my experience with the HCO scanner at Kitt Peak National Observatory, I anticipate that a next generation spectrophotometric instrument on a 1–m class telescope should easily obtain 10000 counts above background in 10 seconds for 10A wide resolution elements for a 5th magnitude A0 V star throughout most of the optical region. This indicates a bright limit of about 3rd magnitude. In about 40 minutes it would be possible to observe an 11th magnitude A0 V star. With detectors sensitive to cosmic rays it might be desirable to perform several shorter integrations. Fainter stars could be observed by degrading the resolution of the data, obtaining less accurate data, or integrating longer.

The total time to obtain the stellar fluxes also depends on how the sky measurements are made and whether the entire spectral region can be observed simultaneously. If a separate sky measurement is needed for each stellar measurement, then the above times have to be increased. Simultaneously obtaining the sky measurements can have its problems as known from two channel rotating–grating scanners. Nevertheless, a two–aperture twin detector instrument increases observing efficiency and the amount of data obtained. If the full spectral range is measured in two pieces, then the integration time could be greater than the above estimate. But if seeing fluctuations are not important in degrading the resolution and the number of photons received per second, then the observation times for the brightest stars can be as small as 1 second per exposure. As modern electronic detectors are more sensitive in the red than in the blue, the exposure time for the blue will be several times that for the red for most objects.

As much spectrophotometric work has been performed in an all–sky mode, this type of operation has to be provided. For APT operations, D. S. Hayes is implementing all–sky operation and any problems should be solved in a relatively short time. For many nights which are non–photometric, measurements of relative flux distributions will be possible with an ASPT.

Techniques will have to be devised for finding stars in relatively crowded fields. Observations will be limited by the aperture sizes and placement and the seeing to some extent. If the ASPT uses an intensified CCD camera for finding objects, then the ability to store and transmit the essence of such images would be desirable. When there were any questions, the astronomer could find out which object the telescope found and observed. If it was not the desired one, then instructions could be formulated to offset from the acquired image. The ability to accurately offset from some known object is also desirable for mapping out the energy distribution of an object with a finite extent on the sky such as a nebula or a galaxy.

With an intensified uncooled CCD camera for finding, the ASPT should be able to locate isolated 12th magnitude stars. This limit is not a major problem for most astronomers. But we can count on someone wanting to work on fainter objects. This could be done by cooling the CCD camera and/or using longer integration times.

The amount of data that an ASPT could generate is quite large. To slew to and to observe a 5th magnitude star is expected to take about 3 minutes. Thus 20 such observations are possible in a hour or about 160 per night. If there are 150 photometric nights per year, then about 24,000 such measurements per year are anticipated. This is many times the good spectrophotometric data in the literature. In fact, all accessible stars in the Yale Bright Star Catalog and its supplement could be observed several times in one year. In addition, it should be possible to obtain relative energy distributions on perhaps the equivalent of another 50 nights a year and obtain some 8,000 additional measurements. If the observing selection algorithm takes into account both the sky and stellar brightness, it should be possible to extend slightly the usable observing time by making measurements of the brightest accessible stars in twilight at the ends of the nights.

A major failing of rotating−grating scanner spectrophotometer reductions is the use of mean extinction coefficients. If one determined the extinction according to good photometric practice, it would have taken at least 75% of the observing time. At best with the HCO scanner on the KPNO #1 0.9−m telescope, I could obtain on the order of 40 scans per night of which typically 10 scans per night were of standards. If with the ASPT, 30% of the time were devoted to measuring the extinction and obtaining standard star measurements, then 48 scans per night would be used for these purposes. In an 8 hour night this is 6 scans per hour, which is the lower bound for good all−sky photometry. On those nights the APTs at the same site are performing all−sky photometry, the APT and ASPT observations could be combined to improve the extinction solution.

The adequacy, precision, and accuracy of existing secondary standard measurements for spectrophotometry is a major concern. Initially we will have to use such measurements. The secondary standard star measurements in the all−sky spectrophotometry mode will yield data which could be used to improve the relative flux of the secondary standards. In a comparatively short time there would be more such data than was used to establish the secondary standard fluxes. This consideration indicates we should revise the existing secondary standards as well as establish a sufficient number of additional standards during the first year of operation. Unfortunately Vega, the primary standard, is too bright to be observed without stopping down the aperture. But the need to adequately calibrate the standards outweighs the difficulties of this approach. It would be possible on nights when a staff member of the APT Service is at the site to place a stop over the telescope aperture. On the nights Vega is observed, other very bright stars can be observed as well. Once the revision of the secondary standards is completed we will be able to recalibrate the flux measurements reduced using existing secondary standards. I would anticipate a continued effort through the next few years of operation to improve the secondary standards.

Periodically the data could be re-reduced to reflect any significant improvements in the secondary standard values.

Most rotating-grating scanning spectrophotometers could obtain measurements through the entire optical window (3200-11000A) provided one used appropriate photomultiplier tubes and filters. Their resolution was typically 2 to 100A. Different astronomers selected different bandpasses and bandpass center wavelengths. In automating the observing, it will be necessary for simplicity's sake to take all the data in the same manner. For synthesizing photometric colors from spectrophotometric scans (a good way to check the accuracy of the spectrophotometry; Breger 1976), it is best to have continuous wavelength coverage with the flux sampled at equal wavelength intervals rather than a series of measurements at a discrete set of wavelengths which has often been the practice. The later technique is also subject to missing key wavelengths. It is also good practice to observe the unknown and standard stars with bandpasses having the same wavelength centers and same bandwidths.

Selecting a bandwidth is a compromise. The larger the bandwidth the easier it is to go fainter and to build the instrumentation. On the other hand, as the resolution is increased, there is more science one can potentially do. A resolution of order 10A is a reasonable solution. It is still small enough that one can measure the equivalent widths of the strongest lines, but large enough that the integration times will not be too long. But such a resolution introduces a minor problem as the radial velocities of many stars will significantly shift their continuum with respect to the laboratory wavelength frame. At 5000A, a 0.5A shift corresponds to a 30 km/s radial velocity.

To connect the optical region observations to the ultraviolet and to the infrared, it is desirable to observe as many short and as many long wavelengths respectively as possible. Shortward of 3300A, atmospheric ozone absorption makes it quite difficult to do good spectrophotometry. With a more adequate study of the atmospheric extinction it might be possible to go shortward to near 3200A. Longward of order 1 micron, the quantum efficiency of many of the optical region detectors drop. But for bright stars it should be possible to observe longward to near 1.1 microns. So we have a spectral region of order 8500A to study. Some wavelengths in the red and near infrared are seriously affected by telluric bands. But again, more adequate study of the atmospheric extinction should aid the calibration. If initially it is not possible to properly model the atmospheric extinction in certain wavelength regions, then such problems can be made a subject for future research. Some interesting meteorological data may be hidden in the extinction measurements.

With a resolution of 10A, we will need a detector width of order 850 total linear elements. With a grating instrument, it is not possible to observe the entire wavelength range in a single order. Thus, questions of separation filters, cross dispersing elements, the format of the detector, and whether or not to make simultaneous sky observations arise.

III. ASTROPHYSICAL CONSIDERATIONS

Much of the ASPT science with a resolution of 10A will involve studies of continuum or quasi-continuum features. Only the equivalent widths of the strongest atomic lines can be deduced if the continua are sufficiently smooth. This will be important for the Ca II K line and for the strongest isolated He I lines, such as $\lambda 4922$ and perhaps $\lambda 5875$ (provided its non-LTE effects can be calculated). Objects with variable and/or strong emission lines can also be studied.

Many spectrophotometric studies involve the determination of stellar effective temperatures and surface gravities via comparison with the predictions of model atmospheres. This type of study will continue to be important. Especially if the errors in the calibration of the primary and secondary standards can be reduced, the confrontation between observations and model atmospheric fluxes will provide critical tests of stellar model atmospheres and their input physics. The existence of non-LTE effects in the continua of hot and/or metal-poor stars can be studied as well as effects due to the non-plane parallel nature of extended stellar atmospheres. Observations of stars with solar type and non-solar type compositions will indicate how the abundance anomalies affect the energy distributions. For hot stars the ability to link optical and ultraviolet observations is needed while for cool stars it is the ability to link optical and infrared observations. Stellar flux observations covering very extensive wavelength ranges are also quite useful for discovering unknown binary companions and for population studies via spectral synthesis. For hot stars spectrophotometric measurements longward of 7500A are few. An ASPT could well open this region to study.

An important project will be a catalog of fluxes of typical Population I and Population II stars which includes examples covering the observed range of metallicity. This will be invaluable in population synthesis studies which utilize ASPT observations of globular clusters, galactic clusters and associations, and galaxies. Such studies will likely also involve spectrophotometric flux measurements of the ultraviolet and/or the infrared.

The next generation of optical region spectrophotometric instruments should be able to provide better time-resolved observations. With rotating-grating scanners and a reasonable number of wavelengths, it was hard to get 5 scans an hour with the HCO Scanner. But with an ASPT, it might well be possible to observe a bright star at least once a minute. This is important for objects with physically important short time scales such as RR Lyrae stars, β Cepheids, δ Scuti stars, flare stars, dwarf novae, cataclysmic variables, and some Ap stars. For some of these types we would want to see the effects of stellar pulsations. Hydrodynamic effects should also be present in the energy distributions of Cepheids of Populations I and II, but in these stars, although their periods are of order days, what one is probably looking for occurs only in a very limited part of their period.

Studies of all types of stellar variability can be performed with an ASPT. For Be stars one could look for the start of an emission episode and then trace its development. It would be necessary to develop a quick look algorithm which would compare the data as observed with some canonical values and if significant differences were found the ASPT would begin a preplanned intensive monitoring program. For magnetic Ap stars, ASPT observations would be important in defining the changes in the continua as well as in the broad, continuum features. Continuous wavelength coverage is desirable to define the extent of such features (see Adelman and Pyper 1985 and references therein).

The long term consistency of the data obtained with an ASPT is an important aspect for the study of stellar variability with periods of order one year or more. The development of dust in Mira atmospheres, the variability of solar type stars, and the striking changes of FK Comae stars should be monitored. Such stability is also important in being able to synthesize good photometric colors from spectrophotometric data.

For unresolved binary systems, ASPT observations could be used to place constraints on the components. Multicolor light curves could be generated and solved. Some energy distributions will indicate the presence of hitherto unknown companions. For RS CVn systems, evidence of the dark wave should be found (see, *e.g.* Shore and Adelman 1984).

If the ASPT is programmed to find and subsequently track a slowly moving object, then spectrophotometric studies of solar system objects become possible. For asteroids, one could study their energy distributions as a function of aspect in an attempt to deduce information on inhomogeneous surface compositions and their shapes. Spectrophotometric measurements of cometary comae and tails through various entrance apertures might be useful in understanding the chemistry of these objects. One could also obtain observations of the planets and perhaps the brightest moons, but this would probably require using the CCD camera images and appropriate offsets in a real time mode.

There are also going to be many nights which the ASPT will not be able to obtain absolute spectrophotometric measurements in an all sky mode, but still will be able to obtain useful data. As the stellar fluxes will be obtained simultaneously over a considerable wavelength range, programs which are concerned with the relative strengths of features could be performed. So on nights with thin clouds we can anticipate monitoring programs such as the activity of Be and WR stars.

ACKNOWLEDGMENTS

I thank Russell M. Genet for his comments on the preliminary version of this paper and Donald S. Hayes for discussion of many aspects of spectrophotometric measurements.

REFERENCES

Adelman, S. J., and Pyper, D. M. 1985, *Astron. Astrophys. Suppl.* 62, 279.
Breger, M. 1976, *Astrophys. J. Suppl.* 32, 1.
Genet, R. M., Boyd, L. J., and Baliunas, S. L. 1986, *Automatic Photoelectric Telescopes*, ed. D. S. Hall, R. M. Genet, and B. Thurston, (Mesa: Fairborn Press), p. 15.
Hayes, D. S. 1987, in *New Generation Small Telescopes*, D. S. Hayes, R. M. Genet and D. R. Genet, eds., (Mesa: Fairborn Press), p. 185.
Shore, S. N., and Adelman, S. J. 1984, *Astrophys. J. Suppl.* 54, 151.

SPECTROPHOTOMETRY AND PHOTOMETRY OF A–TYPE FIELD HORIZONTAL–BRANCH STARS

A. G. Davis Philip

Van Vleck Observatory and Union College

ABSTRACT: A program to identify A–type horizontal–branch stars in the general field of the galaxy is outlined. Follow up spectrophotometric APT (1–m) observations are planned for those stars in the magnitude range of V = 7 – 11 and photometric observations are planned in the Strömgren four–color system of stars with V magnitudes down to V = 14. With a reticon spectrophotometer it should be possible to reach stars at about V = 13, which will greatly increase the number of stars which could be observed.

I. INTRODUCTION

In the last decade there has been a resurgence of interest in the study of Population II stars. Evidence of this interest may be found in the proceedings of seven meetings in this time period concerned with globular clusters and Population II stars (See Figure 1). Four of these meetings were IAU Symposia or Colloquia, one was a NATO Study Institute and two were smaller meetings. The titles of these meetings were "Globular Clusters" (Hanes and Madore 1980), "Star Clusters" (Hesser 1980), "Astrophysical Parameters for Globular Clusters" (Philip and Hayes 1981), "Dynamics of Star Clusters" (Goodman and Hut 1985), "Horizontal−Branch and UV−Bright Stars" (Philip 1985a), "Spectroscopic and Photometric Classification of Population II Stars" (Philip 1986a) and "Globular Cluster Systems in Galaxies" (Grindlay and Philip 1987). In the last mentioned meeting, there are many papers in which the results of CCD observations are presented. With CCD techniques it is now possible to construct accurate color−magnitude diagrams of globular clusters which extend well down the main−sequence (for some recent examples of this type of work see Harris and Hesser (1986, 87) [47 Tuc], Penny and Dickens (1986) [NGC 6752], and Richer and Fahlman (1986) [M 13]. An example of the dramatic improvement in color−magnitude diagrams of globular clusters is shown in Figure 2, taken from Hesser (1987). It shows a very tight main sequence for 47 Tuc going town to a V magnitude close to 22 mag. A small insert compares the CCD photometry with older photographic photometry.

An important series of investigations of blue horizontal−branch stars in globular clusters has been made by Buonanno, Corsi, Fusi Pecci and others. They show color−magnitude diagrams with very extended blue horizontal branches reaching to a magnitude V = 18.5 in the case of M 15. A summary paper concerning these results can be found in Fusi Pecci (1986). One problem that they are investigating is illustrated in Figure 3. (taken from the summary paper by Hesser 1987) where color−magnitude diagrams are shown for three globular clusters with similar [Fe/H] but which have quite different horizontal−branch morphologies. The helium abundances and the ages of these three clusters are also similar so the different morphologies indicate that at least one other parameter must vary among these three clusters.

Extensive spectroscopic work is being done to measure the abundances of individual elements in globular cluster stars. An important question under investigation is whether abundance differences detected in some stars is the result of primordial abundance differences or the result of evolutionary changes during the lifetime of the star. Two recent reviews of these problems may be found in Smith (1987) and Norris (1987).

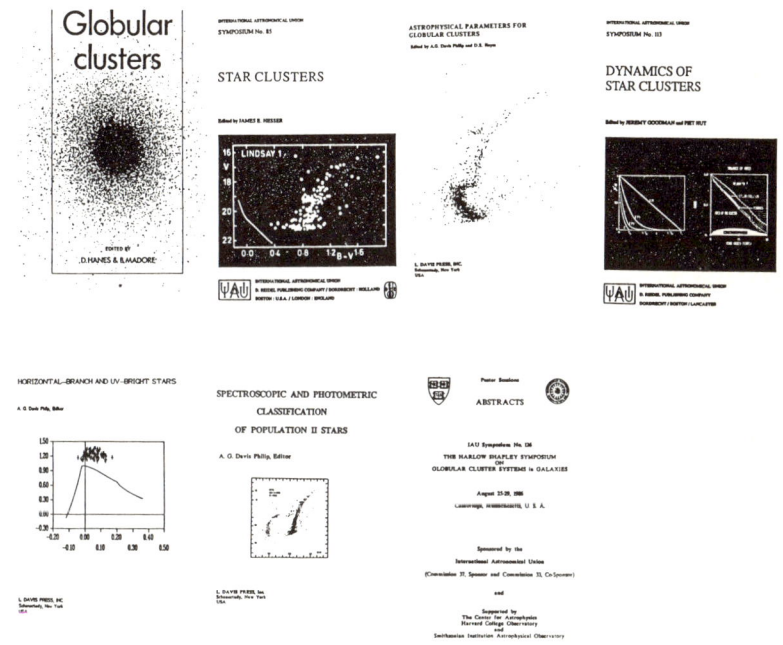

Figure 1. *Meetings in the past decade concerned with Population II Stars.*

At the same time that higher quality data (and data for fainter stars) are being obtained from observations, more detailed evolutionary models have been constructed for Population II stars. VandenBerg and Bell (1985) have published an extensive set of isochrones for globular clusters which can be displayed in color–magnitude diagrams. An example (taken from their paper and shown here as Figure 4) show isochrones for a set of ages and metallicities indicated in the figures. Observational data can be matched to such a series of isochrones and estimates of the age and metallicity of a globular cluster can be made with an accuracy much improved over earlier methods of determining these parameters.

Stars which appear to have the same characteristics as the blue horizontal–branch stars in globular clusters have been discovered in the general galactic field. Some of these discoveries have been made from spectroscopic plates taken in Schmidt object prism surveys (for example see MacConnell et al. 1971); others have been made by a combination of Schmidt spectral surveys and follow up four−color observations (for example see Philip 1985b, Bond and Philip 1973) and others have been made by four−color photometry of stars from various sources (for example see Graham 1970 [measures of the Feige stars], Kilkenny et al. 1975 [proper motion stars and blue surveys] and Stetson 1981 [four−color photometry of high velocity stars]).

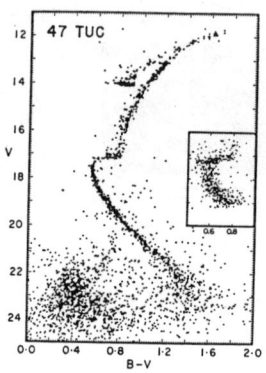

Figure 2. Comparison of the 47 Tuc main−sequence from CCD Photometry (Harris and Hesser 1986) with that (inset) from traditional techniques (Hesser and Hartwick 1977).

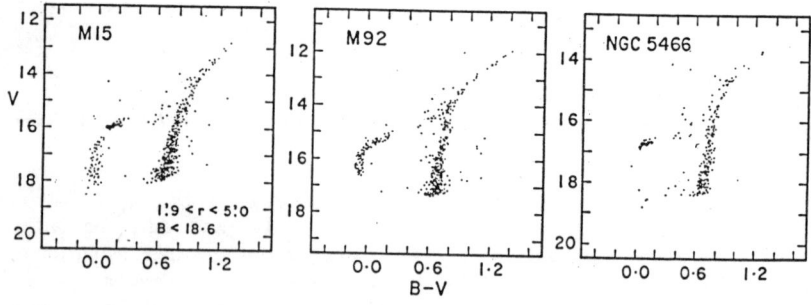

Figure 3. Three clusters, each having [Fe/H] near −2.1 show different horizontal−branches in these photographic CMDs from Buonanno et al. (1985).

Another technique has been used by Pier 1983 [Spectroscopy and UBV photometry of stars in the Preston and Schectman Schmidt survey]. A catalog of A−type field horizontal−branch stars with four−color photometry has been published by Philip (1984). Among the 71 stars in the catalog there are 27 that are in the magnitude range V = 7.0 – 11.9, 38% of the total. Of these 27 stars, 19 have been observed with the Harvard Scanners at Cerro Tololo and Kitt Peak and the Oke Multichannel Spectrometer at Palomar Observatory. Energy distributions between 3450 and 6790 A have been determined (Philip and Hayes 1983, Hayes and Philip 1983). These scans provided additional evidence of the similarity between the field stars and the globular cluster stars because there were pairs of stars, one a BHB star and one an FHB star, whose energy distributions between 3450 and 6700 A were the same to within about 0.02 mag.

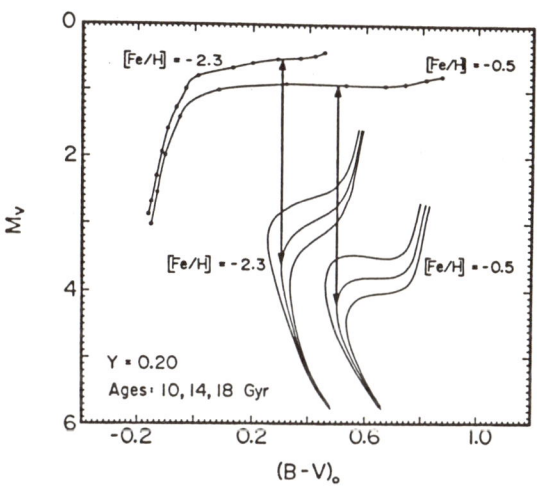

Figure 4. Plot of VandenBerg and Bell (1985) isochrones for the ages and metallicities indicated, along with fully consistent zero–age horizontal branches. Arrow illustrates the definition of the age dependent quantity ΔV_{TO}^{ZAHB}.

Spectroscopic observations have been made for some of the brighter FHB stars from which estimates of [Fe/H] have been made. The three most observed FHB stars are HD 86986, 109995 and 161817. In Figure 5 a set of spectra of these three stars is shown (from Philip and Lee 1985). The range in the values of the [Fe/H] estimates, for stars with many [Fe/H] determinations, is large; over 1 dex in the cases of HD 109995 and HD 161817. The [Fe/H] values are displayed in Table I. These determinations have been made over two decades in 8 different papers and part of the spread in the [Fe/H] determinations for a single star is the result of different approaches in the reduction of the spectroscopic data. It would be useful if all the stars in Table I could be observed spectroscopically and reduced under a common program.

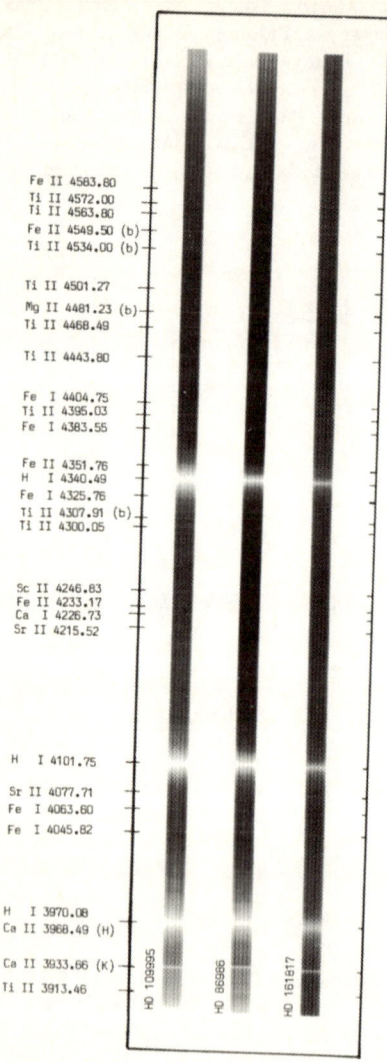

Figure 5. Spectra of three of the four proto—type FHB stars, obtained at a dispersion of 9 A/mm with the Soviet Union's Six Meter Reflector. (Philip and Lee 1985).

II. SCHMIDT OBJECTIVE PRISM SURVEYS FOR EARLY-TYPE STARS

McCuskey (1965) summarized his work on the distribution of stars in the galactic plane. He was an early practitioner of the technique of using Schmidt spectral plates to make surveys of stars of various spectral types in selected regions of the galaxy. At Case Institute of Technology I continued this type of work, but extended it to high galactic latitude regions. Upgren (1962, 63) used spectral plates obtained with the Hamburg Schmidt telescope to classify stars of all spectral types at the North Galactic Pole, following the work of Slettebak and Stock (1959), who classified the stars at the NGP earlier than spectral type F2. Philip and Sanduleak (1968) did similar work at the South Galactic Pole. This survey was followed by a larger SGP Schmidt survey by Slettebak and Brundage (1971). Drilling and Philip (1970), Philip and Drilling (1970) and Philip and Relyea (1971) have published finding lists of early-type stars in regions at galactic latitudes of +30 and −45 degrees. A survey has been made for southern metal-poor stars by Preston and Schectman (1979). Another source of metal-poor stars is the list published by Houk (1986) of weak-lined stars found in her reclassification of the HD stars. The Houk stars found so far are all south of declination −12 degrees.

When follow-up photometric work in the four-color system is done in the areas of these high galactic latitude surveys, A-type stars that have characteristics similar to those of BHB stars in globular clusters have been found. These characteristics are: high c_1 indices for their $(b-y)$ colors, lower than normal m_1 indices (for stars with $(b-y)$ values > 0.1), a velocity dispersion for a group of such stars of over 100 km/s, equivalent widths for the Ca II line at 4226 A which are smaller than the values for normal Pop I stars, and for those stars for which high dispersion spectra have been obtained, [Fe/H] values near −1.0. In Figure 6 c_1, $(b-y)$ and m_1, $(b-y)$ diagrams are shown for FHB stars. In each diagram the solid line indicates the position of Population I, main-sequence stars and crosses represent the positions of the FHB stars (from Philip 1985b).

TABLE I

[Fe/H] Values for FHB Stars

Star	[Fe/H]	Reference
PS 53 II	-1	Kodaira and Philip (1984)
HD 002857	-1	Kodaira and Philip (1984)
	-1.8	Danford and Lea (1981)
	-1.38	Danford (1976)
HD 014829	-1	Kodaira and Philip (1984)
	-1.78	Danford (1976)
BD 01 0513	-0.86	Danford (1976)
4 HLF 2a 44	-1.41	Danford (1976)
HD 060778	-0.5	Danford and Lea (1981)
	-1.03	Danford (1976)
HD 074721	-1.11	Danford (1976)
4 HLF 3 16	-0.70	Danford (1976)
HD 086986	-1.0	Danford and Lea (1981)
	-1.11	Danford (1976)
	-1.7	Kodaira (1973)
	-1.43	Kodaira, Greenstein and Oke (1969)
BD 36 2242	-0.7	Danford (1976)
BD 42 2309	-1.24	Danford (1976
HD 109995	-0.5	Danford and Lea (1981)
	-1.9	Kodaira (1973)
	-1.32	Danford (1976)
	-1.24	Kodaira, Greenstein and Oke (1969)
	-1.2	Wallerstein and Hunziker (1964)
HD 130095	-1	Kodaira and Philip (1984)
SS 206 II	-0.61	Danford (1976)
HD 117880	-1.73	Danford (1976)
1 HLF 5 10	-0.77	Danford (1976)
HD 161718	-1	Kodaira and Philip (1984)
	-0.5	Danford and Lea (1981)
	-1.20	Danford (1976)
	-1.6	Kodaira (1973)
	-1.28	Wallerstein and Hunziker (1964)
	-1.21	Kodaira (1964)
	-0.98	Wallerstein, Stone and Williams (1962)
	-0.54	Aller and Greenstein (1960)
HD 213468	-0.84	Danford (1976)

Positions and finding charts for all non-HD or BD stars in the list above can be found in Philip (1985).

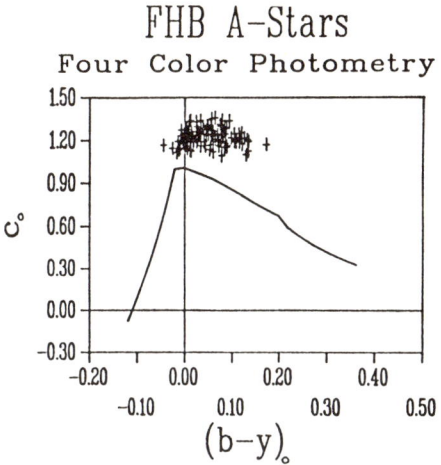

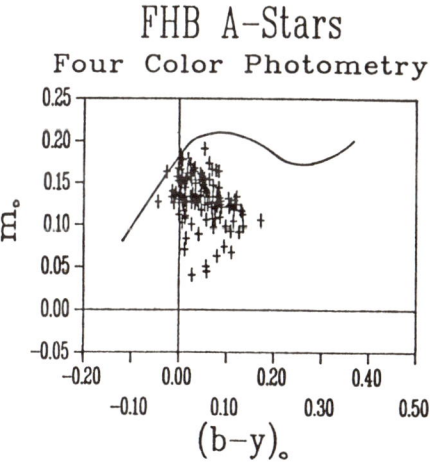

Figure 6. FHB A−Stars plotted in c_0 vs $(b-y)$ and m_0 vs $(b-y)$ diagrams (from Philip 1985b). The solid line represents the Zero−Age Main Sequence. FHB stars are represented by crosses.

A very efficient way to identify FHB stars is that outlined in the preceding paragraphs; namely to take Schmidt spectral plates in high galactic latitude regions, mark all the stars that fall in spectral class A and then measure all these candidates in the four-color system. The A stars which turn out to be normal Pop I stars can be used to determine the interstellar reddening in the region. The reddening is important and is needed when the four-color data are used to determine temperatures and surface gravities for the Pop II stars. The intrinsic color of the star must be used, unmodified by any possible effects of interstellar reddening.

The A stars with c_1 indices 0.2 mag. or greater than the c_1 value for a Pop I star at a given $(b-y)$ index are good candidates for the FHB list. However, all stars with these characteristics are not FHB stars. Population I stars of luminosity classes III to II also fall in this region of the four-color diagrams so additional observations must be made to confirm the four-color classification. An example of a good candidate according to the four-color indices which now is a doubtful case is that of the star HD 57336. The four-color indices placed this star in the middle of the area occupied by FHB stars but UV observations by Huenemoerder et al. (1983) showed that this star had uv absorption lines typical of Pop I stars and not of Pop II stars. A copy of their plot is shown in Figure 7 and one can see a strong absorption line near 1663λ not present in the other spectra.

With the thin prism at CTIO and at KPNO, it is possible to obtain low resolution spectra of stars to V = 17.0 magnitude. The dispersion of the thin prism is 1300 A/mm at Hγ and if one obtains unwidened, 1 hour plates then V = 17 can be reached. Plates have been obtained, using this technique, by Philip (1985b) and by Twarog and Anthony Twarog (1987) of regions at the South Galactic Pole. Pesch and Sanduleak (1986) have been using the Warner and Swasey Schmidt telescope at KPNO to make a thin prism survey of regions near the North Galactic Pole. Hundreds of faint A-type stars have been identified on these surveys, but the follow up four-color photometry has not yet been done. With a 1-meter telescope stars from a thin-prism survey down to V = 14 could be measured in a photometric system such as the Stromgren system.

III. A PROGRAM OF FHB STAR OBSERVATIONS

Among the stars identified as FHB stars (Philip 1984) there are three main groups. There are the "normal" FHB stars which have c_1 indices between about 1.2 to 1.3 at a spectral type of early A. Then there are stars with c_1 indices greater than 1.3 (there are examples of stars with these characteristics among the BHB stars in globular clusters as well, as shown in Philip 1986b.) The third group is a group identified by Jaschek et al. (1985) from ultraviolet spectra obtained with IUE. They found broad depressions at 1600 and 3040 A in several horizontal-branch stars which match features found in λBoo stars. They have no explanation for the cause of these depressions.

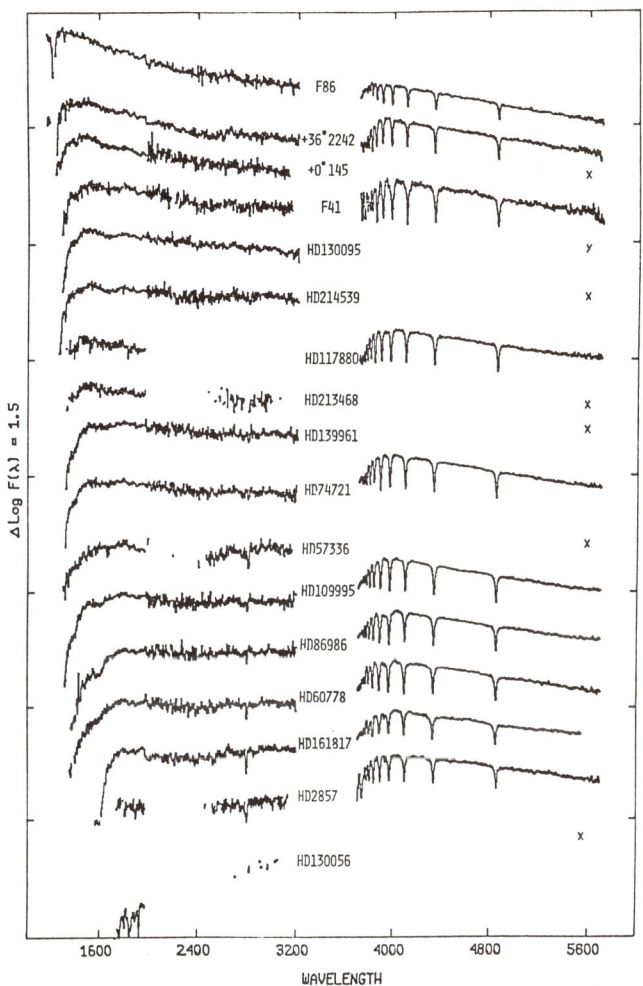

Figure 7. IUE UV spectra of FHB stars from Huenemoerder et al. (1983). HD 57336 shows features near 1663Å not present in the other stars.

When the stars that have good four-color photometry and spectrophotometric energy distributions are divided up into groups, the numbers of the stars (as a function of $(b-y)$) become too small to determine the mean relations for each group. What is needed is a program in which many FHB stars could be identified and then followed up with photometric and spectrophotometric observations. If the database of such stars could be increased by an order of magnitude then meaningful statistical studies could be made of stars in each group and their distribution in the four—color and derived diagrams (such as log g vs T_{eff}) as a function of $(b-y)$ color or temperature could be determined. Mean relations, in the theoretical HR diagram, can be compared with evolutionary tracks and in such a confrontation of observed data with the predictions of theory much can be learned about the parameters of possible models for FHB stars. By matching the mean relations of temperature and gravity for FHB stars with the theoretical tracks, certain combinations of input parameters (such as mass, helium abundance, age and metal abundance) can be eliminated since they predict positions in a theoretical HR diagram that are outside the range of the observed parameters.

As an example of the use of energy distributions a table (Table II) taken from Hayes and Philip (1987) shows temperature and surface gravity estimates made from spectrophotometric measures. An example of some energy distributions from this paper shows the type of data needed for such work. The stars shown in Figure 8 are the four proto—type FHB stars (HD 2857, 86986, 109996 and 161817). The color excesses were derived from measures of Population I A stars in the same general area as the FHB star. In Figure 9 sets of similar energy distributions (of BHB and FHB stars) are shown. Between λ = 3500 and 6600 A the energy distributions in each group do not deviate by more than 0.02 mag. This is further evidence that the FHB stars are similar to the BHB stars in globular clusters.

The temperatures and gravities were determined by extracting the colors (3636—4037) and (4037—6790) from the energy distributions and plotting the de—reddened colors in a diagram in which a grid of fully—blanketed ATLAS model atmospheres (for [Fe/H] = −1) are also plotted. The location of each point is graphically interpolated in the grid of model atmospheres to yield the best value of T_{eff} and log g. See Hayes and Philip (1987) for a full discussion of the method. An example of such a plot is shown in Figure 10 (from Hayes and Philip 1987).

The stars that are bright enough to be observed with a spectrometer on a small telescope (0.5—m or smaller) have been observed already; the number of such stars that has been identified over the entire sky is small. In order to make good progress in this program a 1 m telescope, or larger, should be used. Then stars down to V = 11 could be observed with a conventional spectrophotometer and with a reticon spectrometer stars down to V = 13.2 could be reached. Hayes (1987) describes a reticon spectrophotometer to be used on a fully automated telescope for which it is estimated that the limiting magnitude for acquisition will be V = 13 mag. It is an instrument of this type that this project plans to utilize. Adelman (1987) has presented ideas concerning the next generation of spectrophotometric instruments and one of the projects which he suggests should have high priority is to make a catalog of fluxes of typical Population I and II stars. The project outlined in this paper is an extension of the study of Population II stars to fainter stars.

TABLE II

Temperatures and Surface Gravities of HB A–Stars

Name	(b−y)	E(b−y)	T_{eff}	log g
HD 002857	0.119 mag.	0.014 mag.	7400 K	2.84
HD 014829	0.033	0.009	8750	3.60
HD 057336	0.087	0.04	7680	3.12
HD 060778	0.068	0.01	7880	2.90
HD 064488	0.013	0.01	8240	3.67
HD 074721	0.016	0.01	8640	3.55
HD 086986	0.063	0.02	7880	3.12
SS 287 I	0.037	0.00	8550	3.66
HD 109995	0.046	0.00	8350	3.36
HD 117880	0.042	0.01	8470	3.80
HD 130095	0.02	0.04	8480	3.64
HD 161817	0.115	0.01	7580	3.00
HD 202759	0.139	0.05	7310	2.81
1 HLF 2				
18 21	0.154	0.047	7100	2.58
17 17	0.116	0.047	7460	2.82
17 24	0.074	0.047	8040	3.63
S 14	0.130	0.047	7100	2.60
17 136	0.104	0.047	7680	2.91
S 70	0.062	0.047	7990	3.35

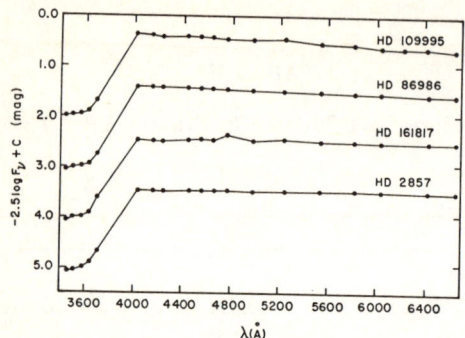

Figure 8. Energy distributions for the four proto-type FHB stars (from Philip and Hayes 1983).

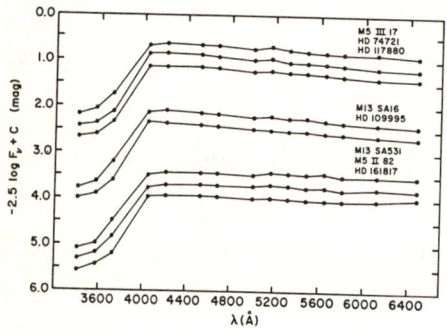

Figure 9. Three sets of energy distributions which are nearly identical for groups of FHB and BHB stars (from Hayes and Philip 1983).

To obtain an idea of the number of FHB stars to be found in regions at high galactic latitudes we can use the data contained in two studies made at the SGP and NGP. In a 100 square degree area in which all A—type stars was measured in the four—color system at the South Galactic Pole (Philip and Drilling 1987) 28 out of 98 A stars measured in the four—color system were classified photometrically as FHB stars. Two of these stars were 11th mag. or brighter, 3 were in the interval V = 12 − 13 and 4 were between 13.0 and 13.2. A similar sort of survey has been done at the NGP (Philip 1968) in a smaller area (30 square degrees) and there were one star at V = 11, 3 between V = 12 −

13 and 3 more between V = 13.0 and 13.2. If we average the results from these two regions then we find that in a 100 square degree region at high galactic latitudes we could expect to find one FHB star brighter than 11th mag., 1 1/2 more if the search is extended to V = 13.0 and 6 more if the search is extended to V = 13.2. At V = 13.2, a survey could turn up about 3 FHB stars per Schmidt plate (a four plate survey covers approximately 100 square degrees).

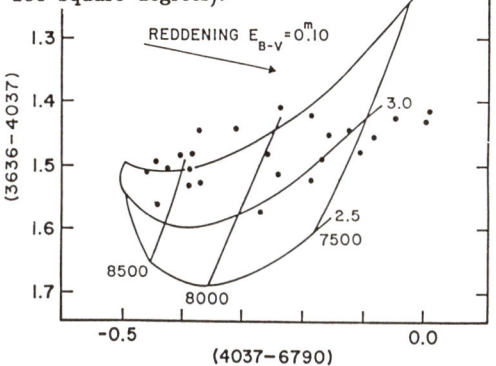

Figure 10. The color indices (3636-4037) vs (4037-6790), obtained from energy distributions such as those shown in Figures 8 and 9, with gravity and temperature lines derived from Kurucz model atmospheres. The points represent measures of BHB and FHB stars (Hayes and Philip 1987).

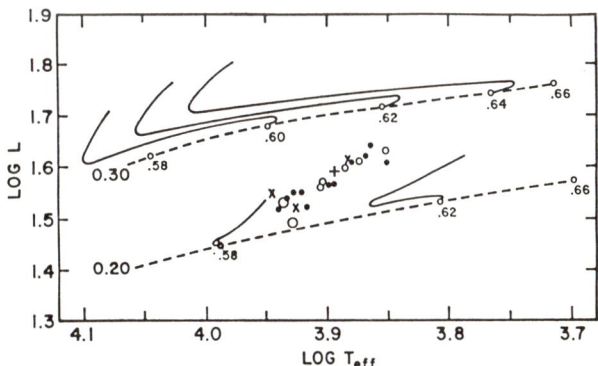

Figure 11. Log L versus log T_{eff} for FHB and BHB stars discussed in Hayes and Philip (1987).

The program of obtaining spectrophotometric scans of FHB stars should start with stars already located. There are 13 A stars in Houk

(1986) which have declinations north of −30 degrees. These weak−lined A−stars are very good FHB candidates and they are relatively bright since each one of them is an HD star. The fourth volume of the Michigan Spectral Catalog is about to be published and there should be some additions to the list of 'weak−lined' A−stars from this volume. In Philip (1984) there are 10 FHB stars brighter than V = 13.2 at the NGP and 15 FHB stars in other high galactic latitude regions for which no spectral scans have been made. In the Philip survey areas there are Schmidt spectral plates on which candidate stars have been marked but which have not been measured in the four−color system as yet. The survey made by Pesch and Sanduleak (1986) should be a good source of new FHB stars. It may be possible to acquire lists of other possible FHB stars from surveys made by other groups, for example the Preston Schectman (1979) survey.

Additional plates could be obtained with the Warner and Swasey Schmidt telescope at Kitt Peak National Observatory in areas adjacent to the NGP survey made with the Hamburg Schmidt telescope. If energy distributions could be obtained for approximately 100 FHB stars, then sufficient data would be available to study the mean relations of surface gravity with temperature for stars in the three groups of FHB stars mentioned at the beginning of this paper. For those stars which have been located by means of complete surveys of all A−type stars in a given region, then the relative number of stars in each group represents a true relative abundance of the stars as they appear in space, out to the limit of the volume surveyed.

If we can increase the number of Population II, early type stars with accurate energy distributions then diagrams such as the one shown in Figure 10 (from Philip and Hayes 1987) can be plotted for the various types of HB stars and compared to the predictions of atmospheric models. In Figure 11 log L is plotted against log T_{eff} and compared to various models with different helium abundances and core masses. Such studies of the brighter FHB stars and comparisons made of these data with the predictions of atmospheric models will enable much to be learned about the nature of the much fainter BHB stars.

REFERENCES

Adelman, S. J. 1987 in *New Generation Small Telescopes*, ed. D. S. Hayes, R. M. Genet and D. R. Genet. (Mesa: Fairborn Press), p. 157.
Aller, L. M. and Greenstein, J. L. 1960 *Astrophys. J. Suppl.* 5, 139.
Bond, H. E. and Philip, A. G. D. 1973 *Publ. Astron. Soc. Pacific* 85, 332.
Danford, S. C. 1976 Thesis, Yale Univ.
Danford, S. C. and Lea, S. M. 1981 *Astron. J.* 86, 1909.
Drilling, J. S. and Philip, A. G. D. 1970 *Bol. Ton. y Tac.* 5, 307.

Fusi Pecci, F. 1986 in *Spectroscopic and Photometric Classification of Population II Stars*, ed. A. G. D. Philip, (Schenectady: L. Davis Press), p. 83.
Goodman, J. and Hut, P. 1985 *IAU Symposium No. 113, Dynamics of Star Clusters*, (Dordrecht: D.Reidel).
Graham, J. A. 1970 *Publ. Astron. Soc. Pacific* 82, 1305.
Grindlay, J. E. and Philip, A. G. D. 1987 in *IAU Symposium No. 126, Globular Cluster Systems in Galaxies*, ed. J. E. Grindlay and A. G. D. Philip, (Dordrecht: D. Reidel), in press.
Hanes, D. and Madore, B. 1980 *Globular Clusters*, (Cambridge: Cambridge University Press).
Harris, W. E. and Hesser, J. E. 1986, in preparation
Harris, W. E. and Hesser, J. E. 1987 in *IAU Symposium No. 126, Globular Cluster Systems in Galaxies*, ed. J. E. Grindlay and A. G. D. Philip, (Dordrecht: D. Reidel), p. 61.
Hayes, D. S. 1987 in *New Generation Small Telescopes*, ed. D. S. Hayes, R. M. Genet and D. R. Genet, (Mesa: Fairborn Press), p. 185.
Hayes, D. S. and Philip, A. G. D. 1983 *Astrophys. J. Suppl.* 53, 759.
Hayes, D. S. and Philip, A. G. D. 1987, in preparation.
Hesser, J. E. 1980 *IAU Symposium No. 85, Star Clusters*, (Reidel, Dordrecht.)
Hesser, J. E. and Hartwick, F. D. A. 1977 *Astrophys. J. Suppl.* 33, 361.
Houk, N. 1986 in *Spectroscopic and Photometric Classification of Population II Stars*, ed. A. G. D. Philip, (Schenectady: L. Davis Press), p. 19.
Huenemoerder, D., de Boer, K. S. and Code, A. D. 1983 *Astron. J.* 89, 851.
Jaschek, M., Baschek, B., Jaschek, C. and Heck, A. 1985 *Astron. Astrophys.* 152, 439.
Kilkenny, D. and Hill, P. W. 1975 *Mon. Not. Roy. Astron. Soc.* 173, 625.
Kodaira, K. 1964 *Ztschr. Astrophys.* 59, 139.
Kodaira, K. 1973 *Astron. Astrophys.* 22, 273.
Kodaira, K., Greenstein, J. L. and Oke, J. B. 1969 *Astrophys. J.* 155, 525.
Kodaira, K. and Philip, A. G. D. 1984 *Astrophys. J.* 278, 208.
MacConnell, D. J., Frey, R. L., Bidelman, W. P. and Bond, H. E. 1971 *Publ. Astron. Soc. Pacific* 83, 98.
McCuskey, S. W. 1965 in *Galactic Structure, Vol V of Stars and Stellar Systems*, ed. A. Blaauw and M. Schmidt, (Chicago: Univ. of Chicago Press), p. 1.
Norris, J. 1987 in *IAU Symposium No. 126, Globular Cluster Systems in Galaxies*, ed. J. E. Grindlay and A. G. D. Philip, (Dordrecht: D. Reidel), p. 93.
Penny, A. J. and Dickens, R. J. 1986 *Mon. Not. Roy. Astron. Soc.* 220, 845.
Pesch, P. and Sanduleak, N. 1986 *Astrophys. J. Suppl.* 60, 543.
Philip, A. G. D. 1968 *Astron. J.* 73, 1000.
Philip, A. G. D. 1984 *Contributions of the Van Vleck Observatory* No. 2, p. 1.

Philip, A. G. D. 1985a *Horizontal-Branch and UV-Bright Stars*, (Schenectady: L. Davis Press).
Philip, A. G. D. 1985b in *Horizontal-Branch and UV-Bright Stars*, ed. A. G. D. Philip, (Schenectady: L. Davis Press), p. 41.
Philip, A. G. D. 1986a *Spectroscopic and Photometric Classification of Population II Stars*, (Schenectady: L. Davis Press),.
Philip, A. G. D. 1986b in *Spectroscopic and Photometric Classification of Population II Stars*, ed. A. G. D. Philip, (Schenectady: L. Davis Press), p. 67.
Philip, A. G. D. and Drilling, J. S. 1970 *Bol. Ton. y Tac.* 5, 297.
Philip, A. G. D. and Drilling, J. S. 1987, Submitted.
Philip, A. G. D. and Hayes, D. S. 1981 *IAU Colloquium No. 68, Astrophysical Parameters for Globular Clusters*, ed. A. G. D. Philip and D. S. Hayes, (Schenectady: L. Davis Press).
Philip, A. G. D. and Hayes, D. S. 1983 *Astrophys. J. Suppl.* 53, 751.
Philip, A. G. D. and Lee, J. 1985 in *Horizontal-Branch and UV-Bright Stars*, ed. A. G. D. Philip,(Schenectady: L. Davis Press), p. 57.
Philip, A. G. D. and Relyea, L. J. 1971 *Bol. Ton. y Tac.* 6, 69.
Philip, A. G. D. and Sanduleak, N. 1968 *Bol. Ton. y Tac.* 4, 253.
Richer, H. B. and Fahlman, G. G. 1986 *Astrophys. J.* 304, 273.
Preston, G. W. and Schectman, S. A. 1979 *Annual Report of the Director, Hale Observatories*, (Washington, D. C.: Carnegie Inst. of Washington).
Slettebak, A. and Brundage, R. K. 1971 *Astron. J.* 76, 338.
Slettebak, A. and Stock, J. 1959 *Hamberger Sternwarte V*, No. 5.
Smith, G. H. 1987 *Publ. Astron. Soc. Pacific*, in press.
Stetson, P. B. 1981 *Astron. J.* 86, 1882.
Twarog, B. and Anthony Twarog, B. A. 1987, personal communication.
Upgren, A. R. 1962 *Astron. J.* 67, 37.
Upgren, A. R. 1963 *Astron. J.* 68, 194.
VandenBerg, D. A. and Bell, R. A. 1985 *Astrophys. J. Suppl.* 58, 561.
Wallerstein, G. and Hunziker, W. 1964 *Astrophys. J.* 140, 214.
Wallerstein, G., Stone, Y. H. and Williams, J. A. 1962 *Astrophys. J.* 135, 459.

MUSINGS ON A SYNTHETIC PHOTOMETER

Nathaniel M. White

Lowell Observatory

The proven value of filter photoelectric photometry is diminished only by the inefficiency of measuring one or a few colors while the rest of the collected photons are filtered out. The photoelectric photometer is a unique combination of a very sensitive, high−gain, low−noise detector presently requiring no extraordinary technical needs with a simple imaging system and the stability of bandpass definition by optical filters. Can a spectrophotometer be designed that will provide a multiplex advantage yet maintain the precision and reproducibility of filter photometry?

Ideally a synthetic photometer would be a spectrophotometer−computer−software combination that would input starlight and output fully reduced colors and magnitudes on any photometric system. The value of such an instrument is not in time−saving alone. Color indices could be measured during less than photometric conditions. There are numerous monitoring programs where simultaneous coverage is an important advantage such as wavelength dependent variability.

A nonscanning spectrophotometer designed for absolute photometry is characterized by four interdependent requirements. These are the wavelength coverage, spectral resolution, the image stability on the detector, and the need for a high−resolution array detector. A wavelength range of .3 to 1.1 microns with a spectral resolution of 10A (.001 microns) would enable the approximation of most photometric systems using photoelectric photomultiplier tubes. Photometric stability relies on the imaging system to produce a stable color to pixel relation for normal seeing fluctuations and image motion. The detector must be sensitive over the entire wavelength range with sufficient spatial resolution and size to gain the full multiplex advantage.

Although a grating easily provides sufficient spectral resolution and high throughput at specific wavelength intervals, only a prism can produce a non−overlapping spectrum with nearly constant high throughput for the entire .3 to 1.1 micron range. A 40mm, 60 degree crown prism, and a 52mm camera focal length yields a dispersion of from 2A per 25 micron pixels at .4 microns to 6A per pixel at .6 microns. At 1 micron, the dispersion is 30A per pixel. This arrangement does not meet the 10A resolution requirement over the entire spectrum, but does meet the requirement in the spectral region where most narrow−band photometric systems are defined and the spectrum may be most cluttered. For wavelengths longward of .7 microns, most photometric systems are either broadband or measure molecular bands both of which can be effectively synthesized with the lower prismatic dispersion.

The resolution of a spectrophotometer is degraded if there is image motion on the detector. Variable resolution will to some degree reduce the accuracy attained by a synthetic photometer. This problem is solved in a spectrograph with the use of an entrance slit, but the unknown vignetting cannot be tolerated in a spectrophotometer. The problem is a difficult one with three possible solutions. The first is to use a combination of collimator−to−camera focal ratio and high dispersion so that the input image motion is reduced to an insignificant amount at the detector. For our desired wavelength coverage, this would require too long of a collimator focal length and a very large detector. A second method might use a circle−to−rectangle light pipe. Geometries of this type produce very fast exit beams which require large optics and reduce the practical range of collimator−to−camera focal ratios. Fiber bundles which go from a circular shape to a linear array suffer from a variable fill−in factor. A third method might use the Fabry image of the entrance pupil as the entrance slit of a conventional spectrograph. This suffers from the restriction of a fast exit beam similar to variable geometry light pipes. The problem of image motion on the detector is a difficult one and it is not clear what the best solution will be.

Only two detectors meet both the requirements of wavelength sensitivity and high spatial resolution. The diode array and the CCD both have their weak and strong points. The diode arrays are inexpensive and electronically simple. The cost and technical expertise required to put together a photomultiplier system with computerized data acquisition would be sufficient to put together an effective diode array system. For bright work the diode array would be the detector of choice while the more expensive, liquid nitrogen cooled CCD would be required for faint objects.

A PRELIMINARY DESIGN FOR A SPECTROPHOTOMETER TO BE USED ON A FULLY-AUTOMATED TELESCOPE

D. S. Hayes

Fairborn Observatory

ABSTRACT. The design for a low-resolution spectrophotometer using a reticon as a detector is presented. The design is matched to the characteristics of a fully-automated telescope of 1-meter aperture.

I. INTRODUCTION

For many problems in astrophysics, low-resolution (roughly 10A) spectrophotometry using small-to-medium-sized (0.4- to 1.0-m) telescopes is necessary to give more information than can be obtained from conventional filter photometry. Many of these problems require the

New Generation Small Telescopes, ed. D. S. Hayes, R. M. Genet, & D. R. Genet.
© 1987 Fairborn Observatory.

routine observation of a large number of stars for their successful investigation, and this mode of observation can now be done most efficiently with a fully-automated telescope.

Fully-automated telescopes, which operate under computer control without an observer in attendance, have already been successfully applied to differential photometry of variable stars with single-channel photoelectric photometers by the Fairborn Observatory (Boyd, Genet and Hall 1985c; 1986). Such telescopes are operated by the Automatic Photoelectric Telescope Service (APT Service; Boyd, Genet and Hall 1985a,b). The APT Service is a joint operation of the Fairborn Observatory (equipment and operations) and the Smithsonian Astrophysical Observatory (site and facilities) (Baliunas, et al. 1985). The APT Service operates three telescopes, ranging from 0.25 to 0.4−m aperture, at an excellent site on Mt. Hopkins, Arizona.

In this paper a preliminary design for a spectrophotometer to be used on an APT is described. Spectrophotometric observations are similar to filter photometry in the manner in which they are carried through, but a spectrophotometric system must be different from the present APT systems in three major respects. One is that spectrophotometry normally is done in the all−sky mode; the development of an all−sky capability for the APTs on Mt. Hopkins is not particularly difficult (Hayes and Crawford 1986), and is planned for the near future. The second difference is that spectrophotometry requires larger telescope apertures for a given brightness of star because of the narrower bandpasses used. A 0.75 m telescope is currently under development by the Fairborn Observatory; for the present discussion an aperture of 1−meter is assumed, but the instrument described here would also work on a 0.75−m telescope.

The other extension to the current practice of the APT Service is in the method of acquisition of stars. The APTs acquire stars by making an open−loop motion to a calculated position, and then executing a square−spiral search while monitoring the output of the photomultiplier. This method works without problems for bright stars in uncrowded fields. To find fainter stars, particularly in the presence of possible confusion, an acquisition system based upon a television−type system using a CCD will be more appropriate. Such a system would be used only for acquisition, so the detector would not have to be cooled, and would not have to have special characteristics; it would have to be intensified in order to have a sufficiently faint limiting magnitude. The APT Service has started the development of such a system, and one can expect that it would be available by the time that the instrument being described here would be completed.

II. THE TELESCOPE

For the purpose of designing the instrument, the telescope is assumed to be a fully-automatic computer-controlled classical Cassegrain of 1.0-m aperture. The Cassegrain f-ratio of the telescope is fast-f/3.5. This is a result of a safety constraint: the APT Service telescopes must not impede the closing of the roll-off roof even when pointed at the zenith. It could happen that the telescope has a malfunction during operation on a night when it clouds up quickly and starts raining. The telescope will, as a result, have an f/2 primary, and a large, low power, closely-spaced secondary which gives the low overall f-ratio. The mount is also very compact, resulting in constraints on the size and shape of the instrument at the Cassegrain focus. The spectrophotometer described here will not have any problem fitting within these constraints.

With an aperture of 1.0-m and an f-ratio of f/3.5, the scale would be 60 arcsec/mm. All-reflecting optics would be used, so the full wavelength range would be available. Essentially only on-axis observations would be made. Acquisition and guiding would be done with an intensified uncooled CCD TV camera. The camera would be operated at TV frame rates, but with integration over several frames. As a result the limiting magnitude for acquisition would be about V=13.0 mag.

III. INSTRUMENT DESIGN PARAMETERS

Several objectives which constrain the design of this instrument must be stated at once: 1) The instrument must be simple and inexpensive to build and operate, and be as efficient as possible, consistent with the other design objectives. This means that versatility will be sacrificed, if necessary. 2) The instrument should be quite compact. 3) The dispersion should be no lower than 5A/pixel in the second-order blue. 4) The full wavelength range of 3200-10000A should be available with no more than a change in filter. When a star is observed over the full wavelength range, the combination of two exposures, one in the second order blue and one in the second order red, should give the entire useful wavelength range.

With a 1-m f/3.5 telescope, the scale will be about 60 arcsec/mm, so good seeing will produce a star image about 17 microns in diameter at the focal plane. This image must project to considerably less than a pixel at the detector, so that image blowup, poor seeing and guiding errors do not affect the resolution or wavelength scale. Since a typical reticon has pixels about 30 microns in width, and over ten times this in height, we see that in good seeing the image projects to only one-half of a pixel along the dispersion. Because of the likelihood of poor seeing, a further demagnification would be helpful. In the design presented here, a demagnification factor of two is used. This results in a rather fast camera; the alternative is a larger instrument—that is, there is a trade-off between total instrument size and camera speed.

In Table I are listed the parameters of the spectrum projected on the chip. The wavelength ranges to be projected in the first and second orders are selected to give some overlap so that full wavelength coverage is achieved. The choice of wavelength limits arises from the characteristics of the $CuSO_4$ filter to be used for order separation in the second order, and from the desire to cover the entire wavelength range of 3200-10000A in two orders.

TABLE I

Parameters for spectrum on the detector

Wavelength range	Order	Dispersion		Filter
2600-5600A	2	3.2A/pixel	110A/mm	$CuSO_4$
5200-11200A	1	6.4A/pixel	221A/mm	lowpass at 5000A

Since both orders will be overlapped on the chip, order separation filters will be necessary. A simple filter slide driven by a small motor under computer control will be all that will be needed to select the order to be observed. Thus, a full-wavelength observation will require two exposures. The worst cost of this second exposure will be the read time, since the exposure times in the second-order UV-blue can be expected to be roughly a factor of ten longer than those in the first-order red (because of the higher dispersion and the lower sensitivity of the chip). This cost in observing time can be expected to be small compared to the extra complication of cross-dispersion.

The layout of the spectrometer is shown in Figure 1. It is shown reduced in scale, but at full scale the portion shown occupies a square about 8 inches on a side, so the result is a compact instrument. All components of this system should be straightforward to produce and integrate except the camera, which is very fast and must work over a wide wavelength range. On the other hand, the demands on the image quality are not severe.

IV THE RETICON DETECTOR

A reticon appears to be a better choice than a CCD for this application. This is because of its superior sensitivity (particularly in the

UV) and greater dynamic range. Additionally, the cost of the chip and the system for operating it is lower. The well−known presence of much higher read noise is not such a problem in high−S/N observations. The use we are proposing generally involves high−S/N survey observations of relatively bright stars. That is, the faintest stars commonly observed would be about 11th mag; with a 1−m telescope one could observe to 14th at lower S/N. In our case, the larger telescope aperture makes possible higher S/Ns and shorter integration times.

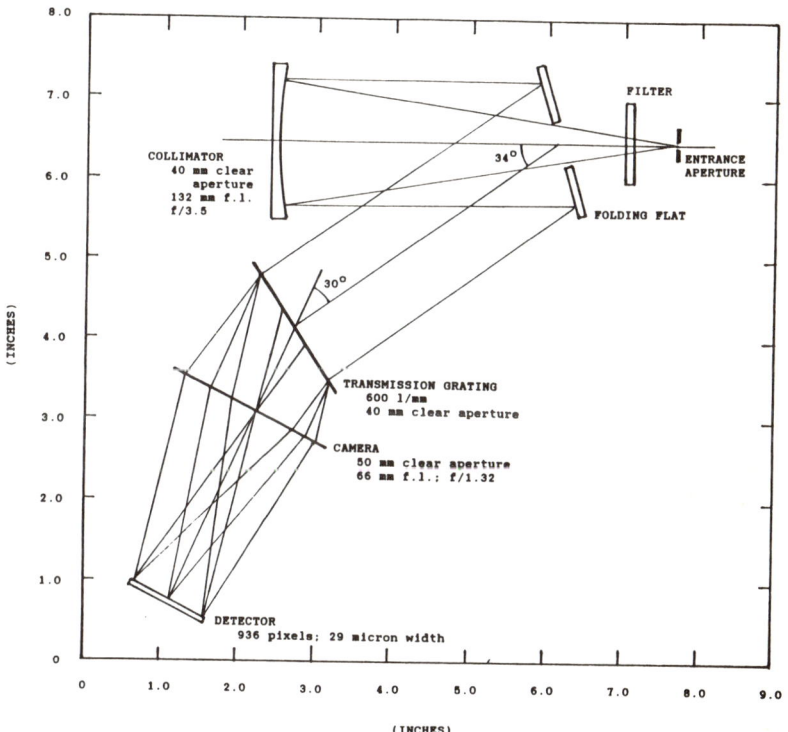

Figure 1. Optical layout of the spectrophotometer.

The use of an array detector generates some problems, as well as opportunities. Two problems come to mind which are peculiar to the operation of the instrument in the fully−automatic mode in a remote location. These are: 1) cooling the detector, and 2) storing the data for extended observing periods. In the first case, the use of cryogens is impractical, since it would be necessary to store large quantities and somehow deliver limited flows to the dewar. In the case of a reticon, where relatively bright objects are being observed, a multistage thermoelectric cooler will suffice. In the second case, we observe that the amount of data produced will be about 100 times that currently being produced by a single−channel photometer using a photomultiplier tube. On the other hand, the data rate would not strain the system; the only issue would be storage. An observation of a star (with simultaneous sky measurement) will produce about 10k bytes of data, compared to less than 100 bytes for *UBV* photometry. A typical night's observation will include a few hundred star−observations, or a few megabytes of data. Thus, a hard disk with a capacity of 70−100 Mb will suffice for several weeks of observation; such capacities are now readily available for IBM−PC−AT clones for less than $3000. (Several ads in the January, 1987, issue of Byte give discount prices of: $1789 for 70 Mb; $1349 for 80 Mb; $1195 for 85 Mb; $1695 for 86 Mb; and $2996 for 140 Mb).

V. SYSTEM PERFORMANCE

It is desirable to estimate the integration times which would be required for stars of various types and brightnesses for the purpose of selecting scientific projects to propose for its use. Such estimates are presented here, although they must be qualified as being very rough at this stage. The estimates are based upon the author's experience with the Intensified Reticon Scanner (IRS) at Kitt Peak on a 36−inch telescope. The fact that an intensifier was used means that it is the quantum efficiency of the photocathode, rather than the reticon, which determined the count rates obtained. The estimates given here assume that the quantum efficiency of the bare reticon is about 3 times higher than that of the photocathode of the IRS in the near UV and blue (the photocathode in the IRS is not a particularly good example). The ratio grows to much higher values in the red, of course.

The efficiency of the system in the near UV and blue is the most significant, for two reasons. One is that these regions are important scientifically for many of the projects likely to be done with this instrument. Another is that this region will require the longest exposures because of the lower quantum efficiency, the larger atmospheric extinction, and the low intensity of most stellar spectra in the near UV. When the two exposures are made, one in the "blue" and one in the "red," the "blue" one will be by far the longer except for very bright or very blue stars. For the bright stars, both exposure times will be short, and scintillation noise will be dominant. The scintillation noise spectrum is such that integration times under a second will not be appropriate even when this gives more than enough counts. Thus, the "blue" and "red" exposure times will be about the same length. But since both will be very short, and since the performance on fainter stars is of primary

SPECTROPHOTOMETER: A PRELIMINARY DESIGN

interest for this discussion, the "blue" exposure time will be emphasized in this discussion.

Count rates for the IRS have been adjusted to the present conditions. The size of a star image on the pixel is about the same for the two instruments, a fact which simplifies the estimates. For a star with the monochromatic magnitude (per unit frequency interval) = 10.0, the count rates are estimated to be about 9.3 and 31 counts/sec/pixel at 3500 and 4000A, respectively. Thus, for such a star it would take 5 1/2 min to obtain a total of 10^4 counts at 4000A. In order to prevent the measurements of the background sky from taking up additional observing time, the star and sky measurements will be made simultaneously. This will require using a dual parallel array and two diaphragms. Since the sensitivities of the two sides of the detector array will, in general, be different, the star will have to be "beam switched" from one side to the other so that observations can be made in each. Trends in sky brightness and small-scale extinction changes can be nearly eliminated if a symmetrical ABBA sequence is used, where one diaphragm is called "A," and the other "B." Thus, a complete observation requires four exposures, and 10^4 counts/pixel at the K-line could be obtained in 20 min. at a magnitude of 11.5. For A-stars, the flux at 3500A is about one magnitude fainter than that at 4000A, so a 20-min ABBA sequence would give 10^4 counts/pixel at 3500A for an A-star with a V magnitude of about 9.0. For spectrophotometry with a passband of 30A = 10 pixels, one could get the same precision at 11.5 mag. Thus, to reach 13th mag would require about 1 2/3 hours per star, a time which would be best split into five 20-min ABBA sequences.

An APT such as that considered here can be expected to acquire and center a star in less than 2 min. With the figures above, for spectrophotometry with 30A passbands of A-stars, one could expect to observe about 200 stars/night at a V mag of 6.5, 150 stars/night at mag 9.0, 25 stars/night at mag 11.5, and 5 stars per night at mag 13. An approximate allowance for the "red" exposure has been made. These numbers can be compared with the requirements of the scientific programs described in other papers in this Symposium.

VI. SUMMARY

Spectrophotometry with a reticon detector on a 1-m telescope is well-matched to fully-automatic observations of relatively bright stars, where surveys of large numbers of stars with high S/Ns are desired. As shown above, it is possible to design such a system which will give full wavelength coverage, with sufficient resolution, and which will be sufficiently compact to fit in the limited space available. Further, the cost of the system would be small. Most of the important concepts and characteristics of the system and proposed mode of observation are already well understood. Those aspects which require some extrapolation into new territory, such as scaling up the APT concept to 1—m aperture, and observing with an array detector and storing megabytes of data per night, are still not particularly demanding. In all aspects automated spectrophotometry requires only modest extrapolations of current practice, while promising great increases in the rate at which information is gathered.

REFERENCES

Baliunas, S.L., Boyd, L.J., Genet, R.M., Hall, D.S. and Criswell, S. 1985 *IAPPP Comm*. No. 22, 1985.
Boyd, L.J., Genet, R.M. and Hall, D.S. 1985a *IAPPP Comm*, No 19, 1.
Boyd, L.J., Genet, R.M. and Hall, D.S. 1985b *IAPPP Comm*, No 12, 59.
Boyd, L.J., Genet, R.M. and Hall, D.S. 1985c *Sky and Tel.* 70, 16.
Boyd, L.J., Genet, R.M. and Hall, D.S. 1986 *Publ. Astron. Soc. Pacific* 98, 618.

AN ALGORITHM FOR CALCULATING ATMOSPHERIC EXTINCTION FOR SPECTROPHOTOMETRY AND NARROW−BAND PHOTOMETRY

Donald S. Hayes

Fairborn Observatory

Paul C. Schmidtke

Department of Physics, Arizona State University

ABSTRACT. The determination of the coefficient of atmospheric extinction is one of the key steps in the reduction of photometric data. The coefficient is generally determined by observing one or more stars at least two times each at different airmasses and applying a least−squares solution to the resulting data. Several approaches to this process may be found in the literature for photometry with filters. They all incorporate one conspicuous omission in the procedure: no method is used to estimate the error in the coefficient determined by the least–squares solution. We present a method of solution which incorporates an explicit calculation of this error.

I. INTRODUCTION

Photoelectric photometry involves the apparently simple process of measuring the brightness and colors of stars by use of "linear" detectors and a number of filters. Spectrophotometry differs only in the use of more numerous and narrower passbands. This simplifies things in that the "color terms," which result from wide filter passbands, are eliminated; the passbands used in spectrophotometry are sufficiently narrow that they behave nearly monochromatically. In both filter photometry and spectrophotometry, a number of physical effects influence the results, including the effects of the earth's atmosphere on the starlight and the collective characteristics of the telescope, photometer and filters. As a consequence, the reduction of the data involves several steps directed at determining the magnitude of these effects. We will discuss the step in which the effects of the earth's atmosphere on the intensity and color of the light are determined.

It is important to emphasize that maximizing the effectiveness of measuring the effects of atmospheric extinction and, as much as is possible, correcting the data for them, depends upon observing procedure as well as upon the method of reduction. Observing procedures and techniques have been discussed in the literature (Hardie 1962, Hayes and Crawford 1986, Rufener 1984, 1985, Young 1974), as have methods of reduction (Harris, *et al.* 1981, Popper 1982, Manfroid and Heck 1983, 1984, Rufener 1984, 1985, Schmidtke 1983a, 1983b, Young 1974). The physical nature of atmospheric extinction has been discussed by Hayes and Latham (1975). We will not consider these topics here.

The effect of the extinction by the atmosphere is generally characterized by a linear equation relating the apparent magnitude to the airmass; the constants in the equation are the atmospheric extinction coefficients. At least one star is generally observed a number of times over a range of airmass; the result is one or more linear equations which can be solved for the coefficients. If more than two observations are made at distinctly different airmasses, the system of equations will be over–determined, and as a result a technique like the least–squares method can be applied. Such approaches are common, and may be found in the references cited above. All of these discussions are lacking, however, in that they do not take full advantage of the least–squares method. Once the method of solution is established, it takes little additional effort to calculate a standard error of the extinction coefficient from the solution.

CALCULATION OF ATMOSPHERIC EXTINCTION 195

In any scientific measurement, the error in the determination is as important a quantity as the value itself. Photometric measurements are not exceptions to this rule, and the determination of errors is a vital part of the reduction process. There are many contributors to the errors in the final results, and atmospheric extinction is just one of them. Photometrists use a number of techniques to estimate the errors in their data, but a common one is to compare the observed and standard−system values for standard stars observed during each night. A quantity called the residual (*i.e.*, the difference between the observed and standard−system value) is calculated. The standard deviation of the distribution of the residuals is a measure of the precision of the data. Plots of the residuals against relevant quantities such as airmass or time can be an effective diagnostic of the observations.

The standard−star residuals include the contributions of many sources of error to the data, including errors in the standard−system values themselves. Unless the various contributions can be separated, one can only conclude that a given data set contains "good" or "poor" photometry, but not why it is poor. When properly functioning equipment and sensible observing procedures and reduction techniques are used, atmospheric extinction is probably the major contributor to errors in the final data for all−sky photometry. Thus, the ability to determine the precision of the extinction coefficient can make an important contribution to understanding the quality of photometric data.

In the section to follow, the least−squares solution for the extinction coefficient is given, along with the additional equation giving the error in the coefficient. Note that these equations are written for the "nearly monochromatic" case where the passbands are "narrow." This algorithm was originally developed for the spectrophotometric investigations reported by Philip and Hayes (1983) and by Hayes and Philip (1983), but it could be used for Stromgren *uvby* photometry without change. The equations are written for magnitudes as the measured quantities, but they could be rewritten for color indices with only a change in notation. Similar equations for broadband filter photometry, where the passbands are "wide" and color terms must be included, could be developed, but have not been, to the authors' knowledge. Strategies for avoiding the necessity to solve for the first order and color terms simultaneously are suggested by Schmidtke (1983c). The algorithm to be discussed in the next section assumes that a number of stars are each observed a number of times over a significant range of airmass. For this to be effective in measuring the extinction, the observations should be spread through the night and include stars which are rising and setting. Further discussion of the strategy of measuring extinction may be found in Hayes and Crawford (1986).

II. THE ALGORITHM

Suppose we observe J stars, K_j times each at I different wavelengths. For observations in which the coincidence correction is already applied and the sky readings have been subtracted, the observed magnitude of the kth observation of the jth star at the ith wavelength is given by:

$$m_{ijk} = m^o{}_{ij} + a_i \cdot X_{ijk} \qquad (1)$$

where $m^o{}_{ij}$ is the magnitude outside the atmosphere (*i.e.*, for zero airmass), a_i is the extinction coefficient, and X_{ijk} is the airmass. Observations at the I wavelengths for a given star are made simultaneously using spectrophotometers with array detectors, so X_{ijk} is taken to be independent of i. With grating—scanners, the scan is done up and back (or down and back), so the combined scan may normally be treated as if the observations at the I wavelengths had been made simultaneously. This will not hold true at airmasses so large that the time rate of change of airmass cannot be approximated as a constant. The equation makes no assumption regarding the standard—system values M_{ij}, which is related to the $m^o{}_{ij}$s by the transformation zero points. The determination of the zero points will not be discussed here.

For simplicity, the is can be suppressed, remembering that the set of equations given below is for one of the I wavelengths; the algorithm must be repeated for each wavelength in the photometric system. Since there are no color terms, the equations for the different wavelengths are independent of each other.

If just one star were observed K times, where K is greater than 2, then the least—squares method could be applied to solve for both the extinction coefficient, a, and the extra—atmospheric magnitude, m^o, simultaneously. This is the process of fitting a straight line to the observations of one of the stars whose data are shown in Figure 1. If a number of stars were observed, the same process could be used and a value of a would be obtained for each star; a weighted mean would be calculated to give the extinction coefficient for the night (Schmidtke 1983a). The method adopted here is a **global** solution; we assume that the same value of a pertains to each star. Thus, all multiple—observed stars are included in the solution at the same time. This might seem to be impossible, since the magnitudes m_{jk} and $m^o{}_j$ are not the same for the different stars. Rewriting the first equation will clarify the situation:

$$(m_{jk} - m^o{}_j) = \text{constant} + a \cdot X_{jk} \qquad (2)$$

The quantities on either side of the equation are independent of which star is observed. In Figure 1, this process is equivalent to shifting the lines until they lie on top of one another. Instead of shifting one line to lie on the other, we shift **both** lines until the magnitude at zero airmass is zero; that is, the constant is **defined** to be zero. One consequence of this approach is that one cannot also solve for the extra-atmospheric magnitude, m^o_j, at the same time that the extinction coefficient is being determined. The standard procedure is to follow the determination of the extinction coefficient by a series of substitutions of the value of a into the K equations like Eqn. (1) for each of the J stars and calculate weighted means of the m^o_js.

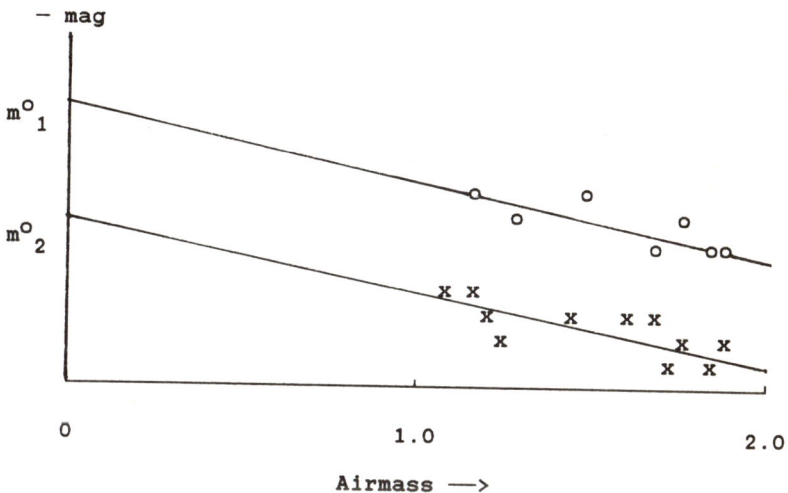

Figure 1. *Observed magnitudes as a function of airmass for two stars. See text for further details.*

Applying the least-squares method (Beers 1958) and assuming all the error to be concentrated in the magnitudes (*i.e.* the airmasses are measured perfectly), the extinction coefficient is:

$a = A/B$, where

$$A = \sum_{jk}^{JK_j} m_{jk} X_{jk} - \sum_j^J K_j^{-1} (\sum_k^{K_j} m_{jk} \sum_k^{K_j} X_{jk}) \quad (3b)$$

and

$$B = \sum_{jk}^{JK_j} X^2_{jk} - \sum_j^J K_j^{-1} (\sum_k^{K_j} X_{jk})^2 \qquad (3c)$$

These equations are similar to the conventional least–squares solution for a simple linear equation except that the normal "single sums" over k have been replaced by the "double sums" over j as well as k. Setting J=1 gives the conventional solution.

The next step is to determine the error in the extinction coefficient. The standard deviation, s_a, in the value of a is given by:

where

$$s_a = \left[\frac{\sum_j^J \sum_k^{K_j} (\delta m_{jk})^2}{\left[\sum_j^J K_j - (J+1) \right] \cdot B} \right]^{1/2}$$

and

$$\sum_j^J \sum_k^{K_j} (\delta m_{jk})^2 = C - A^2/B,$$

$$C = \sum_j^J \sum_k^{K_j} m^2_{jk} - \sum_j^J K_j^{-1} (\sum_k^{K_j} m_{jk})^2$$

The calculation of s_a needs only one new set of sums, denoted "C," and these are similar to the sums in A and B. In the denominator of the equation for s_a is the quantity: $\Sigma K_j - (J+1)$. In a least–squares solution for a single star this would read: $\Sigma K_j - 2$. The quantity "2" counts the number of constants in the solution, which includes the slope and the intercept. In our solution there are J stars; a line of arbitrary slope can be placed through a single point, so at least two observations of star j are needed if the observations are to contribute to the solution for a. This means that for J stars the constants in the solution include the J intercepts plus the common slope a, or a total of $J + 1$.

ACKNOWLEDGEMENT

This approach was inspired by the use of program SCANNER, which was constructed by David W. Latham. The algorithm was derived independently some years later in order to benefit from the advantages of the global solution and the explicit calculation of the error in the extinction coefficient. The details of program SCANNER have not been published, to the authors' knowledge.

REFERENCES

Beers, Y. 1957, *Introduction to the Theory of Error* (Reading: Addison–Wesley Publ. Co.), p. 38 ff.
Hardie, R. H. 1962, in *Astronomical Techniques*, ed. W. A. Hiltner, (Chicago: U. of Chicago Press), p. 178.
Harris, W. E., FitzGerald, M. P. and Reed, B. C. 1981, *Publ. Astron. Soc. Pacific* 93, 507.
Hayes, D. S. and Latham, D. W. 1975, *Astrophys. J.* 197, 593.
Hayes, D. S. and Crawford, D. L. 1986 in *Automatic Photoelectric Telescopes*, ed. D. S. Hall, R. M. Genet and B. L. Thurston, (Mesa: Fairborn Press), p. 87.
Hayes, D. S. and Philip, A. G. D. 1983, *Astrophys. J. Suppl.* 53, 759.
Manfroid, J. and Heck, A. 1983, *Astron. Astrophys.* 120, 302.
Manfroid, J. and Heck, A. 1984, *Astron. Astrophys.* 132, 110.
Philip, A. G. D. and Hayes, D. S. 1983, *Astrophys. J. Suppl.* 53, 751.
Popper, D. M. 1982, *Publ. Astron. Soc. Pacific* 94, 204.
Rufener, F. 1984 in *Advances in Photoelectric Photometry*, ed. R. C. Wolpert and R. M. Genet, (Fairborn: Fairborn Press), p. 156.
Rufener, F. 1985 in: *IAU Symp. No. 111: Calibration of Fundamental Stellar Quantities*, ed. D. S. Hayes, L. E. Pasinetti and A. G. Davis Philip, (Dordrecht: D. Reidel), p. 253.
Schmidtke, P. C. 1983a, *IAPPP Comm.* 9, 95.
Schmidtke, P. C. 1983b, *IAPPP Comm.* 12, 74. (*Erratum* to 1983a).
Schmidtke, P. C. 1983c "IRAF Filterphot Package: All–Sky Reduction Equations, Differential Reduction Equations, Methods of Coefficient Calculation," (Tucson: Kitt Peak National Observatory).

SECTION 2b – INTRODUCTION: AUTOMATED SPECTROSCOPY

Michael A. Seeds

Pennsylvania Astronomy Research Consortium
Franklin and Marshall College

INTRODUCTION

The feasibility of automatic photometric telescopes (APTs) has been established by a number of instruments which now produce good quality data with only routine assistance from an observer. These telescopes have shown their value primarily in synoptic programs which require regular, long term observations over a number of seasons. But many research problems in astronomy require spectroscopic as well as photometric observations, and synoptic spectroscopic observations are difficult to compile because of the way telescope time is allocated at the major observatories with suitable facilities. An automatic spectrographic telescope (ASGT), with simultaneous photometric capabilities, could provide such observations.

An ASGT could consist of a meter–size telescope capable of locating and photometering stars automatically. A fiber–optic cable

---

New Generation Small Telescopes, ed. D. S. Hayes, R. M. Genet, & D. R. Genet.
© 1987 Fairborn Observatory.

would allow the telescope control system to direct light to the spectrograph located in an environmentally controlled room separate from the telescope. A separate computing system, when strobed by the telescope control system, would operate the spectrograph, record the CCD image, and then return control to the telescope control system. Though an ASGT is more complex than an APT, there appear to be no serious problems barring an ASGT.

It is now feasible to build and operate an ASGT. Present technology in the form of microcomputers, CCD detectors, and data storage media are sufficient to control the instrumentation, detect the spectra, and record the data. Those problems that do exist seem tractable. The papers in this session describe some of the problems unique to ASGTs and the proposed solutions. Such problems include control algorithms, CCD data handling and storage, routine verification of spectrographic image quality, and the acquisition of and guiding on faint stars.

It seems very likely that one or more ASGTs will be placed in service in the coming years. Synoptic, spectrographic observations are a powerful tool for the study of many classes of stars from Be stars to long period variables. The value of such observations far outweighs the modest cost of an ASGT.

THE FEASIBILITY OF AN AUTOMATIC SPECTROGRAPHIC PHOTOMETRIC TELESCOPE

Michael A. Seeds

Pennsylvania Astronomy Research Consortium
Franklin and Marshall College

I. INTRODUCTION

An automatic spectrographic telescope (ASGT) is capable of locating and observing stars spectrographically and photometrically without the assistance of a human observer. Thus an ASGT resembles an APT (automatic photometric telescope) with the addition of a spectrograph and the associated data handling system. Such a system would be ideal for synoptic studies of several classes of stars.

The Pennsylvania Astronomy Research Consortium (PARC), a group of nine astronomers at five private institutions in south-central Pennsylvania, has made a year-long design study directed toward developing an ASGT for use in a major research program concentrating

---

New Generation Small Telescopes, ed. D. S. Hayes, R. M. Genet, & D. R. Genet.
© 1987 Fairborn Observatory.

on chromospherically—active stars. Though the study has isolated a number of problems unique to the ASGT, none of these appear to be insurmountable.

II. ASGT OVERVIEW

An ASGT would resemble an APT with the addition of a spectrograph. the spectrograph would be fed by an optical fiber, which would allow the spectrograph to be located off the telescope in an environmentally controlled housing (Ramsey and Huenemoerder 1986). Thus the spectrograph would not be limited by space constraints on the telescope and could be built on a quality optical bench for stability (Bopp 1986).

During routine operation, the ASGT would work from a starlist in the telescope control computer (TCC), moving from star to star and obtaining photometric data as indicated in the starlist. Certain stars in the starlist would be flagged for spectrographic observations, and when the TCC selected such a star, it would center the star and obtain photometry before moving the beam to the telescope end of the optical fiber. Once the starlight was directed onto the fiber, the TCC would hand over operation to the spectrograph control computer (SCC) and then supervise guiding while the SCC proceeded to obtain a spectrum.

The SCC would be responsible for controlling the spectrograph and managing the data flow from the system. Exposure times based on magnitudes in the starlist or derived from current photometry would be set by the SCC. When an observation was complete, the SCC would be responsible for handing control back to the TCC and then recording the CCD image on suitable data storage media.

Thus an ASGT could operate automatically at a remote site on a large number of nights during the season. A human would be needed for periodic support such as the exchange of data storage media, the modification of the starlist, and maintenance of the CCD cooling system.

III. DESIGN PROBLEMS: TELESCOPE

Problems in the design of an ASGT can be classified as problems with the telescope, problems with the spectrograph, problems with operational control, or problems of management. The telescope problems arise from the constraint that the ASGT under consideration support the PARC research project in chromospherically active stars. In order to do so, the system must be able to record the spectrum of a 9th magnitude star at a resolution of 1 to 0.5 Å with an exposure of about one hour.

The constraints imposed by the research project require a 1—meter class telescope. In addition, the mirror must be relatively light—weight to permit a light—weight telescope with a low moment of inertia that can be moved easily and positioned accurately by stepper motors (Baliunas, Boyd, Genet, and Guinan 1985). Thus the PARC design study suggests an "egg—crate" or slumped—thin mirror of meter class.

A second design problem argues for a mirror slightly smaller than 1 meter diameter. A short focal ratio is most economical in terms of mounting and housing the telescope at a pre—existing site. But a 1 meter mirror of very short focal ratio would be prohibitively expensive even if it could be constructed. A longer focal ratio or a smaller diameter dramatically lowers the cost of the mirror.

The PARC design study proposes a mirror of 0.8 meter aperture. Such a mirror can be constructed economically from thin sheet pyrex using a slumping technique and can be housed in a light—weight, compact telescope suitable for automation. Most of the loss of light in reducing the aperture from 1 meter to 0.8 meter can be recovered by careful design of the fiber coupling head at the telescope end of the fiber as well as by more careful attention to the mirror coatings.

Because the ASGT must be capable of observing fainter stars, acquisition and centering is more difficult. Current APTs routinely acquire and center stars as faint as 8th magnitude. An ASGT suitable for the PARC research project must be capable of photometry to 12th magnitude, and the sky (at a galactic latitude of 40 degrees) contains 45 times more stars per square degree of 12th magnitude or brighter, compared to stars of 8th magnitude or brighter (Allen 1976). Put another way, if an APT can detect stars of 8th magnitude or brighter, then on the average, stars are separated by about 54 arcmin. An ASGT capable of detecting stars of 12th magnitude or brighter must be able to locate stars whose average separation is only 8 arcmin.

Clearly the mounting and control system of an ASGT must be as precise as possible, but is unlikely that an ASGT will be able to slew large distances across the sky and acquire 12th magnitude stars without error. The solution to this problem is to use brighter acquisition stars near the program stars (Boyd, Genet, and Hall, 1985). The telescope can locate the acquisition star and then offset to the fainter program stars. This requires that the identification algorithm include magnitude information, but it avoids the complexities of pattern recognition.

In order to obtain a spectrum, the ASGT must be able to place the star image on the telescope end of the optical fiber. The fiber in the PARC design has a diameter of 4.4 arcsec projected on the sky. Placing the image on a fiber of this size should not be difficult if the system can first center precisely using the photometer and then offset to the fiber. Designs using a flip mirror have the disadvantage of one additional reflection and the repeatability in the positioning of the flip mirror, though new techniques in microstepping stepper motors may at least partially resolve the latter problem.

The centering problem can be largely resolved if a closed—loop system is used for guiding. Certainly, it is necessary to keep the star image centered on the optical fiber during spectrographic exposures that could be as long as an hour. If the telescope control system is capable of detecting the location of the star image with respect to the optical fiber, it can generate a correction to move the star image back to the fiber. Thus centering and guiding are improved by a closed loop system.

It seems possible for an intensified CCD to observe reflected starlight from the plate holding the optical fiber just as human spectroscopists guide by watching light reflected from the slit decker or a conventional spectrograph. Another design proposes interrupting the exposure for 1 second at regular intervals to check the location of the star image. The former solution is preferable because it eliminates the repeatability problem in positioning a mirror and because it would permit the CCD to integrate for short intervals to guide on faint images. Designs which use beam splitters (Heacox 1986) are much less desirable because they steal photons from the spectrograph and because they are likely to have brighter limiting magnitudes than a decker reflection system.

IV. DESIGN PROBLEMS: SPECTROGRAPH

The PARC design study for the spectrograph calls for a cross dispersed echelle using a refrigerated CCD as a detector. Here we will consider three problems with the automation of such an instrument.

One of the most difficult problems is the cooling of the CCD chip to liquid nitrogen temperatures without requiring human support more often than once a week. Four options are available, but none are ideal. Thermoelectric cooling systems are the most convenient, but they cannot get as cold as other systems without using a running water bath to remove heat. Given the availability of water on mountain tops, this would require a closed cycle water circulation system.

A second solution is closed-cycle refrigeration systems, which require no maintenance and can operate for roughly 1000 days without attention. In a laboratory setting, such refrigerators can cool to a few degrees Kelvin. However, such systems must pump coolant through the cooling system, which can introduce serious vibration communicated to the CCD chip by the cold finger. Of course, vibration in the spectrograph appears in the data as a loss of resolution. Also, some refrigeration systems require water cooling.

Joule-Thomson cooling systems can attain temperatures down to 77 degrees Kelvin by venting clean, dry nitrogen or argon through fine pores in a plate (Little 1984). Experiments at Lick Observatory developed problems with ice clogging the pores. Although this problem may have been related to contamination in the nitrogen, such systems seem presently unsuitable for use at remote telescope sites because of a lack of reliability and safety.

A liquid nitrogen (LN_2) bath is not acceptable because it must be topped off a least once a day. However, automatic LN_2 refill systems have been used in laboratories (Landis, Madden and Goulding 1986) with no valves in the supply line. The freezing of valves is the principle source of trouble with such systems. A 50-liter supply dewar refilling a 5-liter detector dewar requires fresh LN_2 about once a week. Thus if LN_2 is available at the site, and if weekly support is available, an automatic refill system is feasible.

In addition to cooling, an automated spectrograph could require periodic attention for alignment and focus. Experience with an existing fiber-coupled echelle spectrograph shows that once adjusted and focused, the spectrograph is quite stable if the environment is controlled. However, the adjustment of such a spectrograph is delicate and probably cannot be automated using present detectors and computing technology without tremendous design effort. Such adjustments, though rare, will require the presence of a human.

A third problem with the automation of a spectrograph is the volume of data produced. A single 512 × 512 CCD image can require roughly 0.5 megabyte (Mb) of storage. On any night when spectra are recorded, roughly 20 flat fields and 20 bias fields will also have to be recorded, so a minimum of 21 Mb is needed per night. Averaging flat fields and bias fields can reduce this minimum to 2 Mb per night, but that is still a large amount of data to record. Another paper in this session (Karshner 1987) will describe solutions to this problem.

V. DESIGN PROBLEMS: OPERATION

A number of design problems are related to the operation of an ASGT. For example, photometric data flowing from an APT must be monitored nightly to assure that equipment failures have not degraded the accuracy of the data. Should an equipment failure temporarily destroy the value of spectra recorded by an ASGT, large amounts of telescope time could be wasted until the problem was detected. Thus, the quality of the spectra recorded by the CCD chip must be verified as often as possible. This would seem to require that a sample of the spectrographic data be examined as often as daily. Such a verification system could be set up using modems and telephone lines if some preprocessing could be done on site to exact specific portions of the spectrum for examination. Transferring entire CCD images over telephone lines would be very time consuming.

Another problem relates to the recovery of data should an error occur during a night. A single power fluctuation during the exposure of a flat field could degrade all spectra on that night. Obviously bad flat fields or bias fields could be detected automatically and omitted from the averaging process, but this would not detect all possible errors. To recover from such errors, the astronomer must have access to the original CCD images. Thus the designers must resist the temptation to extract linear spectra automatically at the telescope and minimize storage capacity by discarding the CCD images themselves. Individual CCD images must be recorded for later analysis at least until the long term reliability and stability of the system can be verified.

Because an exposure could be as long as an hour, spectrographic observations could easily dominate the observing program unless great care is taken to balance the starlist and the selection algorithm. The starlist must include, at a minimum, an object ID, coordinates, magnitude, a list of filters, a spectrum flag, an observation probability, and "must observe" dates. Obviously coordinates and magnitudes are used to locate and identify the right star. The filter list maximizes the efficiency of the system by preventing it from observing through more filters than are necessary for a given star.

The remaining items in the starlist are related to the operation of the spectrograph. The spectrum flag identifies the star as a candidate for spectrographic observation. The observation probability specifies the likelihood that a spectrum is to be recorded on a given night. Thus we can prevent the system from recording the spectrum of a particular star more often than necessary and the randomness in the data spacing introduced by the use of a probability will prevent any unfortunate aliasing that could be produced by observing the star at precisely regular intervals. The "must observe" dates permit us to assure that a certain star will be observed at a critical point in its variation or at a time when collaborating astronomers are also observing.

Because the selection algorithm is complex and spectrographic observations are time intensive, it is critical that the starlist and selection algorithm be simulated before being placed into service or modified. Otherwise an inadvertent combination of stars and selection criteria could result in some objects never being observed.

Some current APTs now operate with a starlist that is modified quarterly, but the complexity of an ASGT and its potential require easier access to the starlist. The complexity of the control system and selection procedure means that errors may occur which require quick correction to prevent the loss of valuable telescope time. Modifying the starlist quarterly is not sufficient to avoid lost time. Also, the potential of a one-meter class ASGT is so great that some provision must be available to permit observation of unexpected events. The flaring of a cataclysmic variable, the sudden change in period of an intrinsic variable, or the appearance of a nova or supernova could call for quick modifications to the observing list.

On the other hand, modifications to the observing list must not be too easy. One can imagine how a member of the observing team could use a modem to inadvertently disable the observing program by making what seemed to be a minor change in the starlist.

VI. DESIGN PROBLEMS: MANAGEMENT

A small set of problems inherent in the design of an ASGT can be classified as management problems. While an ASGT could be developed and operated by any individual or institution, the potential of such a facility is well suited to a consortium of astronomers from small institutions. A consortium brings together people with a wide range of talents ranging from digital electronics to cryogenics. Being a long-term project, an ASGT may not appeal to astronomers at major institutions where the fruitfulness of a line of research is often judged in months rather than years. A consortium of astronomers at smaller institutions may be better able to manage an ASGT for a number of years to pursue research that larger institutions cannot consider for a variety of reasons.

A consortium of astronomers may take one of two paths toward the utilization of an automated telescope. The N astronomers in the consortium may each claim 1/N the of the telescope's capacity for their particular observing program. Thus the facility becomes a common observing facility on which individuals work on separate research projects. As an alternative, the members of the consortium can cooperate on a single research project to utilize all of the telescope's capacity. Given the diversity in interests among astronomers, this almost certainly asks some to sacrifice their research interests.

The PARC study has taken a middle path. The members of a group developing an ASGT could agree to work as a team on a single major research project which would use 70 to 80 percent of the ASGT's capacity. The remaining capacity could be apportioned to individuals in the consortium for secondary research projects based on observing proposals which they submit. This provides a balance between making the instrument as fruitful as possible and maintaining the interests of a diverse group of astronomers.

VII. CONCLUSION

The PARC design for an ASGT described here is research driven. That is, characteristics of the design are fixed by the specific major research project which PARC wishes to pursue. A smaller instrument with lower dispersion might be adequate for bright stars, while a much larger ASGT would be needed to obtain quasar red shifts. Nevertheless, the PARC design would be adequate for a wide range of research projects.

The age of the general purpose observatory capable of making a wide range of observations in support of almost any avenue of research may be drawing to a close. Dedicated, automated instruments are technically and economically feasible. Many of the components of an ASGT can be reused in new system designs, and, compared to a major general purpose observing station at a good site, an ASGT is a bargain.

Should research trends 10 years from now lead in new directions, it may be less expensive in terms of design effort, development time, and actual funds to build an entirely new ASGT designed to maximize the return for the new line of research.

ACKNOWLEDGEMENT

This study was supported in part by funds from the Central Pennsylvania Consortium of Colleges. The author would like to acknowledge the contributions from the members of PARC, in particular Robert Boyle and Anthony Nicastro, who studied refrigeration systems. The PARC institutions are Bucknell University, Dickinson College, Gettysburg College, Franklin and Marshall College, and Lafayette College.

REFERENCES

Allen, C. W., *Astrophysical Quantities*, (London: Athlone Press), 1976.
Baliunas, S. L., Boyd, L. J., Genet, R. M. and Guinan, E. F. 1985, *IAPPP Comm.* 22, 32.
Bopp, B. W. 1986 in *Automatic Photoelectric Telescopes*, ed. D. S. Hall, R. M. Genet, and B. L. Thurston (Mesa: Fairborn Press), p.123.
Boyd, L. J., Genet, R. M. and Hall, D. S. 1985, *Sky and Telescope*, 70, 16.
Heacox, W. D. 1986, *Astronomical Journal*, 92 219.
Karshner, G. 1987 in *New Generation Small Telescopes*, ed. D. S. Hayes, R. M. Genet, and D. R. Genet (Mesa: Fairborn Press) p. 211.
Landis, Madden and Goulding 1986, *IEEE Transactions on Nuclear Science*, 33, 399.
Little, *1984 Review of Scientific Instruments*, 55, 661.
Ramsey, L. W. and Huenemoerder, D. P. 1986, *Proc. S.P.I.E.*

METHODS OF IMAGE STORAGE FOR AUTOMATED TELESCOPES

Gary B. Karshner

Physics Department, Gettysburg College

I. THE PROBLEM DEFINED

The first automated telescopes were used for photometry, in which the data gathered was condensed down to one point per observation. As the instruments evolved, the need to store more and more data grew dramatically. As an example, one and two dimensional CCD detectors presently gather thousands to millions of points of data in a single observation. This increased need to inexpensively store enormous amounts of data is examined here.

Typical linear−CCD devices have 128 to 2048 detectors in a row which make them an obvious choice for a spectrum scanner or a multi−channel narrow band photometer. Two−dimensional detectors have arrays of 100 X 100 to 1024 X 1024 detectors covering a two−dimensional surface. Since each detecting element or pixel generates one data word, it can be seen that each observation can generate up to a million words of data. Thus, in the course of a single, night many millions of data words

New Generation Small Telescopes, ed. D. S. Hayes, R. M. Genet, & D. R. Genet.
© 1987 Fairborn Observatory.

need to be stored. By the end of a month's time of observations with the telescope, this number would swell to hundreds of millions or even to billions of words that need to be stored and later archived.

For example, suppose the image size is 512 X 512 pixels and each pixel needs 16 bits (or 2 bytes) to represent its analog value. Then:

bytes stored per picture = $512 \times 512 \times 2$ = .5 Megabytes

A night's worth of data might include 20 pictures or 10 Megabytes of storage per night. Thus, if the telescope is to maintain itself for one month unattended, then 300 megabytes of system storage will be required.

Modern computers are very capable of this kind of storage but have not been standardly configured for our specific problem. A roll of 9−track tape can store 120 megabytes of data, for instance, and even larger capacity replaceable magnetic disks of 300−megabyte capacity are made. But most of these devices are designed for high−speed systems where speed of access to the data is very important and this speed, which our situation doesn't warrant, comes at high cost. Another drawback to present configurations is that, with the exception of 9−track tape, most of these devices are designed to work directly with the expensive high−speed super−mini and main−frame systems. They would therefore require modifications in order to be used with the low−speed microcomputers that better suit the rest our problem.

Speed and ease of access is not at all critical for our application. The data is gathered over many minutes and, once stored, need not be accessed until it is to be processed, hardly a high speed problem. A 16−bit microcomputer is ideally suited to this computational problem and these systems are inexpensive and very flexible. Because the recent development of large Winchester disks that are fixed and not removable for storage, the industry has developed inexpensive archival storage. These devices are ideally suited to our task, and will be examined in some detail.

II. A DETAILED LOOK AT MASS STORAGE METHODS

Mass storage methods offer a large number of possibilities. The considerations here are the number of units (as opposed to their storage capacity per unit) necessary to do the task, the cost and reuseability of the storage media, and the long term availability of each kind of media. That is, will it become a 'standard' or a 'dinosaur'? Four different options are discussed here:

a) *Magnetic tape storage*

This is the least elegant solution, and is also the easiest to implement as tape storage systems are old−style technology. It makes no

IMAGE STORAGE FOR AUTOMATED TELESCOPES 213

special demands on the operating system and media cost is low and the media are reuseable. Aesthetically we would be building something that would look like a large business computer with many tape drives standing around to do the task. The drives would not be in an ideal environment, and would suffer more from the environment than the other solutions given here.

There is a new magnetic tape media designed for archival storage with a capacity necessary for our task. The least expensive types are quarter inch tape cassette and half inch tape cartridge. The cartridge will store in the neighborhood 100 megabytes, and cost about $20.00 each, thus making it the least expensive media at present for large mass storage. The main disadvantage is the low capacity of the drives relative to the other solutions discussed here. Nonetheless, this solution would reduce the physical size of the system significantly and the media cost is extremely low. However, there is no indication that this kind of drive will become standard in the future.

b) Hard disk storage with tape backup

In this solution, one tape drive and a number of very large Winchester disk drives would provide the total storage needed. There are several advantages to this system: the technology is relatively mature, therefore down−time should be minimal; the system's physical size is greatly reduced (modern disk drives are much smaller than 1/2 inch reel tape systems); it would be much less susceptible to environmental damage; the high redundancy of parts means that swapping can be used to keep the system running in event of problems. There are two disadvantages with this system: many DOS systems will not address this volume of disk space; and the routine maintenance will require some time to down−load all the data from disk to tape.

c) Optical disk storage

Optical disk storage technology is a very new mass−storage technique that offers hundreds−of−megabyte data−densities that are of the magnitude useful for us. These disks range from very large archival devices, comparable to the large magnetic disk, to smaller ones ideally suited to our task. They are the simplest to implement in that one drive could do our entire job. Although they have a greater capacity, they also lack the erasability of the magnetic disks. Some of the drives are Read Only (RO), but others that can be Written Once and Read Mostly are the useful ones for us and go by the acronym (WORM). WORM drives have capacities of 100's of megabytes to gigabytes and can physically be the size of floppy−disk drives.

Also to be considered is that the cost of the media is currently very high, especially considering that it can only be used once! If one picks a drive that becomes standard, the media costs should drop. The non−standard ones will rise in cost or become unobtainable. Even the eventual standard disks will cost more than comparable magnetic media for some time, and it must be kept in mind that they may only be used once. In addition, although it would seem the media lifetime should be indefinite, manufacturers are only guaranteeing ten years.

III. PROPOSED SOLUTIONS USING CURRENT TECHNOLOGY

a) Solution A: System requiring once−a−month maintenance

In our system it is necessary to store a month's worth of data and be highly reliable during that time. In addition it must be flexible enough to allow it to control the spectrograph as well. For these reasons, we have chosen a VME−bus−based computer system with a 40−Megabyte hard disk. The disk would store backup software and a night's worth of data which would, during the day, be streamed to an available cartridge tape drive. Two or three such drives will be needed for a month's worth of data, depending on the exact number of pictures to be stored and the use of data compression algorithms. At current costs it breaks down as:

VME−bus 68010 computer with 1 Megabyte of RAM, power supply and chassis to hold additional boards.	$4000.00
Image−processing hardware and software	$4500.00
SCSI−bus interface for disk and tape drives	$2500.00
120−Megabyte 1/2" tape−cartridge tape drive ($1900 each)	$5700.00
Total	$16,700.00

This version of the system is designed for high reliability, so it will run unattended for a month at a time.

b) Solution B: System requiring daily maintenance

A low cost version that would be serviced once a day could be made for much less. We can use an IBM−PC−compatible computer which is very cost−effective due to its being in mass production.

IBM–PC compatible with terminal hardware and 20–Megabyte hard disk.	$1200.00
40–Megabyte tape–backup unit.	$800.00
Image processing board.	$1500.00
A/D converter board	$400.00
Total	$3,900.00

This computer can be used to control the telescope; the image processing hardware is mainly to reduce the data and check the alignment of the hardware.

SUMMARY

In responding to the increased need to inexpensively store enormous amounts of data during astronomical observations, I have offered two solutions which can be implemented using existing technology at moderate expense. The need for high–volume data storage should not stand in the way of adapting imaging systems to automated telescopes.

216   PEOPLE

Louis J. Boyd (left) discusses his approach to keeping automation simple. Douglas Hall (middle) and Eric Pearce (right) listen.

AMATEURS AND THEIR PLACE IN PROFESSIONAL SCIENCE

Robert A. Stebbins

The University of Calgary

I. INTRODUCTION

In the wider community, the thought that amateurs might contribute anything other than, perhaps, money and goodwill to professional science is only slightly less than preposterous. Science, according to the popular conception, is a highly technical and oftentimes abstract undertaking mastered only by men (the conception completely overlooks women in science) with a unique bent for intellectual esoterica and a passion for such cloisters as the library and the laboratory. Conceived of thus, the scientist is a special strain of humanity who develops into a social curiosity after years of specialized education and unstinting dedication to the solutions of problems so arcane that the average citizen can only marvel at their incomprehensibility. This is the public's image of science and the profession of scientist.

That some people might try from time to time to enter this lofty realm purely for the fun of it, for leisure, is even more inscrutable than

New Generation Small Telescopes, ed. D. S. Hayes, R. M. Genet, & D. R. Genet.
© 1987 Fairborn Observatory.

science itself and the professionals who work there. And when some of these leisure-seeking "eccentrics" indicate that they occasionally contribute something new to the science they are pursuing, the man in the street is more likely than not to disintegrate in utter disbelief. Science is for the spectacled, half-bald, wild-eyed genius, not for the ordinary being who lives next door.

But, notwithstanding these stereotypes, amateur scientists abound. Many of them contribute new knowledge to the science they are pursuing, although their professional counterparts sometimes depreciate those contributions. This paper, while it centers chiefly on the science of astronomy, draws on what sociology can tell us to date about the process of contribution engaged in by amateurs to all sciences, both physical and social (see Stebbins 1978, 1980). Our concern here is with the social aspects of this process rather than with the technical products that result from it.

II. INITIAL INTEREST

Professionals in science find enormous self-fulfillment and self-enrichment in their work, more than enough to offset any misgivings they might have about being seen as an "egghead," as "weird," or as antisocial. The sociological evidence on the matter is that amateurs find similar rewards (Stebbins 1978, 1982). The professional's interest in science as a vocation is usually sharpened during the years of undergraduate studies to a point sufficient to spur him or her on to graduate work and a master's or doctoral degree. From what has been learned so far (chiefly from research on archaeologists and astronomers), very few of these professionals were once amateurs.

The vast majority of amateurs, by contrast, enter their science as an expression of their intense curiosity about local phenomena. Local historians ordinarily live or once lived in the area about which they write, and they write from an interest in a past whose influences are still being felt in the present. Amateur ornithologists first gravitate to bird watching in their part of the country and then on to some form of data collection about certain species that have caught their fancy. Even studying the sky can be conceived of as a local activity, inasmuch as it is part of the local environment. (This conception is, to some extent, naive, since certain astronomical processes and objects [*e.g.*, lunar occultations and binary stars] have a different appearance when viewed through a telescope from different parts of the world.)

There are two additional factors that bear on this initial interest. One is that it is often mediated by friends or relatives who are already involved in science and who stir the interest of the neophyte to the point where he or she wants to join in. The neophyte's interest is there in latent form. All that is needed is interpersonal encouragement to pursue it and information about how to do so.

The second factor is that only a small minority of amateurs enter their science with any significant amount of formal education in it, say a bachelor's degree or several university-level courses related to it. Amateurs do commonly take a few courses to advance their interest once it is established. Many of these, however, are of the noncredit variety. The abstract and general nature of university degree courses in the sciences appears not to encourage the conduct of local science. Why this is so is taken up later in this article.

III. TYPES OF AMATEUR SCIENTIST

Drawing on a distinction used among amateur astronomers, it is possible to categorize scientists as either <u>observers</u> or <u>armchair participants</u>. The observers directly experience their objects of scientific inquiry; the armchair participants pursue their avocation largely, if not wholly, through reading. The latter hold to their approach either because they prefer it over observation or because they lack the equipment, time, opportunities, or physical stamina to observe.

Amateurs as observers vary much more than their professional counterparts in their levels of knowledge and their degrees of willingness and ability to contribute original data to their discipline. Observers can be described as one of three subtypes: apprentices, journeymen, and masters.[1]

Scientific <u>apprentices</u> are learners. They hope to learn enough about their discipline, its research procedures, and its instrumentation to function as journeymen and eventually, perhaps, as masters. As their knowledge about their science grows, apprentices may select a specialty, becoming learners here as well. The scientific apprentice, unlike his other opposite number in the trades, is normally independent; there is no formal association with a master over a prescribed period of time. Even at this stage apprentices have the freedom of a kind of avocational entrepreneur, though they are still incapable of contributing anything original to their field. Archaeology is an exception; its apprentices must be supervised lest they do irreparable damage to the excavation site.

The <u>journeymen</u> are knowledgeable, reliable practitioners who can work independently within one or a few specialties. They have learned enough to be able to make an original contribution to science. It is a matter of self-definition whether or not an amateur has reached this level of sophistication. The amateurs with whom I have had contact, however, are typically modest, even humble, about their attainments. They seem to sense when they are effectively apprentices, when they have much to learn, when they need supervision in excavating a site or they need more

---

[1] This is no attempt to liken amateur scientists to tradesman. In everyday usage these three terms are applied to any field where extensive knowledge and ability must be developed before independent practice is possible.

experience before producing a valid set of observations. Even journeymen may feel "inadequate" when they compare themselves with the local professionals with whom they have frequent contact. Like all scholars, the journeymen are always learning; expanding their grasp of the discipline as a whole and absorbing the new developments that pertain to their specialties. The same holds for the masters.

The <u>masters</u> actually contribute to their science. This they do by collecting original data on their own which advances the field. The master is aware of the major knowledge gaps in the specialty. The master knows how to make the observations that could conceivably close those gaps, he or she systematically collects the relevant data, and publicizes them through talks, reports, and journal articles. Any amateur may contribute through serendititious means, such as the chance discovery of a new celestial object, but masters systematically seek new data through a program designed by themselves (*e.g.*, writing a local history) or coordinated with someone else (*e.g.*, working as one of a team of lunar occultation observers spread across the continent).[2] Validation of one's status as master emerges from this activity: amateurs and professionals alike acknowledge the master's research when properly conducted, journal articles are accepted for publication, and speaking invitations are received.

There are two further observations to make about this typology. First, the passage of an avocational scientist from apprentice to journeyman is an inexact process. It is based on the acquisition of knowledge, experience, and personal confidence. Such development takes place gradually. Second, it is doubtful that any avocational science society contains a high proportion of masters; fifty percent is probably the upper limit. Though many members are capable of being masters, family and work obligations and other leisure interests drain away the time and energy available for steady original research. Furthermore, some amateurs dislike the regimentation and systematization required of the true master. Other, more easily gained, rewards, such as the fun of looking at the heavens, exploring for new minerals, or participating in the social life of amateurism, hold appeal for many amateurs. Different types of avocational scientists realize different returns from their leisure investments.

IV. AVOCATIONAL SCIENCE

At least four conditions encourage the emergence and persistence of an avocational interest in a science: (1) There are believed to be many undiscovered phenomena of scientific significance. (2) Many important characteristics and behaviors of known phenomena have yet to be

---

[2] All amateurs, even apprentices, are aware of the possibility of accidentally discovering something new. But even apprentices observe or do field work chiefly for other reasons, realizing how rare such discoveries are.

systematically studied. (3) These phenomena, characteristics, and behaviors are observable in their natural environment with the naked eye or with relatively inexpensive specialized equipment. (4) The number of professionals is too small to complete this work alone.

The dominant orientation in avocational science is active research, not armchair reading. But it is usually the observational research of exploration rather than the controlled research of experimentation. Amateur scientists tend to lose ground to the professionals as the problems to be dealt with grow more abstract. Some amateur geologists outdo their professional colleagues at on—the—spot recognition of minerals (Desautels 1969, p. 220), but I found that most amateur archaeologists needed and wanted supervision in classification of their materials. Hypothesis formation and theory construction are rarely done by the part—time scientist.

Contacts with professionals are cherished, for these scholars are both objects of a great deal of respect and models to be emulated. An amateur scientist soon learns that nothing catches the attention of a professional like research, especially research that is somehow publicized. Consequently, the master amateur enjoys the most intimate association with professionals; this in itself is a strong incentive to develop knowledge and skills. The amateurs appear to be widely accepted in archaeology and astronomy (especially the latter), though there are professionals in these fields who decry their presence.

There are still other reasons for pursuing the serious leisure of avocational science. For the master, and even the journeyman to some degree, there is recognition associated with doing research and particularly with disseminating it. Other rewards include self—enrichment, self—actualization and self—expression (Stebbins 1979). Though of secondary importance, the social and re—creative qualities of avocational science are also attractive.

V. AMATEURS AND PROFESSIONALS

Let us turn to a study of amateur—professional relations in astronomy (Stebbins 1982), relations that, from casual observations by the author, appear to be typical of such relations in many other sciences. From the standpoint of professional astronomers, the main intellectual link between them and their avocational colleagues is founded on the research contributions of the latter. In other words, when master amateur astronomers are writing reports and journal articles, the readers of which are other amateurs and professionals, they are contributing in ways considered valuable by a vast majority of the professionals in the study.

How is it that someone without the advanced training of the professionals is able to further the development of their science? As already noted for astronomy and other sciences with amateur wings, the lack of certain kinds of data and the conditions under which they must be collected invite amateur participation. There are celestial phenomena that require continual scanning of the evening sky if they are to be discovered (*e.g.*, comets, novae, meteors) and objects, though known, that require continual monitoring in order to record their major descriptive features (*e.g.*, variable stars, double stars, variations in brightness of novae, aurorae, positions and movement of the moon as determined by occultations). Sometimes these phenomena and objects are best observed, or can only be observed, where there is no permanent telescope (see Stebbins 1980). At other times a phenomenon or object may be visible from an observatory, but long—range commitments merely to describe or discover something are ruled out on expensive telescopes, which are often tightly scheduled with professional projects. Discovery and long—term observation are costly in time and fit poorly with "publish or perish." As we shall see shortly, they frequently center on peripheral questions and interests and are sometimes complicated by the need for teamwork, as in lunar occultations where a group of observers will spread themselves over several miles to facilitate the accuracy of their measurements.

Thus, a remarkable division of labor has emerged between professionals and amateurs in astronomy. Recognizing that, historically, amateurs are capable of original, high quality work, there is a tendency in this discipline to leave to them the exploration and description of the sort just mentioned. Though the sentiment is by no means universally shared, many professionals have faith in the work of "serious amateurs" (i.e., journeymen and masters).

Yet, despite this benign outlook, none of the professionals who were interviewed saw the amateur as strictly indispensable today, which jibes with the perspective of normal science (Kuhn 1970, pp. 23—34).[3] Astronomy would still develop at an acceptable rate if suddenly all its amateurs were to vanish from the face of the earth. The questions raised in normal science come from professionals and are eventually answered by them. At times, as in research on variable stars and the moon, there is reliance on the amateurs for certain descriptive data that they can supply.[4] When observing variables, they may actually note unrepeated phenomena which will never occur again. Were the amateurs to drop out of sight, professionals or specially trained assistants would have to collect this information. But much of astronomy, especially its theoretical side, operates quite independently of them. As noted later, professionals at work in these areas are scarcely aware of them and their contributions. In sum, there would be noticeably fewer gadflies (discussed in the next paragraph) and discoveries, and such losses would certainly have effects on

---

3 Up to as late as World War II, in the history of astronomy the amateur was often indispensable (for example, see Sidgwich 1958, pp. 14—16). But even during this period, indispensability was probably determined in retrospect.

4 One respondent described a research proposal he had reviewed just prior to the interview. Its author, a theoretically inclined professional, was planning to use amateurs to monitor some long—term but rapid variations in a star. The proposal had been submitted to a well—known funding agency.

the future course of astronomy. But the disappearance of amateurism would go relatively unnoticed in the normal science of the present.

In this connection, some professionals mentioned the presence of locally, and sometimes even internationally, known gadflies: those amateurs who publicize their explanation of some major celestial event as correct and that of established astronomy as in error. Perhaps the most celebrated of these is Grote Reber, an American amateur whose cosmological views are variously regarded by professionals. The remarks of one professional about Reber in particular capture the outlook of some of his colleagues about gadflies in general: "While there are some obvious misunderstandings, we would be foolish to ignore him completely." Another respondent, commenting on the same amateur, expressed the feelings of another segment of professional astronomy toward such people:

> "He is the epitome of the amateur astronomer who goes too far. . . These people don't have the background to give their opinions any substance. . . Nobody's impolite to him or anything like that, but he's not taken all that seriously."[5]

At least one set of gadflies have organized themselves into a group known as the British–American Scientific Research Association. Though such activity is rare among avocational scientists, some of its members work at a theoretical level; they are amateur cosmologists. Their main activities include publication of a quarterly mimeographed periodical entitled BASRA and correspondence with professional astrophysicists on the scientific differences existing between them.

But the amateur gadflies who, in their own way, try to serve as watchdogs (and occasionally succeed at this) are rare individuals, for most avocational scientists are quite restrained in their expressions of what they know about their discipline. Having received little or no systematic training in the field, as professionals do, they suffer a nagging uncertainty as to when their opinions are correct or incorrect (square with or contradict the conventional wisdom). Most avocational scientists are vaguely aware of their major knowledge gaps, but they are unable to clearly identify many of them. The result is lack of confidence, in some degree, in what they do know. This, in turn, discourages the adoption of the gadfly role.

The professionals' conviction that amateurs, except for certain gadflies, neither criticize nor are capable of useful criticism challenges the idea that such assessments reach professional ears and are taken seriously.

---

5 Even from the professional standpoint, Reber also has his respectable side. Until his recent retirement, he was employed as a radio engineer, while in his leisure he was a ham. He is credited with having single-handedly nursed radio astronomy through its infancy (Asimov 1964, p. 584) and with having coined the term "radio astronomy" (Edge and Mulkay 1976, p. 422). In the 1940's, some of his theories attracted a number of distinguished professionals whose calculations initially confirmed his ideas. Later, however, contradictory evidence emerged (Edge and Mulkay 1976, p. 85). He is still active as an astronomer through his own observatory in Tasmania.

So does the finding that most of the professionals interviewed had only superficial contact with amateurs. It is the sort gained when giving a talk at a local society meeting and fielding questions afterward. Four of the respondents had even less than this to do with amateurs. Only a third had continually associated with amateurs, which was done through some combination of regular attendance at Society meetings, frequent presentations at these events, joint research and writing projects, open nights at a local observatory, and telephone question−and−answer conversations.

For many professionals the problem is one of finding common ground on which to interact. They might conceivably find it in the affairs of their local societies. But nearly a third of the professionals had never belonged to one and another third had terminated their memberships, after having joined during their graduate student days. And, as noted earlier, research interests in the two groups are infrequently shared. Then there are the fundamental differences in motivation: the professionals see their work as a livelihood in which they must write about questions defined as important by their colleagues; the amateurs see theirs as captivating leisure which, for some, includes the thrill of contributing to a well−established and highly respected discipline.

VI. CONCLUSIONS

It is clear, that amateurs, in general, have neither the inclination nor the aptitude for professional or advanced empirical and theoretical work in the discipline. That is, seasoned amateurs in astronomy and other sciences are very likely to remain amateurs and not to be recruited to the ranks of professionals. Yet, they continue to contribute to science in ways valued by many of their full−time colleagues. Those ways are description and discovery. In contemporary astronomy they are still welcome.

Looking at the ideas contained in this paper leads to the question of whether astronomy should try to expand amateur involvement in data collection by systematically organizing descriptive projects in which avocational scientists can participate and by recruiting those who are properly trained to work in them. While the amateurs are less appreciated as gadflies (the same holds for professionals who adopt this role), professional opinion suggests the field is ready for a concerted push to involve amateurs more. In this connection astronomy might model its efforts after ornithology, another established science whose amateur wing has figured prominently in its advance (McCrimmon and Sprunt 1978).

REFERENCES

Asimov, I. 1975, *Asimov's Biographical Encyclopedia of Science and Technology* (Garden City: Doubleday).

Desautels, P.E. 1969, *The Mineral Kingdom* (London: The Hamlyn Publ. Group).

Edge, D.O. and Mulkay, M.J. 1976, *Astronomy Transformed* (New York: John Wiley).

Kuhn, T.A. 1970, *The Structure of Scientific Revolutions*, 2nd ed. (Chicago; U. of Chicago Press).

McCrimmon, D.A., Jr. and Sprunt, A., IV 1978, *Proceedings of a Conference on the Amateur and North American Ornithology* (Ithaca: Cornell Laboratory of Ornithology).

Sidgwick, J.B. 1958, *Introducing Astronomy* (London: Faber & Faber).

Stebbins, R.A. 1978, *American Sociologist*, 13, 239.

Stebbins, R.A. 1979, *Amateurs: On the Margin between Work and Leisure* (Beverly Hills: Sage Publications).

Stebbins, R.A. 1980, *Journal of Comparative Sociology* 21, 34.

Stebbins, R.A. 1982, *Urban Life* 10, 433.

SECTION 3: COMPUTER CONTROLLED TELESCOPES

Russell M. Genet

Fairborn Observatory

There are many cases where full automation is not practical or even desirable, but where some degree of computerization is entirely appropriate. Some amount of direct "hands on the telescope" is educational for students and enjoyed by some observers. There are cases, such as the observation of comets (difficult objects!), where a human really needs to be in the loop. Photometry of asteroid occultations can be over in a few minutes, and the flash of high-speed data can be recorded nicely by a microcomputer.

Three of the finest amateur observatories in the world are heavily computerized. The Braeside Observatory of Robert E. Fried is fully computerized, with all the telescope and instrument functions being controlled via computer from a warm room at quite some distance from the telescope. The operator only needs to verify (on the intensified CCD display) that the system is looking at the proper field, and then that the stars are properly centered. Having worked on telescope and photometer computerization for almost a decade now, Bob is one of the pioneers in the field.

New Generation Small Telescopes, ed. D. S. Hayes, R. M. Genet, & D. R. Genet.
© 1987 Fairborn Observatory.

The Big Cottonwood Observatory, designed and built by amateur astronomer Jerry Foote, is computerized. Jerry uses distributed microprocessing for telescope control—having different processors dedicated to performing different functions.

The Limber Observatory is one of the nicest and most modern observatories in the world. It uses two microcomputers, one to control the telescope and the other to control the photometer. As with the Braeside Observatory, operation is completely computerized, except for an occasional check on the field and "touch up" centering. It is one of the few, if not the only, amateur observatory to make regular photometric observations of the active galactic nuclei (AGNs).

Bruce Rafert, Norman Markworth, and their collaborators have evolved several of the most sophisticated and refined computerized photometric telescope systems in existence. Their designs have evolved over time towards the highly functional system represented by the 25-inch photometric reflector. Designed specifically for computerized photometry, this 25-inch telescope is a prime example of the economies and efficiencies that can be achieved with a specialized telescope/instrument combination.

An important component for many computerized systems is an intensified CCD camera. However, getting the camera and intensifier together efficiently and at low cost is not easy. Peter Manly, an expert in this field explains the process. Sometimes an important system element, such as the LeCroy MVL-100 single-chip amplifier-discriminator is suddenly not available anymore, and it takes a sharp designer like Jeff Hopkins to figure out a low-cost work around.

The group at the University of Florida has a long tradition of excellence of small telescope control with microcomputers, including one of the first such systems and the small fully automatic telescope at the south pole. They tell us about a computerized telescope based on the VME bus, and lessons from space-borne equipment that should apply to ground-based automatic systems.

Finally, Mark Trueblood of the Winer Mobile Observatory tells about one of the most critical parts of his system—the computer interface between his photometer and computer that pulls in high-speed asteroid occultation data.

BRAESIDE OBSERVATORY—A NEW BEGINNING

Robert E. Fried

Braeside Observatory

During 1986, we supplied photometric data to seven professional collaborators and one student. They are Hall of Vanderbilt, Herbst of Weslyan, Barden of Kitt Peak, Burke of King College, Howell of Goddard, Griffin of Cambridge, England, Kennedy of Brisbane, Australia, and the student from Vanderbilt, Jim Bruton. Work from previous years resulted in the publication of two collaborative papers, one in *Astronomy and Astrophysics* on BW Vul and one on EZ Peg in the *Publications of the Astronomical Society of the Pacific*.

In February we purchased a Sanyo Model 3800 CCD instrument video camera for the purpose of providing a starfield display to the control room. We then obtained an 18mm second generation image intensifier from Litton Industries. In addition, we purchased a set of Strömgren filters, and bought a ten acre tract of land on which to relocate our facility. In this paper I will describe how we use the camera and intensifier, and what we expect to achieve from the upcoming move to the new facilities sometime this summer.

New Generation Small Telescopes, ed. D. S. Hayes, R. M. Genet, & D. R. Genet.
© 1987 Fairborn Observatory.

The $750 Sanyo camera measures 2×2×4 inches and is equipped with a standard "C" mount for lens coupling and a standard 1/4–20 threaded socket for mounting. All electronics to produce a standard composite video signal are inside its tiny metal case. The array is a Sanyo Frame Transfer CCD with 485 × 572 pixels or 2/3 inch diagonal measure and is spectrally peaked near 7000 A. The whole system operates on a simple 12 VDC 500 ma AC adapter.

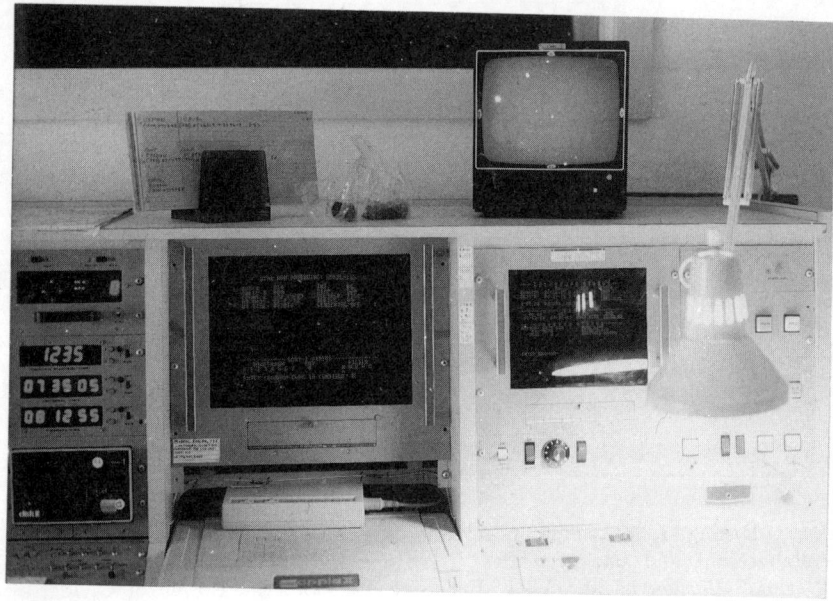

Figure 1. The observer's position showing the CCD camera display. The console display shows the host Apple computer running the observing program while the smaller display shows the complete telescope and dome status.

The Litton intensifier measures just 2 inches long and is 1–3/4 inches in diameter. Its integral high voltage power supply is powered by a 3 VDC AC adapter or the equivalent in batteries. Its flat front face is fiber–optics coupled to the cathode which is broadly peaked at 8300 A. Amplification is provided by micro channel plate (MCP) technology. This output is proximity–focused to the anode phosphor where fiber optics again carry the intensified image to the flat rear face. This image, then, is focused onto the Sanyo camera CCD array by the use of a transfer lens designed especially for this purpose by Litton.

The Sanyo camera will produce a usable image of an 8th–mag K– or later–type star when viewing through the Gort–I photometer we use with our 16" Cassegrain. By adding the Litton intensifier, we are able to gain approximately 5 mag for the same type stars. In tests, we have confirmed sightings of 3C 273 and SS 433 at 12th and 14th mag, respectively.

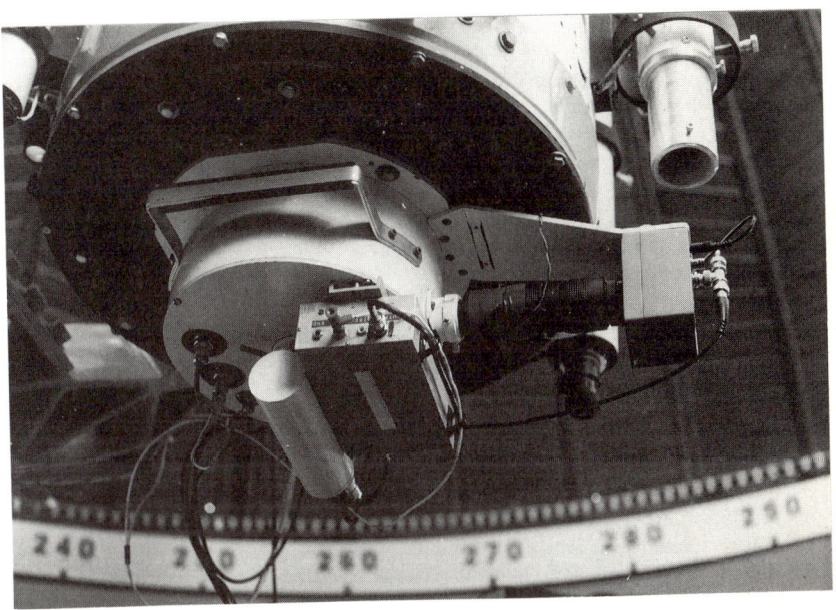

Figure 2. *A view of the PMT, the Taylor pulse amplifier, the Litton intensifier, and the Sanyo CCD camera.*

Now, the purpose of this whole setup is for field acquisition only, so that we can operate the telescope from a remote location. We have considered the possibility of using the camera for auto–guiding and will address this at a future time. We are presently working from a warmed control room separated 30 feet from the telescope. Initial pointing accuracy of the telescope from a cold startup will easily drop the program star onto the 8×10 arcmin screen display. However, flexure induces pointing problems when the telescope is positioned to zenith angles much beyond 30 degrees. We have mapped this flexure in the telescope mounting and found it to be repeatable. Because of this, we expect to complete a correction algorithm to compensate for this flexure.

I have discussed the operation of the GORT−I photometer at past meetings. Little change has occurred in its operation. However, with the purchase of the 1−inch diameter Stromgren filters has come the requirement to redesign the filter wheel to accommodate them. In addition, software changes will be required to permit operator choice in selecting either Johnson or Stromgren photometry together with changes in the reduction algorithm. We have already begun this process.

The last few months of 1986 were busy ones. As already mentioned, in July we purchased a 10−acre tract of level land. It is a quarter mile NW of the U.S. Naval Observatory and is located 5 miles west of downtown Flagstaff. The land is in a Ponderosa pine forest and surrounded by State land and undeveloped forest.

Because this was to be a whole new facility, we had the opportunity to design the place literally from the ground up. But, as with most design work, there are tradeoffs. One of these involved the buildings.

In our present location, the prevailing winds carry the roof boundary−layer air away from the telescope. At times, however, the breeze will come from another direction and wash the warm roof air around the dome causing an obvious deterioration of the seeing image and I felt these seeing problems should be addressed for the new installation.

Because our telescope operates in a remote controlled fashion, the decision was made at the outset to separate the telescope from the telescope control environment. We therefore elected to build two separate buildings, one to house the telescope and one to operate it from. The result is a steel sheathed building 20 feet on a side and 29 feet high. A control building containing a shop, darkroom, bedroom and bath, office space and control room, is located 100 feet north of the telescope building. The trade−off here is the more complicated communications link required in return for better seeing.

The first order of business after acquiring the land was to clearcut a 200 foot circle of trees for the telescope building. At the same time, we began making inquiries into road construction. We were fortunate in being able to obtain the services of a nearby interstate highway construction company. They donated their heavy equipment and material and built us a fine 24−foot wide crushed rock road.

As of today, the Telescope and the Control Buildings have been weathered in. Work is now continuing with the wiring, plumbing and heating. By late spring, we expect to have the buildings completed for installation of the observatory equipment. With the completion of this new facility, we hope to provide added service to our many collaborators. We will continue to offer observing experience to students working under the guidance of their professors. In the meantime, we continue to operate from our original location near Flagstaff.

BIG COTTONWOOD OBSERVATORY: THE FIRST YEAR

Jerrold L. Foote

Big Cottonwood Observatory

I. INTRODUCTION

The major activity at the Big Cottonwood Observatory (BCO) during 1986 consisted of adjustment and alignment of the 0.5 meter f/8.1 Ritchey–Chretien telescope. Construction of this instrument began in 1983 with first light late in 1985. The telescope is controlled by a distributed computer system which uses both single chip microprocessors and microcomputers to handle telescope motion, data collection, reduction and presentation.

In conjunction with the adjustment and alignment effort, the supervisory computer was constructed and the initial station–keeping programs were written. Optimization of the timing algorithms in the individual microprocessors controlling the stepper motors provided improved performance.

II. MECHANICAL MODIFICATIONS AND ADJUSTMENTS

Both the right ascension and declination drive trains are the same, with telescope motion provided by 200 step–per–revolution stepper motors directly coupled to 17–inch, 399:1 worm gears. To achieve the necessary small step angles the stepper motors are microstepped at a ratio of 64 to 1. The 12,800 microsteps per revolution translate into approximately 0.25 arcsec of telescope motion per microstep.

The original design called for 125 oz–in stepper motors (Oriental Motor #268–21) on both axes; however, after careful operational tests it was found that the right ascension motor lost steps under high speed slewing. Switching to a 306 oz–in motor (Oriental # 299–02) corrected the problem. The drive electronics required only a slight adjustment for the increased current that this motor required.

The polar–alignment process consumed a considerable amount of time with adjustments only made after a 1/2–hour observing run. Alignment is such that objects will remain within a 4–arcsec window for over one half hour. Part of this alignment process included fine tuning the timing loops for the microprocessors which control the stepper motors.

As the alignment process was taking place it was found that the microstep multiplexing frequency was particularly annoying at 2 KHz. Changing this to a more tolerable frequency without sacrificing performance was necessary if any long observing sessions were to be contemplated. Trial and error showed that multiplexing at a rate of more than 7 KHz resulted in decreased torque from the motors. A 6.2 KHz pulse from the microprocessor is now used and has proved very good, both from a torque standpoint and for operator tolerance.

III. SUPERVISOR COMPUTER

Both axes of the telescope have identical microprocessors which are used for direct control of that axis stepper motor. Each of these controllers are coupled to the supervisor computer via an STD BUS interface. The supervisor machine is a Z80–based computer running at 6 MHz. The controllers receive primitive commands from the supervisor which select the rate and direction of motion (Slew, Set, or sidereal speed, CW, or CCW rotation). Guiding motions are fed directly to the controller and bypass the supervisor.

The supervisory computer consists of two 8 inch double density disk drives, 128K bank–switched RAM, video–mapped RAM, two serial ports, a system clock and a keyboard. The machine can accommodate a 10–Megabyte hard drive as future storage requirements increase. The machine, power supplies and interfaces are mounted in a rack with wheels, and during development, are located at the telescope.

IV. SOFTWARE

The programs that run in the supervisor computer are written in Turbo Pascal. This language is easy to use, self-documenting and efficient. To date the major programs which have been written comprise the station-keeping routines involving Time, Date and Position, and a 2000-member object catalog with cross-references, callable from the main program.

Continuously displayed and updated are the Local Time, Universal Time, Local Sidereal Time, Julian Day, and Telescope Right Ascension and Declination. Run time access to other future routines is provided for by keyboard interrupt.

V. FUTURE WORK

It is becoming apparent that the 0.5 meter telescope at BCO will be used for more than one task. To this end, a universal instrument platform is under design for the Cassegrain focus. This platform will allow off-axis guiding and a standardized mounting for any instrument mounted to the head. To allow the maximum rigidity for this platform, a moving secondary focus system will be incorporated.

Two specific instruments have been investigated for use on this platform: a photon-counting photometer and a spectroscope. The accumulation of hardware for the photometer has already begun.

REFERENCES

Foote, Jerrold L. 1986, *Automatic Photoelectric Telescopes*, ed. D. S. Hall, R. M. Genet and B. L. Thurston (Mesa: Fairborn Press), p. 123.

Trueblood, M. and Genet, R. M. 1985, *Microcomputer Control of Telescopes* (Richmond: Willman-Bell).

PHOTOELECTRIC PHOTOMETRY AT LIMBER OBSERVATORY

David McDavid

Division of Earth and Physical Sciences
The University of Texas at San Antonio

I. INTRODUCTION

After receiving an M.A. in astronomy from the University of Virginia in 1977, I set off in search of fame and fortune as a saxophone player in a rock−and−roll band instead of pursuing a Ph.D., largely because of the dismal prospects for an astronomical career which prevailed at that time. I remained interested in astronomy, however, and by the time the band folded in 1979, I was teaching physics labs at UTSA and building an observatory/residence in the Hill Country northwest of San Antonio (Figure 1). The rooftop observatory was christened D. Nelson Limber Memorial Observatory in respect for my thesis advisor at the University of Virginia, whose untimely death occurred shortly after I was granted my degree.

New Generation Small Telescopes, ed. D. S. Hayes, R. M. Genet, & D. R. Genet.
© 1987 Fairborn Observatory.

Figure 1. D. Nelson Limber Memorial Observatory, Pipe Creek, Texas.

The observatory was originally built around a 14–inch Celestron, and the photoelectric era began with an EMI Starlight–1 photometer in 1982. It was then that I reluctantly bought an Atari 400 for data reduction, having mastered the concept of the RETURN key a year earlier as part of an indoctrination at McDonald Observatory. The prospect of typing in the raw data and sitting back while the computer did all the work was simply too enticing in comparison with my previous experiences hammering on a pocket calculator and drawing graphs to get photometry results, although I felt a bit uncomfortable about how I was going to be sure the computer didn't make a mistake.

Soon I had bought some chips, wrapped some wires, and written some code that allowed the Atari to control the photometer and log the data through its joystick ports, although I would not allow the numbers to be reduced automatically because I still felt the need to supervise the computer. My system at that time was to log data for each individual integration with an identifier, print out a complete listing later, then select and enter the measurements into a reduction program.

In late 1983 our department at UTSA had adopted the IBM PC for office work, which was the excuse I needed to buy a Compaq Portable and abandon the TV set, cassette recorder, and jumping spiders. About the time I finished the Starlight−1 interface card for the Compaq, Russell Genet came for a visit and introduced me to the world of Fairborn Observatory and the IAPPP. This resulted in the publication of the interface design in *IAPPP Communication* No. 16. The Fairborn Press' publication *Advances in Photoelectric Photometry, Volume 1* gave the details of Petr Harmanec's photometric campaign on Be stars, and, remembering this as one of Dr. Limber's favorite interests, I settled on that concentration for my observing program.

Now the photometry system was operating so well that the weak spot fell in the tracking inaccuracy of the C−14. Bright Be stars presented little difficulty because only short integrations were required, but I also had an interest in Seyfert galaxies and was making numerous observations of two exceptionally bright ones, NGC 1068 and NGC 4151. Thus I was especially vulnerable to Russell's suggestion that I be the first on my block to obtain one of the new computer controlled 16−inch telescopes being planned by DFM Engineering. I knew that DFM (Dr. Frank Melsheimer) had an excellent reputation from my days at UVa when he was brought in to consult on the new 40−inch at Fan Mountain Observatory. Encouraged by my growing success at the computerized photometry business up to that point, I took another cautious step and put in an order.

The months that wore on during the fabrication of the DFM scope allowed me to develop the photometry system I am now using. As outlined in the remainder of this article, the system is not what would be called an Automatic Photoelectric Telescope, because, enjoying the activity of observing and not wanting to attempt a great deal of complexity in automation, I decided to require a human operator for some functions. That is still my philosophy at present, and it is not likely to change in the near future (famous last words).

II. OVERVIEW OF THE SYSTEM

My computer controlled photometry system is unusual because it uses two computers rather than one. The reason is simply that this was the most convenient approach for me, since I had already built a successful interface between the Compaq computer and the EMI photometer before the DFM telescope was available. The DFM telescope control system is designed around an Apple computer and includes in its multitasking capability a programmable "USER" function especially for a purpose such as instrument control and data logging. However, it was more efficient for me to use the existing Compaq as the master control unit for photometry applications and to let it treat the entire DFM telescope system as a peripheral. Thus the main program for photometry runs on the Compaq, acquiring objects by accessing the Apple under the "USER" function. The observer is prompted to intervene in order to

verify and center each object before a counting sequence is executed.
Figure 2 shows the DFM telescope with its Apple control computer.
Figure 3 is from a different angle, showing the Compaq in its place next
to the telescope pier.

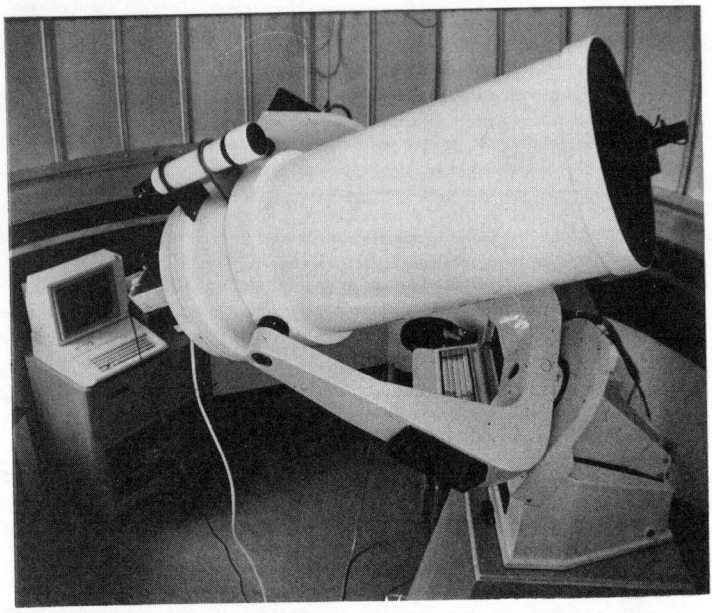

Figure 2. The Limber Observatory 16-inch computer controlled Cassegrain telescope by DFM Engineering, Inc.

The most commonly used programs are for differential photometry data acquisition and reduction. However, there are also programs for all-sky photometry in which observations of standard stars are made for the purpose of deriving extinction and transformation coefficients necessary for the differential photometry reduction.

III. HARDWARE

The photometer is a Starlight−1, made by EMI Gencom Inc.. It has an uncooled EMI 9924A photomultiplier tube and a UBV filter set consisting of a 1−mm Schott UG−1 (U), a combination of 1−mm Schott GG−400 and 1−mm Schott BG−25 (B), and a 1−mm Schott OG−515

(V). The correspondence between this filter–tube combination and the standard UBV system is demonstrated by the results of many all-sky observations of a variety of standard stars, which routinely yield a match with residuals on the order of 0.02 mag (roughly the precision to which the magnitudes and colors of the standard stars are defined). Although some observers have reported numerous difficulties with the Starlight–1, I have none to report except that I must be careful to cap the view eyepiece while counting is in progress because of a light leak.

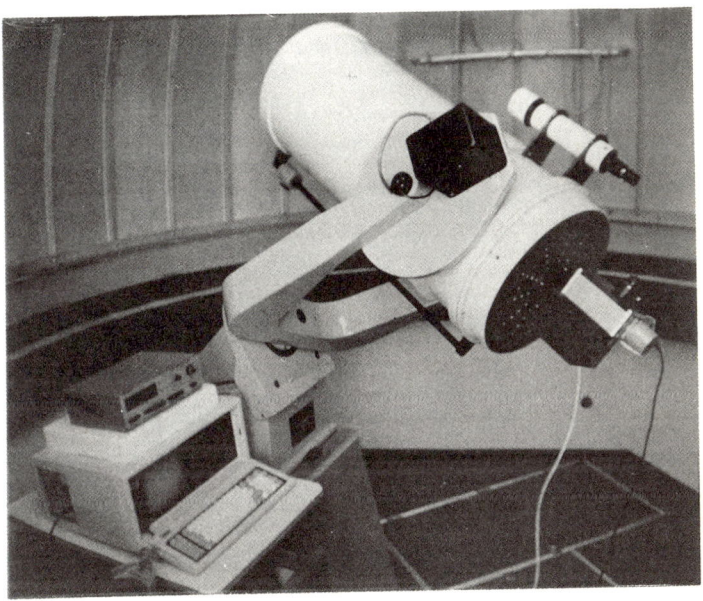

Figure 3. *The EMI Gencom Starlight–1 photometer and Compaq Portable control computer at Limber Observatory.*

The PMT deadtime has been measured as 175 ns to a precision of 18 ns by repeated applications of the "two aperture technique" (analysis of the ratio of counts in two different aperture sizes on a clear daytime sky, using filters to obtain one set of data requiring no coincidence corrections and one set for which coincidence corrections are necessary). A Fabry test (plot of count rate versus position as a star is scanned across the photometer diaphragm aperture) shows no measurable variation in sensitivity over an aperture of 34 arcsec, which is the size used for most observations.

The interface mentioned in the Introduction provides the connection between the Compaq Portable and the counter unit of the Starlight−1 to manage reset, start, stop, and readout functions. It is based on an Intel Programmable Peripheral Interface chip and an IBM Prototype Card which fits into an expansion slot in the Compaq.

The remaining control functions of the Compaq are wired to the standard parallel printer port in accordance with detailed specifications from the *IBM Personal Computer Technical Reference Manual*. An 8−bit output sends coordinates of objects and slew request codes to the Apple. A 4−bit input receives verification codes from the Apple. An additional 4 bits of bidirectional I/O are devoted to filter wheel control, which is accomplished by a garden−variety stepper motor (Superior Electric Type SS25−1159) through driver electronics copied directly from the Braeside Observatory design given by Bob Fried in the Fairborn Press publication *Microcomputers in Astronomy I*. The stepper motor is mounted on a bracket made from quarter−inch aluminum plate attached with screws to existing threaded inserts on the photometer head (Figure 4), so that no modifications to the Starlight−1 were required except removal of the filter wheel knob. When direct coupling between the stepper motor shaft and the filter wheel failed, I solved the problem by moving the motor over and installing a 4:1 gear set.

The number of separate cables in the system is minimized by using a single 25−conductor ribbon cable from the printer port to the small sheet metal box containing the stepper motor interface and power supply. From this box, one 9−conductor Atari joystick cable runs to the filter motor and one 25−conductor ribbon cable runs to the Apple. In other words, the communication line between Compaq and Apple is detoured through the motor driver box to simplify the cabling.

IV. SOFTWARE

The four photometry programs described in this section are all written in COMPAQ BASIC VERSION 2, which has proven entirely adequate for my purposes. Even the filter stepper motor and the photometer counter are managed by BASIC subroutines, so no assembly language calls were required.

1. ASP (All−Sky Photometry Data Acquisition)

The object catalog for all−sky photometry consists of data on 98 UBVRI Standard Stars taken from *The Astronomical Almanac*. After reading this information from a disk file, the computer prompts the observer to initialize the filter wheel, which is done by manually rotating the drive gear to a mark that identifies the "opaque" filter position and then plugging in the power supply. The filter wheel is returned to this position at the end of every star observation before the telescope is slewed to a new position.

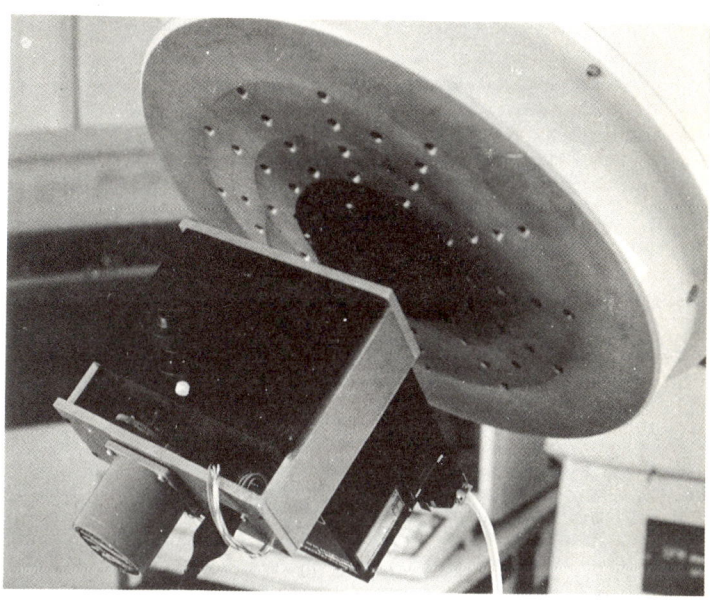

Figure 4. The Starlight-1 photometer head modified for automatic control of the filter wheel.

The next prompt is for the observer to enter a code for the observational data file to be created for this run. A random file is automatically named for the current day, month, and year, but the code entered by the observer is taken as a 3-character extension of the filename, making multiple files possible on the same night.

The computer then performs a sorting of the standard star database to identify 4 stars to be used for determining the extinction coefficients: a red star rising, a blue star rising, a red star setting, and a blue star setting. For instance, to select the red star rising, the series of conditions is (1) star above the horizon, (2) $B-V > 0$, (3) hour angle less than or equal to -40 degrees, and (4) airmass less than or equal to 3. Given the number of standard stars in the database, it is usually possible to find suitable extinction stars for the 3-hour observing sequence. The 4 extinction stars are then placed together in a block which is repeated 4 times, alternating with blocks of 4 program stars each selected for having

the smallest airmass remaining in the catalog whenever a new object is needed.
    Observation of an individual star begins with a prompt to call the "USER" function by pressing the space bar on the Apple. Pressing the space bar on the Compaq then writes an integer to the Apple that breaks a "wait" loop and transfers the coordinates as separate integers for hours, minutes, and seconds of right ascension and degrees, arc minutes, and arc seconds of declination. A subroutine including handshake codes is used for the transmission of each integer. The Compaq then prompts the observer to press the space bar, which sends an integer code to the Apple and initiates a slew. The next Compaq prompt is for the observer to center the star and press the space bar to correct any pointing error (typically on the order of 2 arc minutes) by updating the telescope control system coordinates. One more space bar prompt from the Compaq initiates the counting sequence of 3xU, 3xB, 3xV, followed by another prompt for the observer to offset to the sky for a UBV set of counts. If the standard deviation of each group of three 10−second integrations is less than 10% of the mean, the computer displays "DATA SET ACCEPTED", after which the raw counts together with starting and ending Universal Time and airmass are written to the disk file and the next object is selected. If a statistical inconsistency is found, there are options to abort the run or retry. It may be apparent by now that I find continuous use of the space bar to be a very easy and effective way for the observer to interact with the computers in a dark observatory!

2. RASP (Reduce All−Sky Photometry)

    This program reads a completed ASP observational data file, makes coincidence corrections, converts counts to magnitudes, matches the objects with the standard star data file, then performs linear least−squares fits including a goodness−of−fit parameter to find the extinction and transformation coefficients. An overall evaluation of the fit for each of the transformation coefficients is given as the average of the residuals between the standard values and the values calculated from the derived transformation. The trial fits intended to detect color dependence in extinction have been unsuccessful.
    As a matter of routine, I do an ASP run every few months to check the constancy of the transformation coefficients and to monitor variations in extinction.

3. DIP (Differential Photometry Data Acquisition)

    The differential photometry program is built around the fundamental features of the programs already summarized. After initializing the filter wheel and entering a data file code, the observer is presented with an object menu read by the Compaq from a disk file. The display of current airmass and number of repeated observations so far in the night is included as an aid in object selection. Entering an object number begins a series of Compaq−Observer−Apple interactions via space bar ending when the UBV counting sequence on check star, program star, comparison star, and sky is complete. The observer is in charge of centering each new object before counting begins, just as in the ASP program. One more touch of the space bar provides a quick look at the reduced data using a typical set of reduction parameters built into the program, followed by a dump of the raw data to disk if approved. One complete observation set with 10−second integrations can be made in about 20 minutes.

4. RDIP (Reduce Differential Photometry)

The differential photometry reduction program begins by reading the master file of objects for DIP followed by a requested DIP data file, then displaying the contents of the observational file as a list of records for individual stars. After entering extinction and transformation coefficients, the operator can request the differential reduction of any individual record.

V. CONCLUSION

After about two years of experience with this system, I think it is entirely satisfactory. Some of my recent work has been on period analysis programs written in Turbo Pascal and producing output on a Hewlett−Packard plotter, primarily using an adaptation of the Scargle periodoigram method due to Alex Fullerton of The University of Toronto. I have also written programs to produce data plots against time and phase.

Beyond my beginning attempts at period analysis, the most profitable use of the data will probably come from the international Be star photometry campaign. All the observations to date have been sent to Czechoslovakia for that purpose. In closing, I would especially like to thank Russell Genet for the motivation to expand my involvement in photoelectric photometry. It has been fun and highly successful!

THE SFA 41–INCH MICROCOMPUTER–CONTROLLED TELESCOPE

James B. Rafert

Department of Physics and Space Sciences,
Florida Institute of Technology

Norman L. Markworth

Department of Physics and Astronomy,
Stephen F. Austin State University

I. INTRODUCTION

The Stephen F. Austin State University 41–inch telescope is one of the largest automated telescopes in the world devoted entirely to

---

New Generation Small Telescopes, ed. D. S. Hayes, R. M. Genet, & D. R. Genet.
© 1987 Fairborn Observatory.

photoelectric photometry. Located 11 miles north of Nacogdoches, Texas on the SFA beef farm, the 41-inch represents a novel combination of extremely inexpensive technologies, which together enable the telescope to perform with the abilities of far more expensive, yet less versatile instruments. Presently, the telescope is used for all-sky photometry of the W Serpentis stars (see Wilson, Rafert and Markworth 1984; Rafert and Markworth 1984)-a program which involves the observation of an entire class of eclipsing binary stars, and the analysis of their light curves. As the initial stages of data reduction for each of these stars is performed on the same type computer as runs the telescope, we have come to think of the whole process of observation and reduction as a set of tasks.

This chapter describes the instrumental component of our telescope system-other papers in this Symposium (Markworth and Rafert 1987a, b) deal with our observational program and data reduction software. These activities are linked together by the observational program for several reasons. First of all, we observe each of our target stars which happens to be in its season once each night. As the orbital periods of the Serpentids are relatively long, this frequency of observations yields adequate phase density for most of the stars. But this also means that immediate reduction of the data is required so that the shape of the light curve can be monitored on a daily basis. If some sort of unusual behavior develops on one star or the other we are in a position to alter the observational program accordingly.

Also, owing to the rather long periods of most of the stars, we need to keep track of just which stars were observed on any particular night, so that we can be in a position to have observed all of the stars in the class with a similar, uniform density of observations and to have done so for all stars once we have gone through one orbital cycle of the longest period star (which is well over a year). Sometimes it gets cloudy. On partially cloudy nights not all of the objects which we wish to observe may be seen by the telescope while the skies are still photometric. Those "missing phases" need to be observed on the next clear night because otherwise, a phase gap will develop which cannot be filled for a year or more.

Finally, some of the stars in our program have relatively short (2 week) periods, and need to be observed several times during a single night. The telescope control computer keeps track of this type of data, and allows the operator to select just those objects which most urgently require observation. In this framework, it is easy to see why the processes of observation, reduction, and primitive data analysis must all be performed simultaneously. No doubt, the proliferation of microcomputer controlled telescopes, and the increasingly sophisticated programs which they are performing, will lead to an increase in the number of observational programs in which this approach will be required.

Our telescope system is composed of several major subsystems: 1) optics, including the video imaging system; 2) mechanical components; and 3) electronic components. Each subsystem is described in further detail in this paper.

II. OPTICS

The telescope has a 41-inch primary mirror, built by Tinsley Laboratories, and two secondary mirrors which are mounted in a flip-cage to provide either f/8 or f/14.7 Cassegrain foci. These optics, which were generously donated to SFA by Pan American University, were the initial "condensation nucleus" which got the telescope construction project started. Stored in a departmental laboratory for nearly a decade, the mirrors became "aware" that they were going to be used for something in the fall of 1983, when we crated and shipped them to Denton Vacuum. The primary was aluminized and overcoated, while the two secondaries were silvered and overcoated to improve performance as far as possible on our rather limited budget.

On the day that the crates returned to the science building loading dock, the level of excitement had reached a rather high level. Would the aluminization job be OK? Would the mirror be OK? There certainly was no source of funds to do it again. Of course, the boxes were opened to reveal Denton's normal, perfect job. Final tests in the laboratory indicated, again, that we had an excellent set of optics.

The optical layout, with either secondary mirror in place, is the classical Cassegrain design. Focus is achieved at the focal plane of a Thorn-EMI Starlight-1 Photoelectric Photometer, which is unmodified for our application. Small steppers (described in the next section) serve to rotate the filter, aperture and view/count knobs. The photometer has a set of transfer optics, which divert the light beam through the interior of the photometer, and bring it to a standard 1 1/4 inch eyepiece holder. The focal plane of this arrangement is found 1/4-inch outside the photometer, where we have placed the photocathode of a Varo 25mm Micro-Channel Plate Image Intensifier Tube (#3603-1) with a gain of about 55000. The tube is equipped with its own internal high voltage power supply (2.5V DC required) and has a limited ability to protect itself from bright light sources.

The resulting image on the intensifier's phosphor screen is itself imaged onto the vidicon mesh of an RCA low light level television camera. The video signal is displayed on a standard TV monitor. The transfer lens which we use is an 18mm Criterion eyepiece which happened to be handy during the optical bench layout. A similar arrangement is used on the FIT 25-inch (described elsewhere in this book; Rafert and Oswalt 1987), although a Canon f/2.8 28mm camera lens is used on that telescope and provides better image quality. We are presently designing optics which will optimize both of these video systems to their telescopes, cameras, and sky backgrounds. Because of light losses in the photometer head optics and in the imaging system, the limiting magnitude of objects visible on the control room monitor is about 14.5 under dark sky conditions. This is far more than adequate for our present application.

III. MECHANICAL SUBSYSTEM

Figure 1. A 48−inch Drive Disk is shown during assembly. The disk has just been attached to the south polar plug which uses a 6−inch bearing.

The 41−inch is patterned after the "spare parts" 40−inch at Lick Observatory (see Osborn 1980), the major difference being that we had no spare parts to start with. We did, however, have a pile of scrap steel, several large pieces of pipe which were spied laying in one of the athletic fields, and a machinist who was willing to provide some "on−the−job" training to two astronomers who looked like they could tell the difference between, say, a hammer and a screwdriver.

Figure 2. The Telescope Mount for the 41-inch telescope. Note the relatively large offset of the declination axis on the tube side of the mount (this is required for proper dome clearance). The declination drive disk serves as a large percentage of the required counterweight.

After observing the 40-inch in action, we were impressed by the smooth tracking and ease of construction of its friction drive system. The cost savings that such an approach allow are enormous. The primary components of our drive system are two 48-inch diameter steel drive disks. The disks were turned on a lathe to within 0.001 inch of round, then were hardened to Rockwell 40. Minor warpage was removed by grinding both faces flat, leaving a disk 7/8 inch in thickness. A disk is shown, installed on the south polar axis plug, in Figure 1. In all, the disks cost $2200. This can be directly compared to the $5000 *rental* fee that was required for the *plans* to build a tool to *make* a gear of the same size. Each disk is driven by a small (3-inch) drive roller mounted on a 1-inch steel shaft. We use a Mathis 169 tooth gear and stainless worm to provide the necessary gear reduction so that each motor step will correspond to about 0.1 seconds of arc.

The mounting itself is a modified German Equatorial in which the north end of the polar axis is supported by an A-frame structure and the south end by a 7-foot tall concrete pier. This height is required for the telescope to be able to clear the dome horizon, and to swing completely under the pier for servicing. We used a 12-inch diameter steel pipe with 1/2-inch wall for the polar axis, and an intersecting 9-inch pipe for the declination axis. Construction was largely "cut and weld," although the end plugs were machined carefully to fit surplus 4-inch pillow block bearings (a larger, 6-inch bearing was used on the south end of the polar axis). Two of the bearings were modified to fit within the declination axis; the entire mount is shown in Figure 2. The telescope is mechanically well behaved; we have noticed little flexure (as evidenced by pointing error) for a wide range of hour angles and zenith distances.

The telescope tube is composed of four major pieces. A double bottom mirror cell, scaled from the plans for the Lick 40-inch, supplies support for the primary mirror. Radial compensators composed of aluminum and Delrin provide near-zero thermal expansion control on the mirror's lateral position. The mirror cell attaches to a large tub-like structure, which also serves to attach the telescope to the mounting. The tub was fabricated from cold-rolled 1/4-inch thick steel, and was reinforced with both end rings and longitudinal members. The tub then serves to hold the telescope truss, which was made from 2 1/2-inch diameter steel pipe (hollow with 1/4-inch wall thickness). Finally, the secondary mirrors are held in a flip-secondary housing, which is attached to the end of the telescope truss. The telescope, shown in Figure 3, is relatively light-weight, was easy to build, and was easily assembled within the dome using a single hoist affixed to the dome shutter (we were careful in doing this to avoid lifting a component with the shutter only partly open, or in loading so much weight as to stress the dome).

The photometer is attached to a cylindrical tailpiece at the optical back focal position for the optics. Each of the control knobs on the photometer are turned by surplus Sigma steppers. This circuit is adequate for low speed operation of small steppers, and was designed as part of our effort to minimize costs by doing things ourselves, in the simplest, most straightforward manner. Each circuit is controlled by two bits of parallel I/O so that the telescope control program can control each motor independently.

As a point of interest to the reader, neither of us had any prior experiences in building a telescope of this aperture, in welding, or machine shop work. From the day we cut the first piece of metal, to the day the last bolt was installed spanned a period of only 9 months—much of which was spent dealing with budgeting, procurement and logistical details. We received quite a lot of help, and even more useful advice from the departmental mechanist who likewise, had never attempted such a project. Little by little we became better and better at machining, and in talking to other machinists over the telephone in the language they expected. Entire universes of wondrous products were unveiled for us, from Winfred Berg to Aircraft Spruce. Including the electronics to be described in the next section, our total cost for the project was $31,000 (excluding the optics).

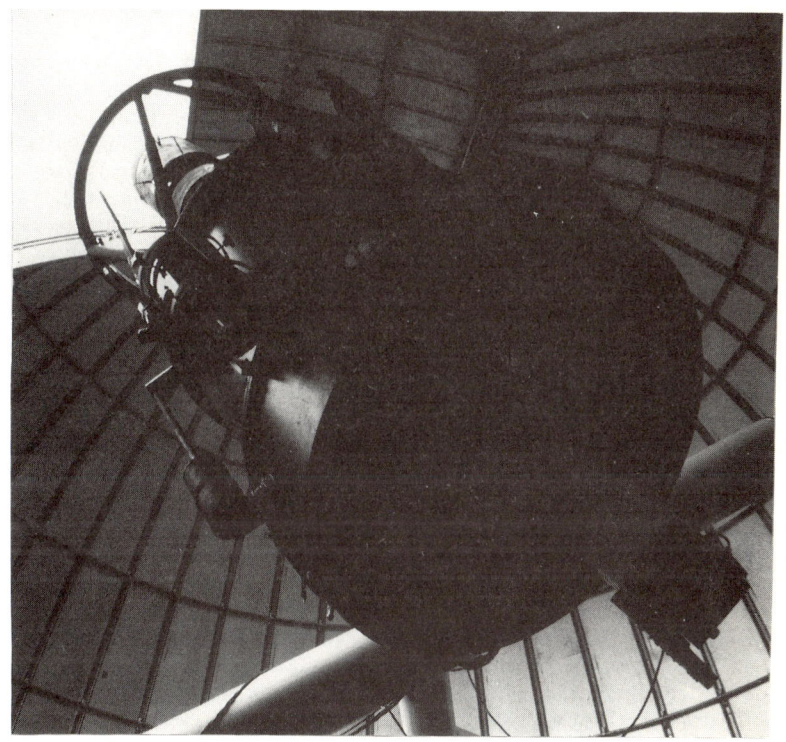

Figure 3. The 41-Inch Telescope, shown after initial assembly and checkout.

IV. ELECTRONIC SUBSYSTEMS

The real heart of the 41-inch telescope is its drive system. As we someday hope to control the telescope using a very sophisticated computer system, we purchased commercially available motion control components which would respond to RS−232 commands. Thus, although our control computers might change, the response of the telescope itself would remain the same.

The telescope is driven by two Superior Electric M172FD306 Stepper Motors. This is the largest Slo−Syn stepping motor and advances 200 full steps per revolution. Under normal operating loads, we have been able to achieve velocities of about 2500 full steps per second. The motors are powered by DRD003 Driver Cards. This driver is a single card which furnishes 200 watts of power. Input to this card is received from the Superior 20 pin bus; both 5V and 12V logic power supplies and 48V motor power supplies are required. All connections are easily connected or disconnected on terminal blocks.

The DRD003 cards are in turn controlled by IMD015 Microstep Indexer cards. With our gear ratio, the telescope has a sidereal drive rate of 207 microsteps/second, and a slew rate of near a degree/second. This rate is reasonable for such a large telescope, particularly if the alternative difficulties of a multiple motor drive system are to be avoided. The IMD015 card is also connected to an external square−wave oscillator, which can be programmed to provide the exact frequency required to achieve (and to adjust) the sidereal track rate. The electronics however, are unable to change the speed of a single motor "on the fly". The cards described in this article require the motors to be ramped to a stop (in hardware) anytime a change in speed command is issued. This presents no great problems for us, as the microcomputer is in control anyway and can easily keep track of the required adjustments, but it does complicate the program coding slightly. Superior Electric now offers another type of card in which speed can be changed on the fly, but that card does not have the ability to be indexed from an external pulse generator, and hence, does not offer the necessary resolution required to exactly achieve the sidereal rate.

Both sets of motors/driver cards/indexer cards are controlled by a single IOD004A RS−232 Interface Card. This card has an on−board MC6802 microprocessor which is used to control the microstep motions, slew rates, acceleration rates, etc., and 2KBytes of RAM for on−board programming. Any RS−232 device can issue commands to this card. The board also comes in an IEEE version, and can simultaneously control up to six indexers/drivers/motors. Each of the cards described here comes with excellent user documentation and detailed wiring schedules.

We use a Commodore C−64 microcomputer to control all of the instruments connected to the telescope, including the telescope drive system. The C−64 possesses the necessary computational power to accomplish essential real−time tasks, as long as not very many numerically intensive demands are placed on it. On the other hand, creation of the necessary system I/O has been something of a challenge: we use the User Port for RS−232 communications (only an adapter cable is required); the Memory Expansion Port is used to control the photometer and small stepping motors which are attached to it or elsewhere on the telescope; the Game Ports are used to service the Telescope Paddle. All of this is very straightforward and follows

Commodore documentation. The only minor difficulty lay in designing extra user ports on the Memory Expansion Interface for the Photometer and small steppers.

Aside from the problem with I/O, the C−64 is not all that poorly suited for the job at hand. It serves as the ideal user node, where printers, disk drives, and color monitors can be easily attached; it has a well−developed screen editor where system software can easily be developed; and perhaps most importantly of all, it is so preposterously inexpensive that several spares (we've never needed one) are always on hand.

V. ACKNOWLEDGEMENTS

We are particularly grateful to the other faculty in the Department of Physics and Astronomy at Stephen F. Austin, who were willing to allocate *all* of the Department's capital equipment budget to this project in 1983. Special credit is due Bennet Montes for his efforts in machining and redesigning telescope parts, and to Ed Michaels who helped us overcome a variety of obstacles. Dozens of other people contributed toward the success of the project and have our thanks as well.

REFERENCES

Osborn, J. 1980 *Sky and Telescope*, 98.
Markworth, N. L. and Rafert, J. B. 1987a in *New Generation Small Telescopes*, ed. D. S. Hayes, R. M. Genet and D. R. Genet, (Mesa: Fairborn Press), p. 355.
Markworth, N. L. and Rafert, J. B. 1987b in *New Generation Small Telescopes*, ed. D. S. Hayes, R. M. Genet and D. R. Genet, (Mesa: Fairborn Press), p. 361.
Rafert, J. B. and Oswalt, T. D. 1987 in *New Generation Small Telescopes*, ed. D. S. Hayes, R. M. Genet and D. R. Genet, (Mesa: Fairborn Press), p. 257.
Rafert, J. B. and Markworth, N. L. 1984 in *Microcomputers in Astronomy II*, ed. R. M. Genet (Fairborn: Fairborn Observatory), pp. 10−1.
Wilson, R. E., Rafert, J. B. and Markworth, N. L. 1984 *I.A.P.P.P. Comm.* No. 16, 1.

THE FIT 25−INCH PHOTOMETRIC REFLECTOR

James B. Rafert and Terry D. Oswalt

Department of Physics and Space Sciences,
Florida Institute of Technology

I. GENERAL CONCEPT

    The Florida Institute of Technology 25−Inch Photometric Telescope is a special purpose reflector in which light losses are minimized by placing the photometer directly at the prime focus. The telescope also employs a wide range of inexpensive technologies similar to the Stephen F. Austin 41−inch and 18−inch telescopes (Rafert and Markworth 1984; Rafert and Markworth 1987; Markworth and Rafert 1984) and currently is operated with a modified version of program TAU (Markworth and Rafert 1987). The telescope, shown set up next to the FIT 16−inch telescope for final tests before installation at our remote site, is shown in Figure 1. Although the 25−inch is similar electronically to the SFA telescopes, it is different mechanically, and possesses several recent additions which have not yet been installed on its "sister" telescopes.

New Generation Small Telescopes, ed. D. S. Hayes, R. M. Genet, & D. R. Genet.
© 1987 Fairborn Observatory.

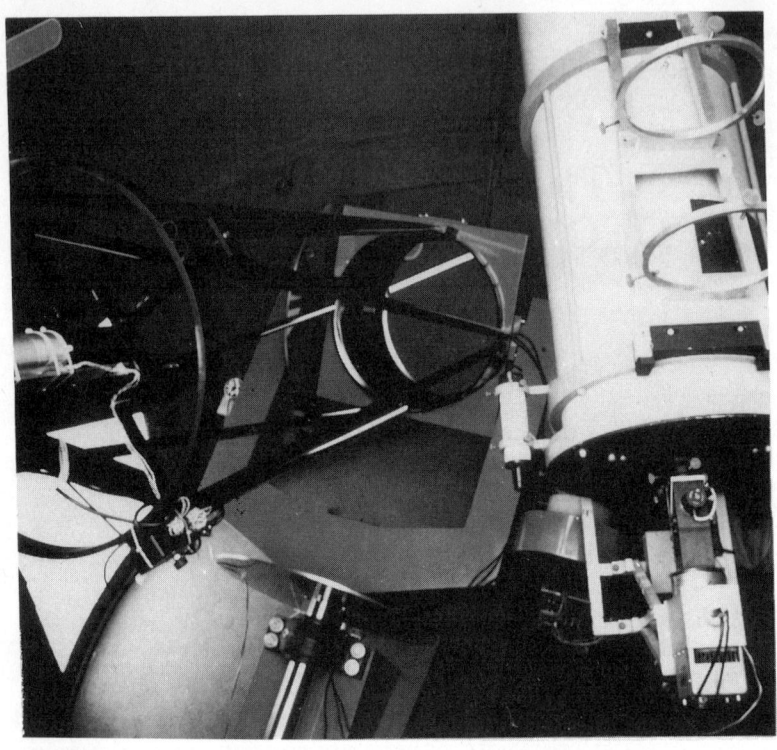

Figure 1. The new FIT 25−Inch Photometric Reflector, assembled next to the FIT 16−inch for testing prior to shipping to our remote site. Our Starlight−1 Photometer will be mounted at prime focus.

The telescope is composed of several major subsystems: 1) Optics, including the 25−inch f/4 primary, photometer head optics, image intensifier and low light level television system; 2) An extremely low−mass mounting, fabricated entirely from 3/8 inch aluminum (except for drive shafts/bearings/etc.); 3) Drive subsystem, built from Superior Electric Modulynx motion control boards; 4) the two computers which control the telescope (a Commodore SX−64 and an Intel System 310/3); and 5) a Macintosh which digitizes the video signal which is also displayed on the control room monitor.

Prior to the start of construction in the fall of 1985, we happened to possess several necessary components, such as the primary mirror and a Starlight-1 Photometer. These pieces would, of financial necessity, heavily influence the type of telescope we would build. We wanted a telescope which could perform both "high" and "low" speed photometry of a wide variety of objects. Not only would the telescope be used to observe eclipsing variable stars, but we also wanted a system which would be able to observe asteroid occultations and which would be able to track rapidly moving objects such as earth satellites. As we possessed no secondary optics and the primary was not perforated, we simply decided to place the photometer at prime focus, avoiding reflected light losses from another surface in the process.

Most of the construction was performed by the authors, using a supply of surplus aluminum sheeting, a bandsaw (with occasional help from a carbide grinding disk; the aluminum was sometimes too large to fit on the bandsaw), a 9-inch metal lathe and an overhead milling machine. Construction proceeded at a slow, but steady pace for the next 15 months. We estimate that the entire project has required about 850 man-hours of effort. Each of the major subsystems will be described in the following sections.

II. OPTICS

As with other telescope construction projects, the "condensation nucleus" for our project was a 25-inch f/4 primary mirror which had been acquired some years earlier by the Physics and Space Sciences Department. The mirror has a focal length of 99.25 inches, yielding a scale of 0.58 degrees/inch at prime focus. Prime focus is at the aperture of a Thorn-EMI Starlight-1 Photoelectric Photometer. We have made no modification to the transfer optics which are normally used in the photometer, other than to redirect the light path so that it comes out the back, rather than the side of the photometer. This is necessary in order to allow the image intensifier and TV camera to be placed behind the photometer so that no additional obscuration of the primary mirror will occur. We hope to construct a smaller cross-section photometer (the Starlight-1 is 6.5 by 7.2 inches, and obscures 9.5% of the total light), similar in size to the 4-inch diameter of the housing for the image tube, transfer optics and TV camera.

The View/Count knob on the photometer retains its normal function: in the "View" position, light is passed through the photometer with the normal relay optics to a Varo Model 3603-300 Image Intensifier, while when in the "Count" position light goes to the photomultiplier tube. This knob and other knobs on the photometer are rotated by surplus Sigma stepping motors, using the circuit of Rafert and Markworth (1987). The image intensifier is a second generation, microchannel-plate device with a maximum gain of about 55000. We use a Canon 28mm f/2.8 wide-angle camera lens to image the phosphor screen of the intensifier onto the videcon of a RCA Model TC2000 low light level camera with an Ultracon tube. This is a relatively inexpensive (<$3000) TV camera which has a variety of user adjustments suitable for optimizing contrast, gain, peak/average response and γ near a low light limit of 0.0015 footcandle at the faceplate.

III. MECHANICAL COMPONENTS

We chose to construct a fork mount as this is an extremely compact type of mounting which is well suited for a friction drive system. Each disk is 34 inches in diameter and was fabricated locally from 3/8 inch thick aluminum scrap using a bandsaw for rough work and an overhead milling machine for the final cuts. Although the declination drive disk introduces an asymmetry to the balance of the mount which must be compensated with some counterweights, the moment of inertia of our mount is very small compared with a mount such as the 41−inch telescope. The fork has an inside width of 36 inches and an outside width of 48 inches. Sufficient length has been allowed for the forks in the event that we eventually choose to perforate the primary mirror and make a Cassegrain focus option.

The polar shaft is a 3 15/16 inch shaft, cut from a surplus propeller shaft of an old minesweeper. The shaft is supported in two large pillow block bearings (see Figure 2) which were generously donated to FIT from SFA (the same bearings are used on the declination shaft of the 41−inch telescope). These bearings provide an extraordinarily sturdy and smooth way to build a mounting for a large telescope; we only apply 1% of the rated thrust load (and that is rated for a continuous duty of the bearing at 10 rpm).

The declination shafts are 2 inch solid steel, and are supported on each side of the fork by two additional self−centered roller bearings. The telescope side of the declination shafts terminate in large aluminum blocks which attach to the telescope "tub" with four 1/2 inch bolts. The tub serves as a point of attachment of the mirror mount and telescope tube truss, and of the telescope tube assembly to the mount. The tub, like all the other structural members was fabricated from 3/8 inch aluminum. The diagonal cuts on the tail piece side serve to make the structure more rigid; they are also required for proper clearance of the drive system.

The mirror mount is of standard design, and provides 9−point floatation to the primary mirror. The mirror mount is attached to the telescope with 4 1−inch bolts, and may easily be removed (with the mirror attached) from the telescope. All points of contact with the mirror are provided with various sizes of soft nylon tipped screws (we get ours from Winfred Berg).

Figure 2. The polar axis is supported by extremely massive pillow−block bearings. The bearings are self−centering, and were kindly donated to the project by SFA.

The telescope truss is also of standard design. Each truss member (1 1/4 inch with 1/8 wall) is welded to a 2 inch by 1/2 inch end ring which was fabricated from bar stock aluminum. The truss provides a very light weight yet sturdy way to support our relatively light weight instrument package.

The spider is a special purpose mounting platform for the photometer, image tube, and TV camera. Focus is achieved with the small stepping motor located at the extreme left of the figure; the entire set of optics has a free travel of 2 1/2 inches provided by four separate roller bearings. These bearings also supply the necessary rigidity so that the photometer does not rotate or move from a fixed optical position as the telescope points to different places in the sky.

After fabrication (including welding), all parts were sanded with 80-grit sanding disks in our automotive shop to remove small scratches, old protective paper, and 10 years worth of aluminum oxide. All surfaces were then covered with aluminum primer, and with Imron (using hardener). The resulting surface is very hard, slippery/smooth, and resistant to Florida weather.

IV. TELESCOPE DRIVE

The large drive disks are driven via frictional contact with a 1-inch shaft which is attached to a Thomas Mathis 224-tooth gear and slipclutch. The worm which engages the 224 tooth gear is attached directly to the shaft of a Superior Electric MO92FD310 stepping motor by a Winfred Berg Wafer-Spring Coupler. The coupler is required to allow for slight (and impossible to eliminate) offsets in the motor and shaft position. As these stepping motors have 200 full steps per revolution, and we use a microstep indexer mode in which there are 16 microsteps per full step, a single microstep corresponds to 0.05 seconds of arc. Under normal conditions it is impossible to observe the effect of a single step. Our drive rate is provided by an external oscillator (see Rafert and Markworth 1987) so the only remaining question is track rate resolution, *i.e.*, how frequently a track pulse is issued. On a Polaris Super C-8 for example, a Sky-Sensor with pulse motors achieves sidereal track rate at 4 Hz. This is observable, even with medium power. You could think of a synchronous motor as supplying track pulses at 60 Hz (although there is smoothing). Our gear train/motor system requires pulses to be given at 282.8661 Hz. This is a rate which is large enough to be able to provide good resolution, and to provide a track rate which is suitable for any type of astronomical observation.

The stepping motors are driven by Superior Electric DRD002 Driver Cards. These are 50 watt bifilar stepping-motor driver cards, which chop at audio frequencies. We prefer to use DRD003 cards (used on the 41-inch telescope), as these cards chop at ultrasonic frequencies (the audible tone of the DRD002's is a little annoying), and will upgrade to those cards once our budget expands. Each DRD002 is in turn controlled by IMD015 Microstep Indexer Cards. These cards allow DIP-switch selection of 5, 10, or 16 microsteps per full step. Up to six microstep indexer cards can be controlled by a single IOD004A RS-232 Interface Card, which has an onboard MC6802 and 2KBytes RAM. These cards are described in greater detail in Rafert and Markworth (1987). Given our moment of inertia (which is very low) and frictional losses, these motors can slew the telescope at 10000 full steps/second. This rate (about 1 degree/second) is satisfactory for almost any application.

All telescope move commands are issued via RS-232 command strings from our host microcomputers. Any type of microcomputer or terminal can be used to run the telescope system. We chose this approach to allow upward development from our Commodore C-64s without needing to change the telescope drive system itself. The Superior Electric Cards are well documented, are meant to be used in an industrial environment, and have a very wide range of programmable functions (such as acceleration, base speed and high speed rates, *etc.*) which operate transparently to the host microcomputer.

V. TELESCOPE CONTROL COMPUTERS

About three years ago we started using the Commodore C-64 to control telescopes and photometers. Since we were operating on a limited budget, we originally decided to put everything we could into hardware for the telescope, and temporarily forego an expensive computer system. Three years later, we are still using C-64 or Commodore SX-64 microcomputers. Why is this?

There are good reasons. First of all, we have not yet been in a position where we could purchase replacements which offered more than an incremental improvement in performance. As was indicated in Rafert and Markworth (1987), system I/O on the Commodore is something of a problem. What would be convenient would be a machine with about a half dozen card slots, into which RS-232 cards, calendar/clock cards with battery backup, or image processing cards could be placed. There are a number of such machines on the market with reasonable prices such as the Tandy 1000, Tandy 2000, IBM-PC and IBM-AT "clones." Yet, we are not talking about just the purchase of a single machine. One is required for each of the telescopes which are presently controlled by Commodores, and also at each of our offices where system development and data reduction take place.

Also, we have invested quite a bit of effort in software and hardware development over those three years. Although we have developed all of the programs which are used in BASIC (not only is it very transportable, but it is the "native" language of the C-64) and are easily modified for use in other machines, it is just not expedient to purchase another machine (or machines), recode the software to accommodate system specific machine—language or I/O calls, **and** to rebuild a lot of interface hardware.

Finally, and most importantly, the Commodore machines possess the necessary computational ability to perform all of our control tasks as long as we are careful not to demand too much in the way of real—time data analysis. Presently, the computers are only used to control the system and store the data once it is acquired. This means that a substantial amount of pre—processing of objects to be viewed and their current coordinates needs to be accomplished before you actually arrive at the telescope. So far, this approach seems to have worked reasonably well.

Yet, there are a number of features that we would like to add to the telescopes. Presently, the FIT 25—inch is run "open—loop." That is, no absolute or incremental encoders are installed. The telescope is presumed to be at a position which corresponds to a known offset (number of pulses issued through a known gear—ratio) from a previously known coordinate. We would also like to be able to read a shaft encoder attached directly to the telescope mount. According to Melsheimer (1983), a system which uses encoders will require about 10 times the computational ability of a system that does not. We are also interested in digitizing the video information at our disposal. The 25—inch possesses a rudimentary video system which is being used at present to display the field of view. It would also be desirable to have on—line access to the SAO catalog (see Rafert and Oswalt 1986), other lists of astronomical objects, or alternative observing programs. The level of computation which would be required for all of these features is beyond that of our present C—64s and SX—64s.

In an attempt to begin developing some of these features, (in the hopes that someday soon we'll be able to purchase the necessary 5 or 6 IBM—ATs) we have added an Intel System 310/3 (in addition to the Commodore SX—64) to the FIT 25—inch telescope. This is an Intel development system which was obtained via an Intel grant program. Although it is a "one—of a kind" system, from our point of view it is quite powerful and is allowing us to begin development of some advanced features. The 310/3 has an 8086 microprocessor and 8087 numeric co-processor, a 10MByte hard disk, and plenty of system I/O in the form of SBX—350 and SBX—351 cards. We are presently writing code which will allow the Intel to take over all telescope control functions, leaving the Commodore in charge of running the photometer.

VI. VIDEO SYSTEM

As stated in the previous section, we are interested in making the best use of our existing video system. The logical thing to do is to acquire a video digitizer or frame grabber board, and to build up pictures from the video signal from the TV camera. At present, several manufacturers have add-on boards for the IBM which carry a price tag of about $2000, and which perform the necessary functions. One such board, which also allows the use of false-color displays is made by Coreco (see Sky and Telescope, Dec. 1986, p. 600). That would be just fine if we happened to have a computer to put it in.

We do however, happen to have several Apple Macintoshes in the department which are used for research projects. A quick scan of the literature revealed "MacVision," a device which Koala Technologies claimed could take any video signal, and with the help of a Macintosh, display or print whatever was being viewed. For $149.00 it seemed worth the gamble. We have noticed that the digitizer works best on point sources, and is capable of displaying anything which is visible on the monitor screen. Presently, we use MacVision to produce finder charts.

There are several reasons why we have not adapted MacVision for "image processing." It takes MacVision about 5 seconds to produce an small image or 15-20 seconds to produce a large image. This is done in the computer with a left to right scan of the screen. Two controls on MacVision control contrast and brightness and there is a pull-down menu option for an adjust mode where the grey scale can be manipulated. Each piece of video data is one byte (8 bits) deep. So far, so good. However, once it comes time to display a picture on the Macintosh screen (which is black and white and hence, only one bit deep) system software generates a "Macintosh pseudo-grey scale" on the screen by controlling groups of pixels. For a very bright scene, all pixels in a group are on; for a very dim scene, all pixels are off; for an intermediate scene, some are on and some off in accord with what was done in "adjust mode." There is no easy way to store the full depth of video information which is available from the TV camera, and the supplied documentation reminds one of the early Commodore era. No program listings; no telephone number to call for user support; no schematics of the circuit. Yet, for the price, it is a very worthwhile way (if you happen to have a Macintosh) to get into the digitized video business.

VII. ACKNOWLEDGEMENTS

We wish to thank Mr. Chuck Inman for his kind assistance in the machine shop, as well as all of the students which have assisted in various aspects of the project. Aspects of this project have received support from Intel Corporation, the Dean's Research Fund, and NSF Grant AST 8543260.

REFERENCES

Melsheimer, D. F. 1983 in *Microcomputers in Astronomy* ed. R. M. Genet (Fairborn: Fairborn Obs.), p. 14.

Markworth, N. L. and Rafert, J. B. 1984 in *Microcomputers in Astronomy II* ed. R. M. Genet and K. A. Genet (Fairborn: Fairborn Obs.), p. 9–1.

Markworth, N. L. and Rafert, J. B., 1987 in *New Generation Small Telescopes*, ed. D. S. Hayes, R. M. Genet and D. R. Genet, (Mesa: Fairborn Press), p. 267.

Rafert, J. B. and Markworth, N. L., 1987 in *New Generation Small Telescopes*, ed. D. S. Hayes, R. M. Genet and D. R. Genet, (Mesa: Fairborn Press), p. 247.

Rafert, J. B. and Oswalt, T. D. 1986 *I.A.P.P.P. Comm.* No. 23, 2 (abstract of paper by D. Kornbluh).

THE SFA TELESCOPE CONTROLLER/DATA LOGGER

Norman L. Markworth

Department of Physics and Astronomy
Stephen F. Austin State University

James B. Rafert

Department of Physics and Space Sciences
Florida Institute of Technology

I. INTRODUCTION

It has long been recognized by professional astronomers and groups of serious amateurs that the automation of data collection at the telescope offers the potential of enhancing efficiency and precision. This is particularly true for photoelectric photometry. Irregularities in the earth's atmosphere require that each observation be of the highest quality possible, and consecutive observations be as closely spaced in time as possible. Concentration on either of these conditions to the exclusion of the other will have deleterious effects upon the outcome.

Once the decision has been made to automate, the observer must sort through a host of decisions, based primarily upon cost, telescope configuration, and skill at computer programming and electronics. The decision tree is extensive enough that it is unlikely than any two installations will have exactly the same mode of operation, even though everyone is pursuing the same goal. The finished product is also strongly shaped by observing philosophy (i.e., my way or the wrong way), so that it seems unlikely that commercial photoelectric software will ever service a large market.

The previous comments notwithstanding, we wish to describe the system (named TAU) we use at the Stephen F. Austin State University Observatory in Nacogdoches, Texas. The system was written three years ago and has undergone extensive field testing and revision. TAU is now used on both the 46−cm and 104−cm telescopes at the SFA Observatory, and the 36−cm telescope at the Florida Institute of Technology Observatory. TAU will also be used on the new 64−cm telescope at FIT (see paper in this Symposium; Rafert and Oswalt 1987).

Telescope control and data logging are traditionally accomplished on separate systems which communicate with one another. The telescope controller is typically small. The first extensive control program for the 46−cm telescope allowed hand paddle control and automatic slew between variable/comparison/check stars and sky. It was able to compute sidereal corrections for the right ascension moves and communicate with the data logger, yet fit comfortably on a 4K−EPROM for a Rockwell AIM−65. The data logger must not only acquire and store the data, but must also schedule the sequence of observations. Since the data logger is the observer's link to the system, ergonometric considerations (user friendliness) generally will make these programs longer. TAU is a hybrid system, in that all of the data logging and the majority of the telescope control operations are accomplished on the single system. TAU computes all telescope moves, but passes the information to a small CPU, which controls the actual stepping of the motors.

A good place to start in the design of a telescope controller/ data logger is to list the requirements of the system, the philosophy of the observations, and the dreams and goals of the designer. Many of the features of TAU were designed around the unique problems of all−sky photometry. In this work many variable stars are observed each night in various parts of the sky. Pointing accuracy is of primary concern, so that the observer need not commit many different star fields to memory. Student observers are used in the project, which greatly influences our philosophy in the program. While TAU offers many features, all options have default values, which ensure that accurate and timely observations can be obtained by observers in all stages of training. By relieving the observer of many of the detailed decisions, greater attention can be paid to the quality of the data. Finally, TAU permits the telescope to be operated remotely. The telescope field is imaged by an image intensifier/vidicon onto a monitor. TAU operates stepper motors on the filter, aperture, and flip mirror controls of the photometer, controls the position of the dome, operates the photon counter, and controls a stepper motor on the secondary mirror housing for telescope focus. The observer is still responsible for centering the star, assessing the quality of the data, and selecting the next target star group (variable, comparison, check, standard). The following sections describe how we accomplished these goals in TAU.

II. SCHEDULING THE RUN

The first set of programs in the TAU package can be used before going to the observatory. These programs schedule the observations to be made that night, and create a disk file for use at the telescope. Figure 1 provides a summary of TAU. TAU offers no way of automatically moving to an object that is not on the observing file for that night. While some may find this to be a disadvantage, we find that it demands that the observer be prepared for the work that night. We maintain several object catalogs on disk file for use with various observing projects (*e.g.*, W Serpentis stars, single variable stars, or deep sky objects for public tours.)

Four different classes of objects are found on the object file. First, and of primary interest, are the variable stars. Second are the check stars for the variables listed. As presently written, TAU defines a group to be comprised of the variable and comparison stars and a sky position. When the check star is to be observed, the user requests a move to it as though it were a variable star group, using the same comparison star and sky position. Since the check star is only monitored occasionally, the program space used for the observing sequences can be shortened considerably. The third class of objects are standard stars, usually selected to be red−blue pairs. The final object class are navigational stars. Since TAU operates the telescope in an "open−loop" positioning mode, field verification and target centering are the responsibility of the observer. Navigational stars are bright objects, strategically placed around the sky, so that moves to these objects are fairly short. Should the observer become confused upon verifying a field, he can move to a navigational star and back again.

The first program in this set is an astronomical calendar program. Given the calendar date, it computes sunrise/sunset, moonrise/moonset, dawn and evening twilight, and the position of the moon at local midnight. These values, along with estimates of barometric pressure and mean nightly temperature (used in the refraction correction), are written to a disk file.

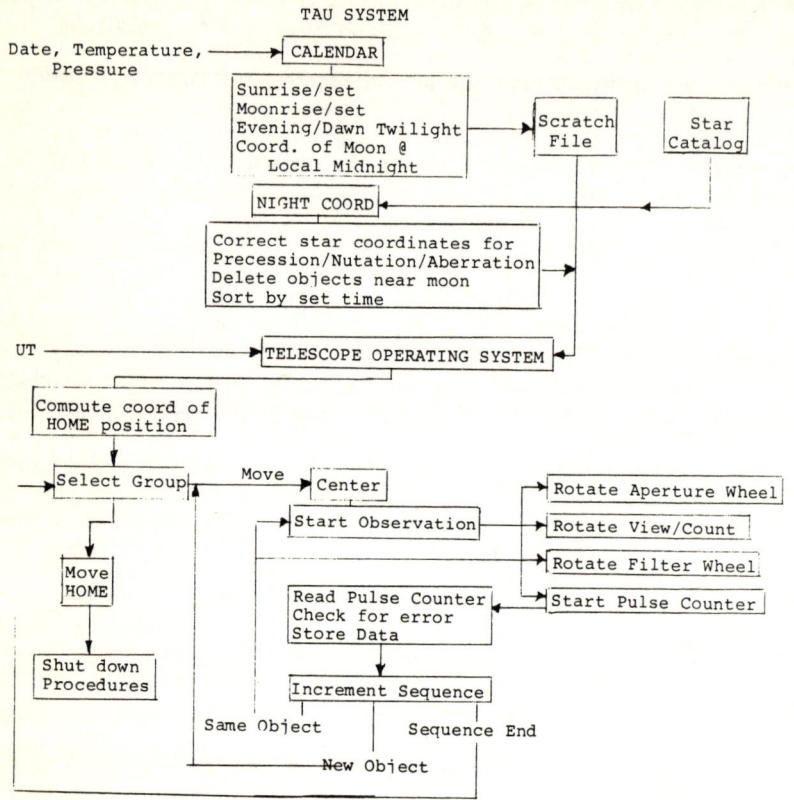

Figure 1. *A software flow diagram of the TAU System.*

The next program in this set acquires the disk file created by the calendar program as well as the object catalog for the night. The object catalog contains the coordinates and epoch of the stars of interest. The program computes the mean coordinates of the stars for that night by correcting for precession, nutation, and aberration. Next it selects those objects which are visible (defined as having an altitude greater than a critical value, usually thirty degrees) between evening and dawn twilight. From this list of possible targets it deletes those within fifteen degrees of the moon. All objects that pass both tests are sorted by their set times and added to the file created by the calendar program. This becomes the observational file for the night.

III. THE OBSERVING RUN

a) Telescope Control

At the telescope the heart of the TAU system is utilized. The observing session begins with the telescope stored in a known "home" position. For us this means on the meridian, fifteen degrees above the southern horizon. Upon user input of the universal time, the system computes the right ascension and declination of "home" and moves the telescope to the first selected object. The observing run is ended by requesting a move to "home." Two general kinds of automatic telescope moves will be considered: 1) Moves from one variable star group to another; and 2) moves within a group. All telescope moves are preceded by the computation of refraction corrections of both the current and destination objects. All moves are entirely precomputed by differencing the refracted coordinates, applying the sidereal correction for the right ascension move (Rafert 1985), and multiplying by the appropriate gear ratio. Since the normal motion of the RA axis is to the west, all east moves provide automatic backlash compensation. A character string is passed to the Superior Electric components containing the acceleration, peak rate, and number of motor pulses for each axis.

We first will consider the moves from one variable star group to another. The telescope is restricted to move only to objects that are currently above the critical altitude (or to the "home" position). This scheme eliminates the need for many hardware "bells and whistles" on the telescope, but does require a fresh observational file for each night. Several of our student observers have "rediscovered" the orbital motion of the earth by using dated observational files, only to find a substantial RA error on the move from "home." Although the objects in the file are ordered by their set time, they may be observed in any order the user wishes. Before moving to the next variable star, the variable star name is displayed along with a request for a decision to either move to this object or skip it. Since the displayed object can be observed, novice users are simply instructed to move to each group in turn. Those users more familiar with the catalog objects are asked to consult with us prior to a run, where orbital period and density of the data points are considered in sequencing the observations.

The success of the moves between groups is a sensitive function of telescope configuration, mechanical alignment, and software. We give suggestions below on how to ensure accurate telescope moves; most of them won through hard experience.

The second type of telescope move executed by TAU is within a variable star group. The move to the group is always to the comparison star. After centering, a keyboard command moves the telescope to the sky position, where observations begin. This is done so that we can center on an object rather than a "non-object." The user may select from a wide range of observing sequences resident within the system. A step in the observing sequence consists of a filter code and an object code. Upon incrementing to the next sequence step, the telescope automatically moves if the object code has changed, and/or rotates the filter wheel if the filter code has changed. All telescope moves within a group are automatic and require no user response. Commands exist to increment or decrement the observing sequence by one step, or move to the top of the sequence. Observations are begun by a keystroke or paddle button, and will continue automatically, changing the filter as needed, until either the object changes or the observation is aborted. The system can be asked to stop observing at each telescope move for centering purposes, or can continue the observations automatically (which can be implemented for any object or only when the new "object" is the sky.) Some observing sequences cycle indefinitely (useful for single star work), while others execute a fixed sequence of steps before requesting a move to the next group (for all-sky photometry.)

TAU also emulates the manual control of the traditional telescope. A five-button paddle is connected to the game port of the host computer. Guide motions can be acquired by combinations of four of the buttons, representing the cardinal compass points. The fifth button can be used in combination with others to acquire slew motion, or used by itself can start and stop an integration. We also use the illogical combinations (e.g., E-W) to redefine the buttons. This "command" mode can be used to change the track rate, focus the telescope, or compute and display the coordinates of an object nearby a catalog object (off-set mode.)

TAU achieves the track rate by controlling the on-board sound chip of the host computer. Since all automated telescope moves are dependent upon the track rate, we required high resolution in frequency. The 16-bit control register for the sound chip can increment the frequency in 1/20th hertz steps, and is regulated by the 1.02 MHz system clock. We have found the system clock to be accurate to better than 3 parts in ten million, by comparing it to a precision EPUT (Oswalt and Rafert 1987).

b) Data Logging

The minimum set of parameters essential for data logging are intensity, time, filter, object, and integration length. As mentioned in the Introduction, however, the data logger acts as the observer's link to the system. A well-designed display must not only convey the essential elements, but must also advise the observer of the system status, and must do so without confusing or annoying the observer. Figure 2 shows the screen used by TAU. Except for special functions, this screen is displayed throughout the observing run. The screen is divided into three major sections. On the left is the system status board, which displays Universal Time, operation (which may be READY, INTEGRATING, or MOVING), next object, filter, and integration length. The upper half of the right side displays the accumulated data. The data format is exactly the same as will appear on the disk file, and includes UT, filter code, object code, integration length in seconds, and intensity. Eleven data points can be displayed before this section is refreshed. Certain status words connected with the flow of the observing sequence may also appear here, such as ABORT, ADVANCE (increment the observing sequence by one step), or BACKUP (decrement the observing sequence by one step.) The lower half of the right side is reserved for systems messages (*e.g.*, MOVE TO RX CAS WILL TAKE 36.2 SECONDS.)

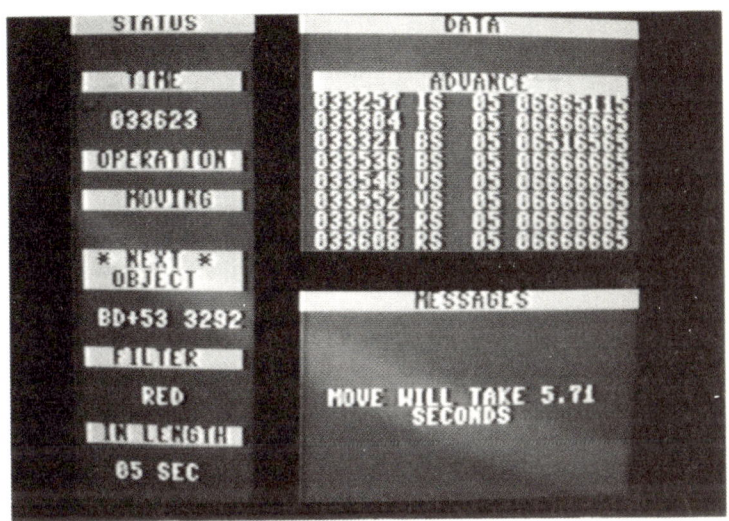

Figure 2. A sample screen used by the TAU System.

Many of the data logging features of TAU are intimately connected with telescope motions, and have been discussed above. A series of observations is begun by a keystroke or hand paddle button. The system will rotate the aperture wheel to a preselected observing aperture (if necessary), rotate the flip mirror to the count position (blanking the viewing monitor), rotate to the filter indicated on the screen (if necessary), and command the pulse counter to count for the time indicated by the integration length. This automatic sequence of events requires about two seconds. At the end of the first integration, TAU will increment the observing sequence, rotate the filter wheel, and continue observations until the object changes. There are two methods of interrupting this process. The observer may abort the observation by a keystroke or hand paddle button, which rotates the flip mirror back into view position and halts the automatic cycling. The second method of halting the sequence is the automatic error—checking routine. TAU stores a matrix of intensity levels for each object in each color. Should the present intensity reading differ from the previous one (same object, same color) by more than a preset amount (we presently use 4%), automatic cycling is halted, the flip mirror is rotated to the view position, and both the previous and present intensities are displayed on the message board. The user is asked to decide whether to store or delete this data point from the disk file. While our present discriminator of 4% is no guarantee of photometric precision in the data, the intent of the error checking is to signal the user of a gross problem (clouds have moved in or the star has drifted out of the aperture.) We find that the error-detection scheme makes the observer much more aware of the quality of the data during the run, which has reduced the amount of wasted effort at the data reduction stage.

When TAU is used in its completely automated form, the observer is only responsible for centering the objects and handling error checking decisions. In our most commonly used observing sequence for all-sky photometry, we obtain 56 five-second integrations in about 7.5 minutes, for an observing efficiency of 62%. This efficiency is very comparable to that of a well-trained professional operating a telescope with manual moves and automatic data acquisition, but TAU allows student observers to operate at this efficiency level. We anticipate being able to improve our efficiency somewhat by changing the host computer. We currently use the Commodore-64 microcomputer. Most of the TAU program is run in compiled BASIC (P-code) with several machine language subroutines where speed is essential. Approximately 19% of the all-sky sequence is spent in program cycling, which we can cut to about 10% using a 16-bit processor. As presently written we can expect a maximum observing efficiency between 70% and 75%. By comparison the Fairborn Observatory is reporting efficiencies of 79% for their automated telescopes. TAU can always be expected to lag behind this efficiency, since we require the observer to center the stars and begin the observational sequence on them (a process which consumes about 15% of the time.) The benefit is that the repeatability of our centering is much better than the APT.

IV. THE DESIGNER'S SKELETON CLOSET

a) The Telescope Environment

Once the telescope controller/ data logger program has been written, the nightmare begins. It is now time to see whether those routines which worked so well in the lab can be used to actually operate the telescope. The programs we have described here are really an integrated hardware/software system. The previous editions of *Microcomputers in Astronomy* (Genet 1983 and Genet and Genet 1984) and the *IAPPP Communications* are filled with interface circuitry, ranging from the elementary to the elaborate. That all of these approaches work suggests that making a microcomputer control device is rather straight forward. The elaborate designs differ from the elementary ones in one critical area, namely: How well does the circuitry function in the real environment of the observatory? Two general areas of environmental concern deserve particular attention.

Because the typical automated telescope uses a wide range of power supplies and electronic equipment, special care needs to be taken in the construction and use of the interfaces to avoid radio frequency (RF) and magnetic interference. Although these precautions have been cited by many authors, it seems that all designers need to discover them independently. The signal cable from the photometer should be very well shielded, and, if the head has no preamplifier, should be as short as possible. All connections should be solid (no cold solder joints, alligator clips, *etc.*) All cables should be constructed as if they were expected to function in combat. All electronic components should be grounded together and connected to a solid earth ground. Computer equipment should be completely isolated from the environment, *i.e.*, all input and output lines should at the minimum be buffered, and preferably opto-isolated from the device they control. Surge protection should be provided for the more sensitive components of the system. Failure to attend to such detail will inevitably cause problems which can require long periods of time to trace.

The second environmental concern at the telescope is the climate. Most electronic components that are used in telescope automation were never intended to live with the range of temperature and humidity found at the typical observing site. Most integrated circuits fail at temperatures below 0 C. Whenever possible, computers and associated sensitive electronics should be stored in a climate-controlled environment when not in use. Humidity protection can be achieved by coating each board with commercially available spray sealant. Heating pads placed underneath computers can be used to maintain reasonable temperatures in cold weather. Precautions should also be taken to minimize intrusion by dust (particularly when using tape or disk drives) and insects.

b) Increasing the Accuracy of the Automatic Moves

Accurate telescope moves begin with accurate coordinates. The source used for object coordinates depends upon the smallest step the telescope can make. Our gear ratios have been selected to achieve a step size of about 0.1 arcsec. We, therefore, require a source of coordinates which is at least this accurate. We prefer to use coordinates from the SAO Catalog whenever possible. Many times, however, a star cannot be identified in the SAO. TAU can be used to measure the offset in coordinates between an object of known coordinates and one of unknown coordinates, using the hand paddle and an onboard timer.

The routines used to compute the night coordinates of the objects must also produce accurate results. We have carefully checked the precession, nutation, aberration, and refraction routines used by TAU over a wide range of conditions to ensure errors no greater than about two arcseconds over the available portion of the Celestial Sphere. Whenever possible, the proper motion of the target stars should also be used.

The computation of the number of motor steps to be issued must properly account for any accelerations used by the system. TAU uses the procedure of Rafert (1985), which is accurate to one part in a million when used to fifth order terms. In the typical telescope move an error of 0.1 to 0.2 arcsec may be introduced here. The telescope moves are completely dependent upon the track rate. A correction factor for the differences in the gear ratios on the two axes is empirically determined for each telescope and is applied to the declination move. The track rate and declination factor are determined by repeatedly moving the telescope between navigational stars separated by at least 30 degrees, refining the parameters during the process. This process can reduce the move errors to less than one arcminute for the stars chosen, but is no guarantee that other moves will be equally accurate.

The causes of residual move errors are mechanical problems, usually requiring long periods of time to adequately model. Markworth (1985) has given a procedure for accurately aligning the pole of the telescope onto the North Celestial Pole. If the alignment is not very good, the telescope moves in a slightly different coordinate system than the sky system. The track rate then becomes a function of hour angle and a declination drift is observed. Telescope flexure can introduce errors of several arcmin, depending on the type of mount used. Modeling flexure is quite time consuming, and can only be accurately done when the other mechanical problems are well−known. Treffers (1985) gives a method for empirically determining and correcting for pointing errors, but his technique requires an independent source of accurate telescope position (good setting circles or optical shaft encoders.) For friction drive systems an additional error is important, *i.e.*, the run−out of the drive disk. This error makes the effective gear ratio a function of hour angle, but a simple measurement with a dial indicator depth gauge allows straight−forward correction.

For those systems where the track rate is not constantly available, care must be taken to ensure that the telescope is always running. When a special function (automatic moves or a data logging function) is not being executed, TAU must maintain the track rate as well as monitor the keyboard and hand paddle. To achieve this TAU issues short, 90 motor−step tracking sequences. After the sequence is sent to the drive electronics, TAU monitors the keyboard and hand paddle. If no commands are being issued, TAU sends another tracking sequence. The length of the tracking sequence is adjusted so that no tracking hesitation results. The problem lies in the fact that the telescope move commands must be sent as character strings. All computers reserve a section of RAM for string variable storage. When this reserved section fills, the KERNAL automatically performs a "garbage collection," during which time no program processing is possible. Garbage collection may take anywhere from milliseconds to minutes, depending upon the speed of the CPU, the amount of available RAM used for string variable storage, and the extent of the collection necessary. TAU solves the problem in two ways. Before every integration, a garbage collection is requested. During normal tracking TAU issues fifty tracking sequences of normal length followed by one sequence five times as long, during which time garbage collection is requested. In this way telescope motion is always maintained. The price to be paid is that 2% of the time, on average, the observer will experience a rather poor response time from the system.

V. CONCLUSIONS

The successful design of an integrated telescope controller/data logger can accomplish the goal of increased observing efficiency, while relieving the observer of most of the tedium of photoelectric photometry. Useful photometry can be accomplished even while training student observers. The designer must become completely familiar with every aspect of the telescope system, but this only makes him a better observer. The ultimate satisfaction is to watch your telescope move successfully and automatically to "your" star.

REFERENCES

Markworth, N. L. 1985, *I.A.P.P.P. Comm.* No. 22, 25.
Oswalt, T. D. and Rafert J.B. 1987 in preparation.
Rafert J. B. 1985, *I.A.P.P.P. Comm.* No. 20, 24.
Rafert, J. B. and Oswalt, T. D. 1987 in *New Generation Small Telescopes*, ed. D. S. Hayes, R. M. Genet and D. R. Genet, (Mesa: Fairborn Press), p. 257.
Treffers, R. R. 1985, *Publ. Astron. Soc. Pacific* 97, 446.

COUPLING IMAGE INTENSIFIERS TO TELEVISION SENSORS

Peter L. Manly

Saguaro Astronomy Club, Metro Phoenix

ABSTRACT. This paper describes the techniques of coupling television sensors to image intensifiers for use in automated telescope pointing and acquisition systems. The emphasis is on applications requiring recognition of star fields and centering of stars of interest with respect to the telescope boresight. The paper describes the methods of coupling television sensors to image intensifiers but does not describe the video processing required to identify the star field. Image intensifier and TV–sensor device characteristics are discussed along with practical engineering applications of these devices. This paper is a companion to one by Boyd and Genet (1987) in this Symposium concerning the recognition of star fields of interest within video signals.

New Generation Small Telescopes, ed. D. S. Hayes, R. M. Genet, & D. R. Genet.
© 1987 Fairborn Observatory.

I. DEVICE CHARACTERISTICS

Image intensifiers are widely used in scientific, security and military applications. Their function is to amplify an image in brightness while retaining the spatial and relative brightness relationships of the scene. They accomplish this by allowing the image at the input end of the intensifier to generate photoelectrons at a photocathode. These electrons are then accelerated across several thousand volts before striking a phosphor screen, yielding many more photons than were in the original image. The trick is to maintain the spatial and brightness relationships of the electrons so that an image scene at the photocathode input is represented at the phosphor output with fidelity.

There are two commonly used approaches to obtaining optical gain while maintaining spatial and brightness relationships. These are the "Generation I" and "Generation II" image-intensifier tubes as shown in Figure 1. In both tubes, the photocathode is held at a negative voltage with respect to the phosphor screen. In the older, first generation tubes, photoelectrons were imaged using a set of electrostatic focusing electrodes, similar to electron beam focusing electrodes used in electron microscopes. A very large voltage was required between the photocathode and the phosphor screen to accelerate the electrons. The focusing electrodes were at some intermediate voltage. Although only one set of focusing electrodes is shown in Figure 1, some tubes have more sets. Electrons strike the phosphor screen with sufficient energy to generate many more photons than the initial photon which started the process. There were a few first generation tubes also made which used large magnetic coils on the outside of the glass envelope to focus the electrons. These tubes were generally large, fragile, and required very high voltages in the 10-30 KV range. In order to keep the number of focusing electrodes small, both the photocathode and phosphor were made curved, thus allowing large spherical aberrations in the electron optics. The input and output planes of the tube thus may be fiber-optic light pipes in order to carry the image from the flat plane to the curved surface. Fiber optics input and output planes also allowed the tubes to be coupled in series.

The second generation tube uses a microchannel-plate electron multiplier, composed of a thin, flat wafer perforated by millions of tiny holes. A bias voltage is applied across the electron multiplier microchannel plate thickness, causing photoelectrons entering the holes to be attracted toward the rear of the plate. Electrons travelling down the holes will strike the walls and generate secondary electrons in a method similar to the classical photomultiplier. Thus, the microchannel plate acts like millions of adjacent photomultiplier tubes. The phosphor screen is proximity focused at the rear of the microchannel plate. The second generation tube shown in Figure 1 has a curved output plane which is useful with simple eyepieces.

COUPLING IMAGE INTENSIFIERS TO TV SENSORS

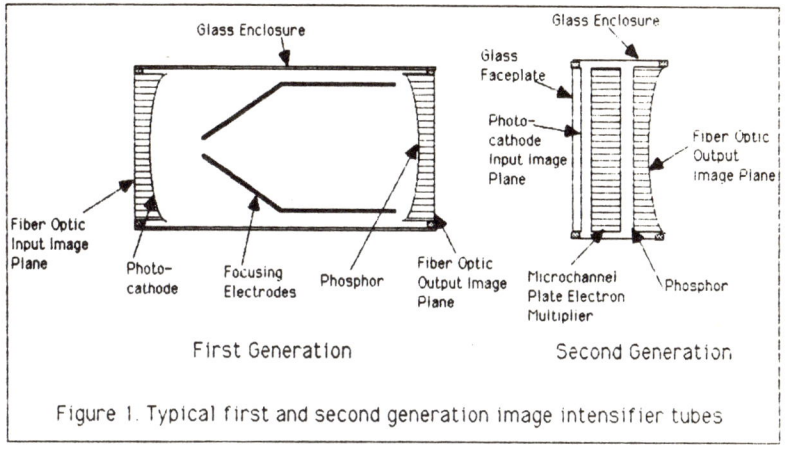

Figure 1. Typical first and second generation image intensifier tubes

Like all other amplifying devices, the image—intensifier tube is not perfect. There are several noise mechanisms which must be either controlled or tolerated. The photocathode has a quantum efficiency identical to conventional photomultiplier photocathodes. Thus, not every photon striking the photocathode is converted into a photoelectron. This is highly dependent on the wavelength of the light. Most intensifiers have an extended red—response for military applications. Most of the usual photomultiplier photocathode materials can be used in image intensifiers. All photocathodes will also give off some electrons even in the absence of any light. This is analogous to the "dark current" in a photomultiplier. The difference in an image intensifier is that the output image resulting from this dark current has spatial structure. In a good tube, this appears like a constantly shifting granular background. In a poor tube, whole areas may emit electrons constantly, resulting in a bright spot somewhere in the field of view. In general, cooling the tube reduces the dark current. Residual gas ions within the tube will also be accelerated and will create flashes or scintillations on the phosphor screen. As some first—generation tubes age they may also become gassy. An aging effect in some second—generation tubes is a gradual reduction of gain due to a change in the characteristics of the microchannel plate hole surfaces. Finally, just as there are no absolutely perfect optics, there are no absolutely perfect electron optics. The resolution of the image intensifier output is always less than the resolution of the input image. Since the intensifier will be used with a TV camera, however, the maximum resolution available in the entire system is about 525 lines by 600 or so picture elements (at best) as limited by the TV camera itself. Almost all new image intensifiers on the market today can easily meet these resolution requirements. All of these noise effects plus photon noise limits—discussed later—are added to the inherent TV camera noise.

II. TV CAMERA CHARACTERISTICS

Television cameras have come a long way from the days of using bulky studio cameras bolted on the backs of telescopes. Modern consumer—grade and industrial—grade surveillance cameras are now small, light weight, and rugged. There are two major categories of TV cameras; tube—type and solid state. The tube—type cameras usually have a vidicon—type image pickup tube in which an electron beam is scanned over a photoconductive surface (called the target) in order to sense the amount of electrical charge resulting from light falling on each point of the surface. Such tubes may have names like Newvicon, Ultracon, Saticon, Plumbicon, *etc.* but they all use the same general principles. The major differences are in the target material's sensitivity to light. For astronomical purposes, it is generally advantageous to avoid color television cameras since they have about 1/3 the sensitivity of comparable black and white (also called monochrome) cameras. Since an image intensifier is to be used in the system and the output color of the intensifier is usually green, no matter what color is input, it is rather pointless to use a more expensive color camera. Most tube—type cameras have a 2/3 format and these will be the cameras we consider. The larger broadcast quality cameras with a 1/1 format are generally too heavy for most small telescopes.

Solid state TV cameras have an integrated circuit on which thousands of individual photodiodes have been arrayed. Light striking these photodiodes will cause each picture element (pixel) to react and measure the picture brightness at that spot. Several methods have been used to extract the data from all of the diodes, one at a time, in a raster—scan sequence. Solid state cameras generally come in two versions; the CCD (Charge Coupled Device) and CID (Charge Injection Device). The two devices differ largely in the method of extracting the data from the integrated circuit which forms the sensor. From the point of view of the user, there is little difference between the two. One defect to be wary of, however, concerns CCD cameras. In some of the early versions, the sensitive area of each pixel was not exactly adjacent to the sensitive area of the next pixel. There were gaps, often as large as the sensitive areas themselves, between pixels. For most extended—area scenes this did not pose a problem but for astronomical use it was disastrous. Finely focused stars could fall between the sensitive areas and not be detected. A star drifting across the field of view would disappear, reappear, disappear, reappear, *etc.* It was like viewing the stars through a picket fence.

Both tube—type and solid state cameras are very good at obtaining time—resolved brightness data. Thus, as soon as the telescope moves, the TV camera detects the motion within one frame time (1/30 second). This makes the devices useful in real—time feedback control of telescope motion. Most tube—type cameras, however, have a certain amount of "lag" built in which smears rapid motion over a few frames. Lag is added to the tube for esthetic reasons to prevent motion from appearing jerky to a human viewer. Some solid state cameras, however, like the Sony XC37 CCD exhibit little or no lag. This also means, however, that intensifier scintillations are displayed in single frames and thus appear at higher amplitudes than in cameras with lag, where the energy from one scintillation event is spread out over several frames.

The noise in TV cameras is composed of components from the input optical signal (both background and stars), pixel noise and readout noise. The noise in the optical input will be discussed later under photon—limited performance of the system. Pixel noise is the thermal variation from frame—to—frame of electron charge per pixel (with or without an optical input). This charge will be read out of the camera as a signal and it cannot be distinguished from a real optical input except by looking at successive frames and determining if the signal changes. Pixel noise typically is decreased at lower temperatures. The readout noise in a tube—type camera is associated with the reading electron beam shot noise and the preamplifier noise. The readout noise in a solid state camera is associated with the charge transfer mechanism and the preamplifier.

III. OPTICAL CONSIDERATIONS

The objective of coupling an image intensifier to a TV camera is to transfer the maximum number of photons from the intensified phosphor output image to the input image plane of the TV sensor. Two common coupling methods are used; fiber optics and relay lenses. In order to use a fiber—optic coupling, both the output of the image intensifier and the input of the TV sensor must be fiber—optic surfaces. The two surfaces are then butted (gently) together and secured with a holding fixture. At present there are few small—format TV sensors with fiber optic faceplates. There are no commercially available solid state TV sensor chips available with fiber—optic inputs although several special installations have been made with varying degrees of success, and a very high cost. While fiber optics are a smaller, more lightweight and more efficient method of coupling intensifiers to TV systems, they are generally not within the budget of most small observatories. It is expected, of course, that costs will eventually come down.

The use of relay lenses in intensified TV systems is an established technology. Relay lenses are basically a conjugate—foci optical system as shown in Figure 2. At the left is the glowing output phosphor of the image intensifier which is transferred to the input image plane of the TV sensor at the right. Adjusting the ratio of the object and image distances allows the user to change the image size. Thus, the image intensifier output and the TV sensor do not have to be the same size. While this may appear to be a simple and elegant solution, there are a few problems encountered when applying it in a practical situation. Using the Lens Maker's Formula, it can be seen that relatively long distances are encountered between the object and the image. This translates to an effectively high f number and results in a decrease in the percent of photons coming off the photocathode which reach the lens. In addition, since both the object and image are in the near field, there will be a field curvature problem for a simple lens. Typically, a multi-element lens with field flattening correction lenses at both ends is employed.

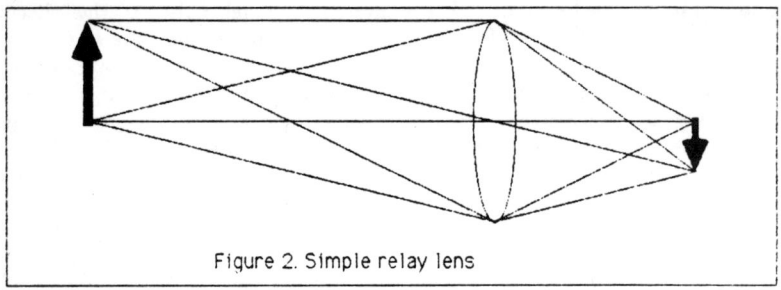

Figure 2. Simple relay lens

There is a trade-off between adding elements to refine the optical aberrations versus using fewer glass surfaces to maximize the transmission of the lens (lowering the F number). While this may seem to be an impossible trade, there is one advantage; any specially designed image intensifier relay lens must work over only a narrow range of wavelengths since the intensifier phosphor output is predominantly green. The lens thus does not have to be color corrected over a broad band. Several relay lenses have been designed specifically for this purpose and they are available from commercial intensifier manufacturers as shown in Figure 3. Such special purpose lenses are generally more expensive than comparable general purpose optics because of their unique nature and the small demand for them. Typically, the lenses cost between $400 and $1,000.

There is, however, another option as is shown in Figure 4. The image intensifier often is delivered with a conventional eyepiece. This is a simple two or three element optical system which collimates the image intensifier output so that it appears at infinity focus. Often the eyepiece lens is adjustable in focus. In addition, most TV cameras are also delivered with an inexpensive objective lens, designed to view objects at infinity focus. If the light from the eyepiece is directed into the objective of the TV camera, then a relay lens has been formed.

COUPLING IMAGE INTENSIFIERS TO TV SENSORS

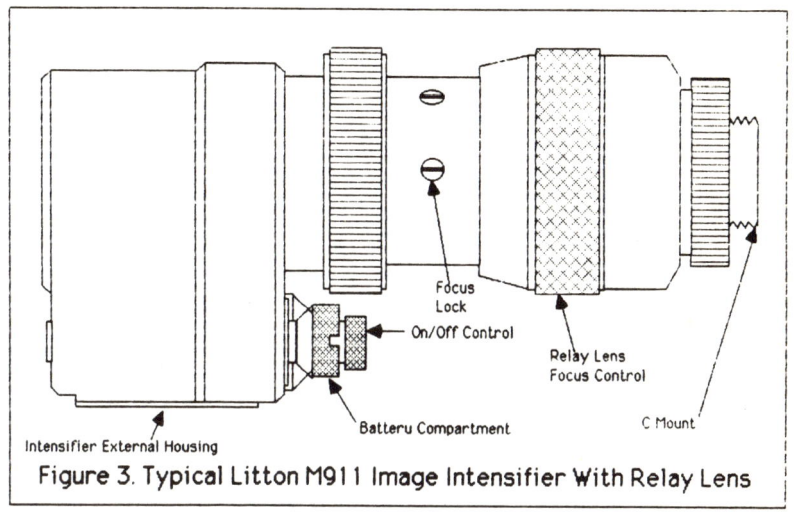

Figure 3. Typical Litton M911 Image Intensifier With Relay Lens

There are some limitations on such a system. The field of view of the eyepiece may not match the field of view of the TV camera objective lens. In this case, either the TV system will be looking at only the center portion of the image intensifier or the entire output of the intensifier may fill only a small portion of the TV screen. It is usually easier to change the focal length of the TV camera objective. For most eyepieces, a 10 mm to 25 mm focal length objective will provide a pretty good match in format size. In most cases, the exit pupil of the intensifier eyepiece and the objective aperture diameter are both about 1 cm. In cases where a gross mismatch occurs in which the eyepiece diameter is much greater than the objective diameter then much light will be lost in the relay lens assembly. Remember, however, that if the gain of your intensifier is 10,000 and you loose 90% of the light in the relay lens, your system still has a gain of 1,000, which is respectable.

If the objective lens which was supplied with the TV camera is used, care must be taken to defeat any auto–iris provisions in the lens, or the relay lens may cut down on the image brightness. In addition, for those cameras with an auto–focus provision, this must also be overridden, lest the camera change focus without warning. The center of the eyepiece lens must be held on the optical axis of the TV camera objective, as is shown in Figure 4. This usually implies an external mechanical support. A light shield should also be added to the set–up shown above so that stray light does not enter the objective lens from around the edge of the eyepiece.

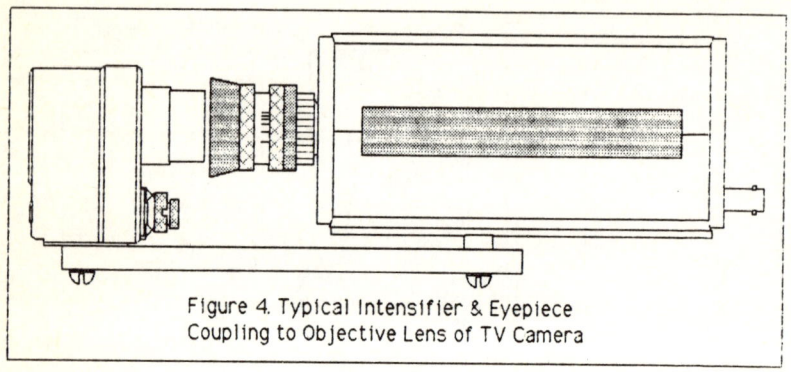

Figure 4. Typical Intensifier & Eyepiece Coupling to Objective Lens of TV Camera

The output of the intensifier may be either clear glass or fiber optic. Typical effective f numbers for most fiber optics are around 1. Any relay lens of reasonable f–number will probably not capture all of the light from the fiber optic output. The best that can be done is to maximize the light transmitted without inventing a physically cumbersome relay lens.

Many intensifier manufacturers use a fiber optic output and then grind the fibers to a spherical surface as shown in Figure 1. This then allows them to use a less expensive and lighter weight eyepiece lens with gross field curvature aberrations which match the curve of the output fibers. If your intensifier has a curved output surface then it is best to use either the eyepiece which came with the intensifier and a standard TV camera objective lens or a complete relay lens made especially for that curved output surface. While some astronomers can design and grind such a special relay lens, the task is not a trivial one.

IV. MECHANICAL CONSIDERATIONS

The image intensifier, relay lens and TV camera form a rather long optical train and they usually require special support, especially on a telescope where the gravity vector can point to almost any position with respect to the optical axis. A typical support is shown in Figure 5. While some light weight TV cameras and intensifiers have been supported off the back of a 20 cm aperture Schmidt—Cassegrain as shown in Figure 6, this is not the recommended practice.

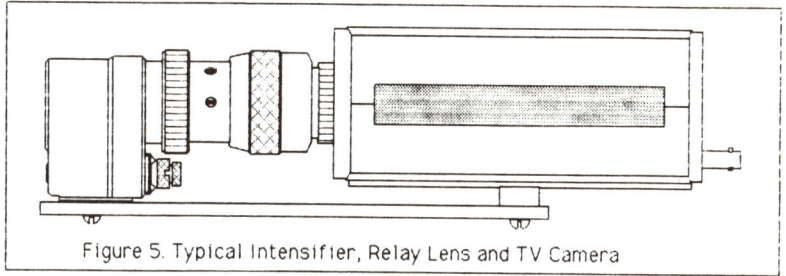

Figure 5. Typical Intensifier, Relay Lens and TV Camera

Such an installation is bulky, prone to being banged into in the dark and can require more counterweights than the telescope manufacturer recommends be used. Ideally, the system should be folded so that it remains closer to the center of gravity of the telescope and is supported along its entire length as is shown in Figure 7.

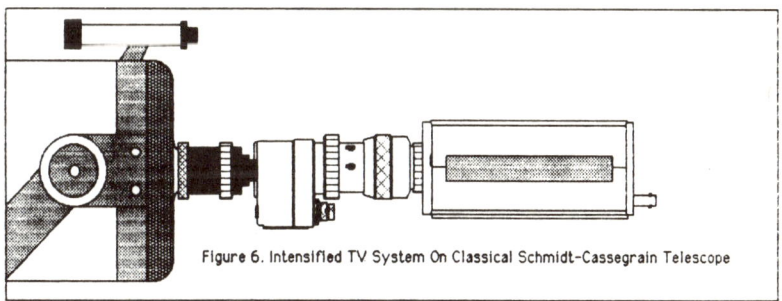

Figure 6. Intensified TV System On Classical Schmidt-Cassegrain Telescope

Provision should be made to allow for small changes in the length of the relay optics when adjusting the focus of the relay lens. For those photometric systems employing a flip mirror, the TV system often sticks out the side of the telescope, and again it requires external support. There should also be provision in the mounting to adjust and align the TV system boresight to the main photometric optics. In this way, the pointing of the data channel may be precisely positioned with respect to the TV acquisition system.

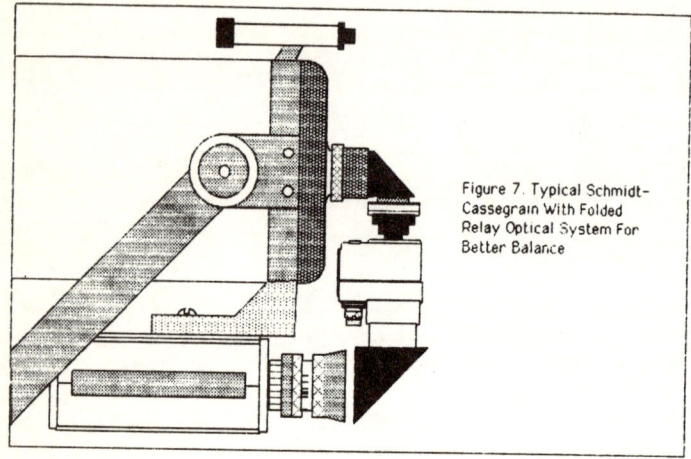

Figure 7. Typical Schmidt-Cassegrain With Folded Relay Optical System For Better Balance

V. ELECTRONIC CONSIDERATIONS

The image intensifier high voltage power supply is usually an integral part of the intensifier itself. Typically, the intensifier power supply input operates on a few volts. The power consumption is so low that most people use batteries rather than fool around with power cords. The TV camera often requires either 12V DC or 115 VAC which is usually no problem in the observatory. The output of the TV, however, may come in one of two forms. One type of output is modulated RF video (on channel 3 or 4) as in most home TV applications. For this type of output, a commercial TV receiver is used for viewing and is tuned to the appropriate channel. The other type of output is called baseband video. The signal contains standard EIA RS–170A video similar to the dubbing output of a video tape recorder. A special baseband display monitor is used. Of the two types, the baseband video is preferred since it does not require the addition of RF modulation to the signal in the camera and demodulation in the receiver. Baseband video will be of higher quality and lower noise than modulated video.

There are two possible sources of electromagnetic interference (EMI). The first is the susceptibility of the TV system to external noise. The second is a possible RF emission from the intensifier power supply. Most TV cameras employ a very sensitive video amplifier at the front end of the system, no matter whether a tube or solid state chip is used. Most good TV camera lens bodies are electrically grounded to the TV camera case via the lens screw mount and thus they shield the front end from external electrical noise. In the event that a substitute lens is used, care must be taken to assure grounding continuity. This is especially so if the substitute lens has anodized threads which may form a nonconducting insulator. One cure for this is to run a separate ground wire from the TV camera case or video ground to the relay lens itself. The TV system may also be susceptible to external EMI from telescope motors and other equipment such as dome slew motors, filter wheel motors and relays, especially if they produce any kind of arcing. The cure for this must be at the arcing source, which is actually good practice for the entire compliment of observatory equipment. A good rule of thumb is that if the interference shows up on a portable commercial TV receiver in the dome then it should be fixed at the source.

The second kind of interference is emission from the intensifier power supply. In general, most intensifiers have an integral high voltage switching power supply. Typically these run at 5 KHz to 40 KHz switching frequency. They usually operate at such low power that they cause no interference. Occasionally a bad power supply in old used units has been known, however, to act up and broadcast at its switching frequency or some harmonic. Since they are usually a sealed unit and are often potted with the intensifier tube, they are difficult to replace. External shielding and grounding is probably the only cure for this.

One final word on electronic problems; remember that the key to image intensification is the acceleration of the electrons through several thousand volts. In some intensifiers the photocathode (input end) may be several thousand volts negative with respect to ground. Similarly, the output phosphor may be several thousand volts positive with respect to ground. While the current capacities of the high voltage power supplies are low, the voltages must be treated with respect. If metallic surfaces are placed too close, then arcing may occur, to the detriment of the gain of the tube, possibly to the permanent detriment of the power supply. A mechanical design which shields these high voltage surfaces from the effects of dust and dew is certainly a plus.

VI. OPERATIONAL CONSIDERATIONS

Image intensifiers are susceptible to damage under bright lights. The damage mechanisms are the same as those encountered in photomultipliers. One difference in the use of intensifiers, however, is that the images of bright stars are focused and concentrated in small areas on the photocathode. In the case of photomultipliers, the image is spread out over the entire tube faceplate. Thus, localized damage may occur in an intensifier at light levels which would not usually harm a photomultiplier. Many image intensifiers have an Automatic Brightness Control (ABC) and/or Bright Source Protection circuit. This is basically an averaged anode or cathode current sensor which decreases one of the tube voltages (and thus the gain of the system) at higher total light levels. This limits the output brightness and protects the tube. The ABC is not designed for stellar scenes. It is possible to image a very bright star or planet and damage components inside the intensifier with localized excess current density without triggering the total current sensor in the ABC circuit. For a 20 cm aperture telescope, stars of V brighter than about 3 mag should be avoided. If brighter stars are to be observed, then a simple 10% transmission neutral density filter inserted in the optical train just in front of the intensifier will suffice to protect the camera. For automated systems, the filter would always be used when slewing and upon initial field acquisition. If the computer doesn't see any bright stars then the filter would be swung out of the way.

A second effect of the ABC circuit in the intensifier is that in high brightness fields of view typical on hazy city nights, the ABC circuit may start to decrease the gain of the system. A similar effect may occur at dawn or with light clouds illuminated by a full Moon. The overall field may still look black on the TV monitor if the haze effect is even. The appearance of the display, however, will be a seemingly unexplained loss of gain. One glance at the sky should explain the problem, but in TV astronomy the observer often sits inside, separate from the elements. It is a good practice to visually check the real sky once in a while when doing video astronomy.

Most image intensifiers have a larger dynamic range than most TV sensors. This implies that if the TV camera sensitivity is matched to the faintest image expected from the intensifier then very bright images will saturate the TV, producing a maximum white level on the video. In bright backgrounds this may mean that the image on the monitor is all white while the intensifier is still operating well within its dynamic range. A second effect is that very bright stars in a dark background will bloom or cover large areas on the TV screen. If the star of interest is adjacent to the bright star, it can fall within the bloomed image. In general, tube—type cameras displaying very bright stars will show the stars as large round blobs as shown on the left in Figure 8. The majority of the star's expanded size is due to charge spreading in the target of the tube. In solid state cameras, however, an overload—level star signal may spread preferentially in either the horizontal or vertical direction as shown on the right in Figure 8. The severity of this will depend on the chip structure. Thus, very bright stars may tend to become a long bright line.

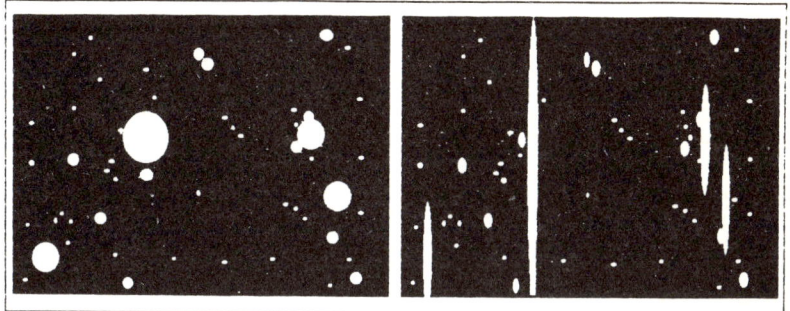

Figure 8. Typical Stellar Scenes Viewed With A Tube Type Sensor At Left And A Solid State Sensor At Right.

One consequence of operating with high—gain image intensifiers is that the arrival rate of photons at the telescope aperture can be quite low and the star will still be detected. Remembering that the TV camera is an imaging device, it is useful to consider that each individual pixel is a separate photometer channel. When the photon arrival rate per 1/30 second frame time is low, then the statistical variation in the rate will cause frame—to—frame differences in the measured brightness of each pixel. This is manifested in the TV display by causing the appearance of broad background noise. In a star, the brightness will vary, and for a faint star, it may disappear entirely in some frames. In general, it is desirable to run the image intensified TV camera system in a background limited mode. This occurs when the dominant noise mechanism is the "sky noise" caused by the statistical variation in the arrival rate of background photons per pixel per video frame.

This is subject, of course, to the operational constraints of the system as regards to the available telescope aperture, desired field of view, etc. For any given system gain there is an f number at which the sky background is barely detectable in the TV signal. This is the optimum operating point from a standpoint of signal—to—noise ratio alone. If the aperture is held constant and the focal length is increased, then there will be no greater sensitivity to point sources while the background noise will go below the threshold of the camera sensitivity. In addition, the field of view will decrease. If the aperture is held constant at the optimum operating point and the focal length is decreased then there will be no greater sensitivity to point sources while the background noise will increase due to the lower effective f number. Since the star signal is constant here and the background noise has increased, then the system signal—to—noise ratio (sensitivity) to stars will decrease. If the focal length is held constant at the optimum operating point and the aperture is increased, then the star signal will increase with the square of the aperture. The f number will also decrease with the increase in aperture and the background light levels as seen at the image plane, and will increase as one over the f number squared. In other words, the background noise will also increase. The result is a nearly equal increase in signal and noise until the background becomes so bright as to saturate the video signal. The user will have to balance his needs in field of view and sensitivity with the capabilities of the available optical system.

Typical limiting sensitivities for common image intensifiers and a "good" low light−level monochrome surveillance camera on a 20 to 32 cm aperture f/10 Schmidt−Cassegrain telescope average around $V = 9 - 11$ mag. At $V = 9$ mag the signal to noise ratio (SNR = mean peak signal above RMS noise) is about 4 to 6. At $V = 11$ mag the signal can still be seen on the display but it has a SNR of 0.5 to 1.0. Most automated video processors require at least a SNR of 6 in order to function properly.

Once the limiting sensitivity of the system has been achieved, the practical designer always asks if there are ways to improve system sensitivity without purchasing a larger telescope or more expensive imaging system. The most common method is to increase the integration time of the TV sensor. The standard 1/30 second TV camera integration time is geared toward providing an esthetically pleasing image. If the integration time were increased to 1/15 second then the system sensitivity might be doubled. This can usually be accomplished by either blanking the read beam in a tube type camera or stopping the clock in a solid state sensor. The technique usually works and the system sensitivity is usually just about doubled. The penalty is a screen that flickers at 15 Hz. Some astronomers can ignore the flicker and others can obtain a monitor with a longer persistence phosphor. The technique usually cannot be extended, however. In most solid state and tube−type systems integration times longer than about 1/10 second run into the limit of charge spreading. This occurs when dense concentrations of charge due to stars tend to diffuse across the target in a tube or silicon array in a CCD. The result is that with longer integration times, the stars don't become brighter but they do become fatter.

There is, however, a second method of increasing the integration time, and that is off−tube integration. The video picture is read out of the TV camera, digitized and stored in a digital computer memory. The next frame is then added to the first and so on. There are several schemes available to accomplish this. The most common is to add the most recent frame to half the value of the previous frames. Thus, rapid motion of brighter stars is preserved and fainter stars can be seen after a few frames of integration. This method also tends to allow transients and scintillations to disappear after a few frames. The disadvantage of off−tube integration as compared with on−tube integration is that every frame of video has its own readout noise. In off−tube integration the readout noises are summed. In on−tube integration, there is readout noise associated only with the frame which is sampled.

VII. CONCLUSION

The concept of using intensified television is a valid one for the purposes of aiding in the identification of stars of interest. The technology is in hand and is sufficiently mature to warrant application as an aid to automated photoelectric astronomy.

REFERENCES

Boyd, L. J. and Genet, R. L. 1987 in *New Generation Small Telescopes*, ed. D. S. Hayes, R. M. Genet and D. R. Genet, (Mesa: Fairborn Press), in press.

CHIPS SET SUBSTITUTE FOR THE MVL 100

Jeffrey L. Hopkins

Hopkins Phoenix Observatory

I. INTRODUCTION

The most sensitive and perhaps easiest to use astronomical photoelectric-photometry system is a photon-counting system. The photon-counting system provides at least 2 magnitudes more sensitivity than a solid state system plus a dynamic range of over 10^7. In addition, the linearity is excellent. With the large dynamic range, no scale switching is needed. This allows very accurate work to be done using a comparison and program star that differ greatly in brightness. Data reduction is also easier as there are no gain factors to worry about. Correction for the system dead time must be made, however.
    The requirements for a photon-counting system are some sort of pulse conditioner, low voltage power supply, and a pulse/frequency counter. The output pulses from a photomultiplier tube are negative going with an amplitude typically in the microvolt region. The pulse

width is typically less than 10 nanoseconds with the signal terminated in 50 ohms. Higher resistance termination will result in wider pulses and correspondingly larger dead time. The pulse conditioner must take these very fast low level pulses and convert them into something that can trigger a counter. Usually a counter capable of at least 10 MHz is used. Bright−star work can result in counts into the millions for a ten second integration. A counter with eight seven−segment LED displays is ideal because it allows displaying up to 99,999,999 counts per integration period and is easily visible in the dark. Also a 10−MHz counter is capable of being triggered with 10 MHz or 100 nanosecond−wide pulses so again dead time will be minimized. Figure 1 shows a block diagram of a typical photon−counting system.

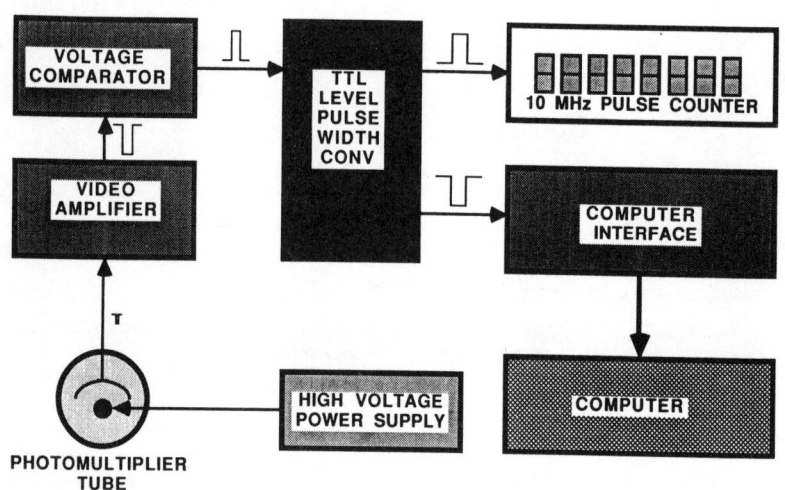

Figure 1. Photon counting block diagram.

Around the beginning of 1980, LeCROY came out with a monolithic integrated circuit, the MVL100, which performed the functions of an amplifier and comparator plus a monostable multivibrator. The price was around $25.00 and it was well suited for use as an amplifier/discriminator in a photoelectric photometry photon−counting system. By merely connecting the output of a photomultiplier tube to a 50 ohm terminated input of the MVL100, and adding a few other fixed components, one would have pulses ready to be counted by most any frequency counter. One minor drawback was that the output of the

MVL100 was emitter coupled logic (ECL). ECL is negative voltage logic and is used for very high speed applications (100 MHz range). To convert the ECL to transistor–transistor logic (TTL) levels, which is 0 to +5 volt levels, a Motorola MC10125 converter could be used. Using these components a photon–counting pulse–conditioning circuit could be built.

In 1985 LeCROY discontinued the MVL100. There are several alternate devices that replace the MVL100 but they are all hybrids and very expensive. Typically an amplifier and comparator hybrid totals over $500. Because I believe photon counting is the best way to do photometry, I set about to develop an inexpensive alternative to the MVL100.

II. PULSE CONDITIONER

Figure 2 shows a schematic of the original MVL100 pulse conditioner circuit. Figure 3 shows a schematic of the newly developed chip set replacement for the MVL100. The following is an explanation of that circuit.

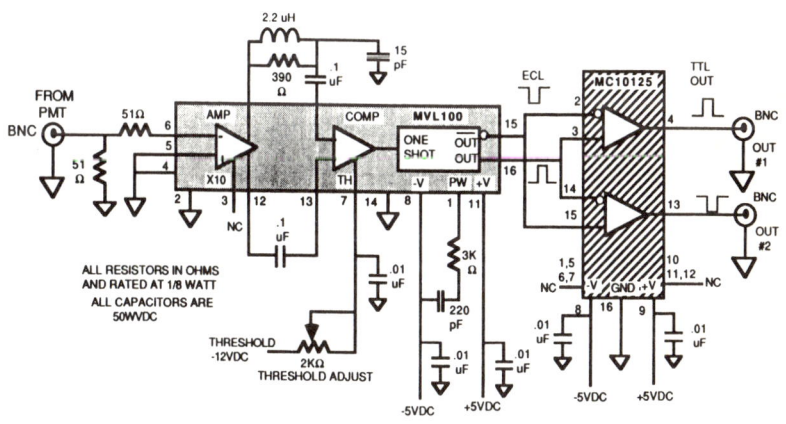

Figure 2. MVL 100 pulse conditioner (MVL 100 low level amplifier/comparator and MC10125 ECL to TTL converter).

Integrated circuit U1 (14 pin DIP) is a high–speed video amplifier (LM733CN) made by National Semiconductor. It has a bandwidth of 120 MHz with selectable gains of 10, 100, and 400. The device operates from +5VDC and −5VDC with a total current requirement of less than 30 mA.

As used in this circuit, the device is programmed for a gain of 100 (specifications indicate the gain will be between 80 and 120). This is accomplished by connecting pin 3 to pin 12. The output pulse from the photomultiplier tube is terminated in R1 (51 ohm). R2 provides protection from reflections of the high speed signal. The "IN 2" (pin 1) input is used as the signal input and causes the output to be inverted (positive going). The "IN 1" (pin 14) is connected to ground. Power is applied to pins 10 (+5VDC) and 5 (−5VDC). The 0.1 uF bypass capacitors are very important and must be used. They must also be close to the power pins 5 and 14. The output of U1 is differential. The "OUT 2" (pin 7) is terminated, through a 0.1 uF capacitor, with a 1 k ohm resistor. The "OUT 1" provides an amplified positive−going pulse to the low−pass filter. The purpose of the low−pass filter is to reduce harmonics in the pulse.

Integrated circuit U2 (14 pin DIP) is a high−speed voltage comparator (LM710CN) made by National Semiconductor. Because not all pulses coming from the photomultiplier tube are due to photons (some are produced by thermal emission) it is desirable to select only the ones due to photons. The pulse produced by a photon goes through all the stages of amplification and attains the highest gain or amplitude. Some thermal electrons also may go through all stages but many are emitted from lower potential stages and produce pulses of lesser amplitude. By using a comparator with an adjustable threshold voltage, the lower level pulses can be eliminated. The other thermal electrons going through all stages produce the measured dark counts. U2 operates from +12VDC (pin 11) and −5VDC (pin 6) supplies. Total current for each voltage is less than 10 mA. Again the 0.1 uF bypass capacitors are very important. The pulse from the low−pass filter is routed to the "+" input (pin 3) of U2. Because of the high input impedance of U2 it is necessary to have resistor R4 (1 k ohms) terminate the signal. The "−" input to U2 (pin 4) is used for the comparison voltage input. Any input pulses to pin 3 that equal or exceed the voltage on pin 4 will cause the output of U2 to go low. The threshold voltage is produced from the +5VDC supply and voltage divider R7 (100k ohms) and multi−turn potentiometer R6 (1k ohm). Typical threshold voltage on pin 4 of U2 is 20 mV. This is a good starting point for the threshold voltage. The output of U2 is on pin 9.

Despite its awesome name, the monostable multivibrator is a very simple and easy to use device. Integrated circuit U3 (14 pin DIP) is a monostable multivibrator, or one shot (74121), made by various manufacturers. The purpose of U3 is to form a constant pulse width for each input pulse and produce two TTL level outputs. The output pulse width is set by resistor R8 (2k ohms) and capacitor C2 (150 pF) and is about 150 nanoseconds. The device is not retriggerable, meaning if a second trigger is applied before the output pulse is finished, it will be ignored. With the input to U3 connected to pin 4 (A2) and pins 3 (A1) and 5 (B) tied to Vcc, the device will produce an output pulse when the input is a negative going transition. Only +5VDC (pin 14) is required for U3 and the supply current is less than 40 mA. The outputs are complementary. Pin 6 (Q) produces a positive–going pulse while pin 1 (Q NOT) produces a negative going pulse. Either of the pulses can be used with a counter because counters count either positive or negative transitions and both outputs produce one of each per pulse.

The printed wiring board on which the above circuit resides is very important. Because of the high frequencies involved the board layout is critical. The board must also have a ground plane, short connections, and proper terminations. Figure 4 shows the parts layout of the top of the board. It is suggested that low–profile sockets be used for the integrated circuits. This makes troubleshooting and replacement easier. Figures 5, 6, and 7 show top, bottom, and drill template diagrams of the printed wiring board. Contact the Hopkins Phoenix Observatory for one–to–one copies of these layouts if you wish to make positives or negatives for etching your own boards.

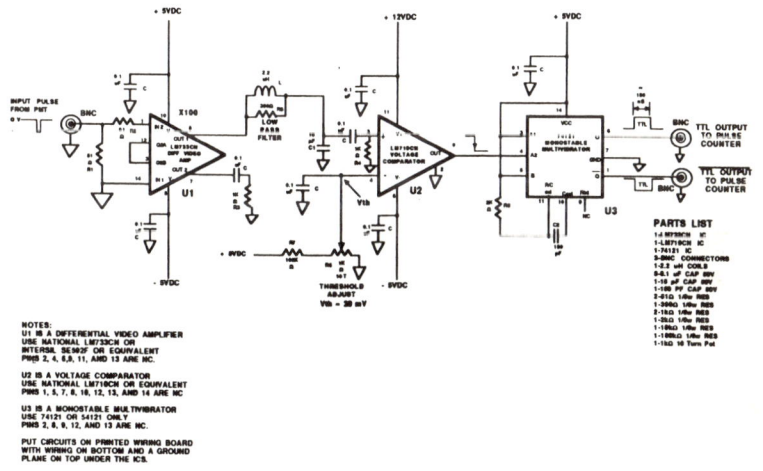

Figure 3. Pulse conditioning circuit for photon counting.

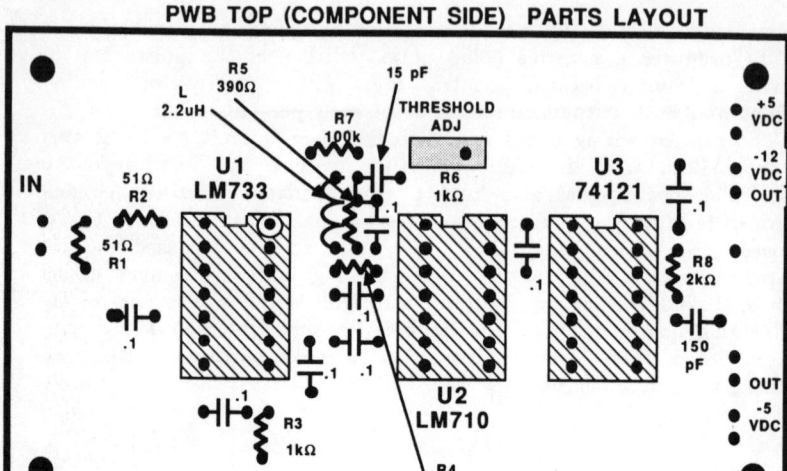

Figure 4. Pulse conditioner PWB parts layout.

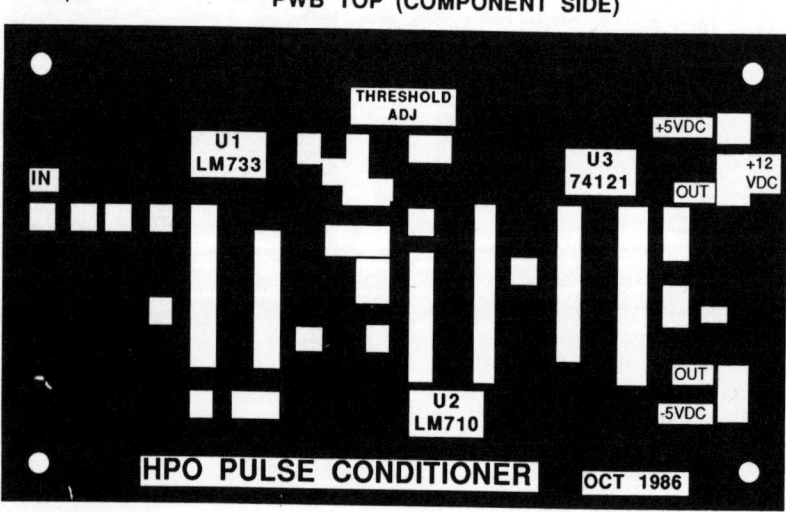

Figure 5. Pulse conditioner PWB top side.

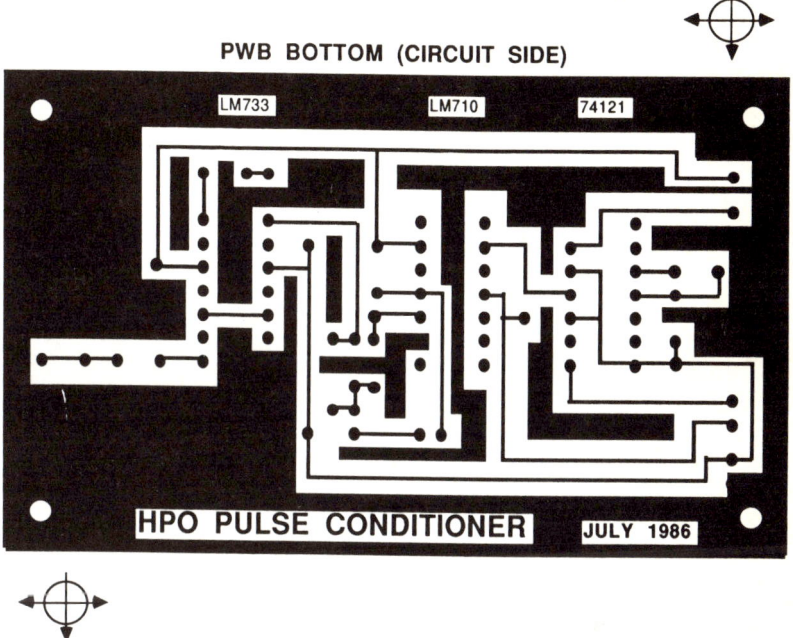

Figure 6. Pulse conditioner PWB bottom side.

To make obtaining this circuit easier, three options are offered by the Hopkins Phoenix Observatory. The first is just the bare two-sided board, tinned and with plated-through holes ($25.00 ppd). The second is the bare board and a kit of all parts ($50.00 ppd). The last is a completed board with all parts, wired and tested ($75.00 ppd). If it is desired to make everything yourself, there are many electronic supply houses that carry these parts. In addition, there are several printed wiring board manufactures that can make the board for you or you can do it yourself. All prices are 1986 and in US$.

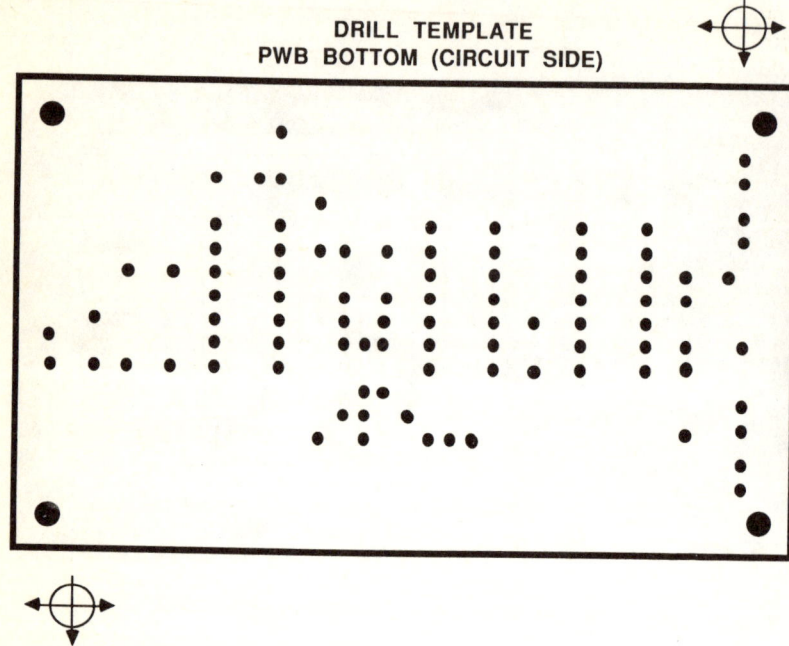

Figure 7. Pulse conditioner PWB drill template.

III. THE LOW VOLTAGE POWER SUPPLY

The pulse-conditioning circuit requires low-voltage power of +5VDC @ 50 mA, −5VDC @ 50 mA, and +12VDC @ 20 mA. The 10 MHz Universal Counter requires +5VDC @ 350 mA. To supply these power requirements the following Low-Voltage Power Supply can be used.

Figure 8 shows a schematic of a low-voltage power supply that can be used to power the pulse conditioner, high voltage power supply, and 10 MHz Universal Counter. Three terminal regulators are used to provide well regulated +5VDC, −5VDC, and +12VDC at up to 1 Amp each. The −12VDC portion may be left out if not needed. It is very important to use heat sinks with these regulators. This may be done by adding special heat sinks to each regulator or using the enclosure they are mounted in as the heat sink. If the enclosure is used, be sure to use heat sink grease and insulating parts (e.g., a mica insulator and insulating feed−throughs and washers) to electrically isolate the regulator from the enclosure. If individual heat sinks are used the insulating parts are not required but the grease is. Parts can be purchased from most electronics distributors and as shown on the schematic. Typically, the regulators are under $2 each and the transformer around $10.

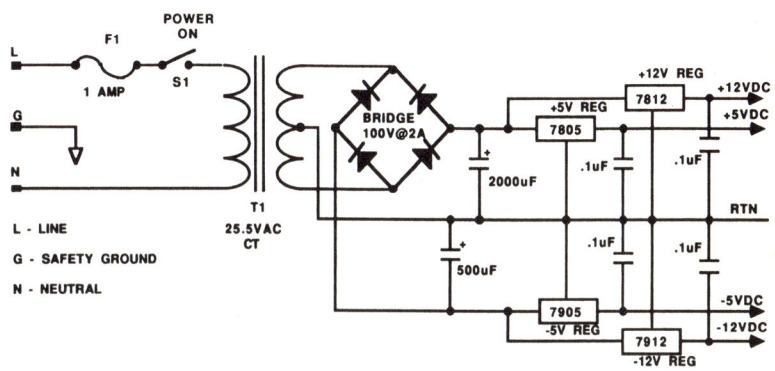

Figure 8. Low voltage power supply.

Most parts can be obtained from any electronic supply house. If there are any particular parts that are hard to obtain, contact the Hopkins Phoenix Observatory for help.

IV. THE UNIVERSAL COUNTER KIT

Figure 9 shows a basic counter schematic of an easy−to−build and inexpensive (under $100) INTERSIL ICM7226 AEV/Kit 10 MHz Universal Counter kit. In addition to buying the kit, it is suggested that eight right angle sockets be purchased so as to allow the display to be positioned at right angles to the printed wiring board. This makes mounting the board and display much easier. In addition to the kit and right angle sockets, you need a +5VDC power supply, mounting hardware, enclosure, plus switches (the board−mounted rotary switches supplied with the kit should not be used as they will make it very difficult to mount the board in an enclosure and still have control of the switches). The switches are used for RANGE (gate time − 1.0 and 10.0 sec) selection, RESET, HOLD, and power. The FUNCTION switch can be replaced with a permanent jumper wire to select FREQUENCY.

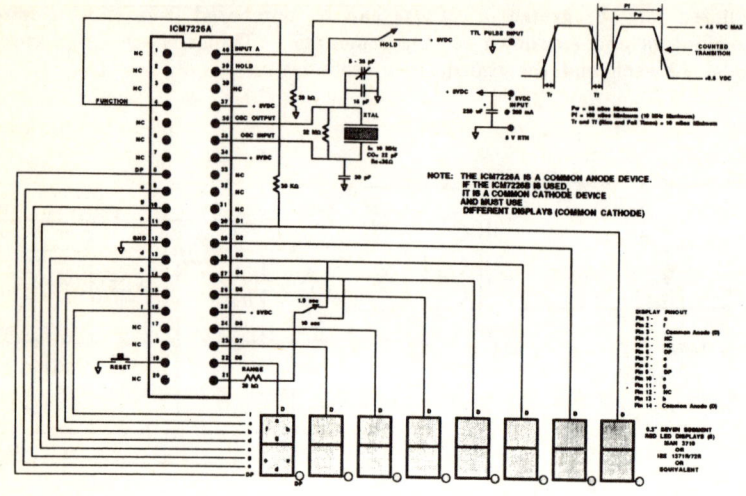

Figure 9. 10 MHZ pulse counter.

The basis for the 10 MHz counter is the INTERSIL ICM7226 A (40 pin DIP) integrated circuit. The device is the only integrated circuit needed to form the counter. It operates from a +5VDC supply and requires a maximum of 400 mA (with all display segments lit). The device allows many different features (frequency counting, period counting, unit counting, frequency ratio counting, and time interval counting) but only frequency counting will be discussed here. It is possible to select gate times of 0.01 sec, 0.1 sec, 1.0 sec, and 10.0 sec. Normally only the 1.0 and 10 sec gates are used in photometry. This allows the use of a simple single-pole double throw switch to select either 1.0 or 10.0 sec gate time.

Another very useful feature is the ability to reset the counter with push button. The count can also be held with another switch, however, this is seldom needed. Usually a 10 sec gate time is used and the RESET button is pushed at the start of the observation. This allows the next readout to be valid and, unlike a popular solid state photometer, you do not need to wait for the second or third readout for a solid count (this is because the RESET always starts the gate time at the instant it is pushed). This can mean a reduction of two or three in the time required to do the observations. What might take two or three hours can be done in one hour. Late on a cold night this can mean the difference between giving up and getting good data. Three to ten readings are usually then taken. The counter holds the current reading while counting during the next gate time. There is then 10 sec in which to make the reading. This is a quite sufficient time to even look around a bit and still copy down an eight digit number.

Another feature of the RESET and HOLD functions is that you can make a remote control box with these switches in addition to having them on the front panel. This comes in handy when the unit is out of reach from the telescope. The switches are just paralleled. Naturally if one HOLD switch is closed at either location, the other one will have no affect; however, the RESET push buttons can be pushed at either location to reset the counter. The input to the counter must be TTL. Other pulse levels can be used but require some leveling circuitry. The schematic shown in Figure 9 is the basic circuit for the counter and does not show all that is optional with the purchased kit. The kit comes complete with all parts (except the +5VDC power supply, external switches, and right angle sockets).

The ICM7226AEV/Kit can be obtained from DIGI−KEY Corporation, P.O. Box 677, Thief River Falls, MN 56701, 1−800−344−4539 (price $86.20).

The right angle display sockets can be obtained from MOUSER ELECTRONICS, 11433 Woodside Ave., Santee, CA 92071, (619) 499−2222 (stock number 535−14−810−90R and price $1.58 each).

A completed kit, with right angle sockets, wired and tested (less switches, power supply, and enclosure) is available from HOPKINS PHOENIX OBSERVATORY for $180.00 ppd.

A LOW−COST AUTOMATIC TELESCOPE AND PHOTOPOLARIMETER

F. Giovane, D. Ely, A. Weisenberger, G. Eichhorn,

Space Astronomy Laboratory, University of Florida

J. Rilum* and B. Soderberg

Royal Institute of Technology, Stockholm

I. INTRODUCTION

During the summer of 1985 a modest amount of money became available from NASA to modify an existing telescope/photometer system

---

* Pres. Addr.: Optical, Materials and Devices Lab., Dept. of Elect. Eng., Univ. of Southern California.

New Generation Small Telescopes, ed. D. S. Hayes, R. M. Genet, & D. R. Genet.
© 1987 Fairborn Observatory.

to make observations of the coma of Halley's Comet at the time of the ESA's GIOTTO space probe rendezvous. This telescope/photometer system was originally developed to act as a test bed for optical space instrument concepts. It included a 32 cm f/16 classical Cassegrain, a sturdy equatorial mount, an optical bench, and a modular photometer which could be configured as a photopolarimeter. As a test bed for space instrumentation, it was configured to be completely automated, although the wiring and computer interface had not yet been implemented at that time. The system was also transportable, and could be relocated to Hawaii to make observations from the 10,000−ft. Mount Haleakala station of the University of Hawaii, an excellent observing site for sky radiance measurements and suitably located for the post−perihelion apparition of the comet. Thus, the existing telescope and photometer were ideal for the intended comet application.

The need for a microprocessor control of the proposed system was obvious from the beginning. The observing program called for a precise repeatable raster scanning of the coma of Comet Halley during the time of the GIOTTO penetration. This was needed in order to correlate the ground based observation's line−of−sight measurements with the results of Halley Optical Probe Experiment (HOPE) on board GIOTTO and the other in−situ space experiments. In this way, the probe observations could be related to future ground based work. It was essential that the observations be accurately positioned with respect to the comet so that knowledge of the specific area being measured could be obtained, and these observations repeated in other colors. It was thus necessary to have an instrument that could be positioned accurately on stars and then moved precisely to the comet to make the coma measurements and then back to the stars to reconfirm the comet positioning. Since a raster scan of the comet had to be reproducible, the telescope mount had to compensate for the non−sidereal motion of the comet. Further, the photopolarimeter required synchronization of the data collection with mechanical operation of the photometer as well as with the rastering operation. The complex motion of the raster, the non−sidereal tracking requirements, and the photopolarimeter synchronization alone necessitated the use of a microprocessor controlled system.

Limitations in both time and money, required a straight forward and low−cost solution to the telescope/photometer positioning. This took the form of using a semi−open loop, where fiduciary positions were obtained by means of inexpensive magnets and Hall sensors. The exact position of the various driven systems was determined by monitoring the Hall sensors and by keeping track of the steps made by the stepping motors which were used for all system motions. The use of worm/worm gears allowed precision motion to be achieved at low cost. This total system of position control was functionally accurate and very low in price.

In this paper we will consider the details of the system installation, the advantages and limitations of the worm/worm gear system using Hall Sensors, and the general nature of the automatic telescope/photopolarimeter developed employing this position sensing system. The microprocessor control system is described in detail in a separate paper in this Symposium (Eichhorn and Giovane 1987).

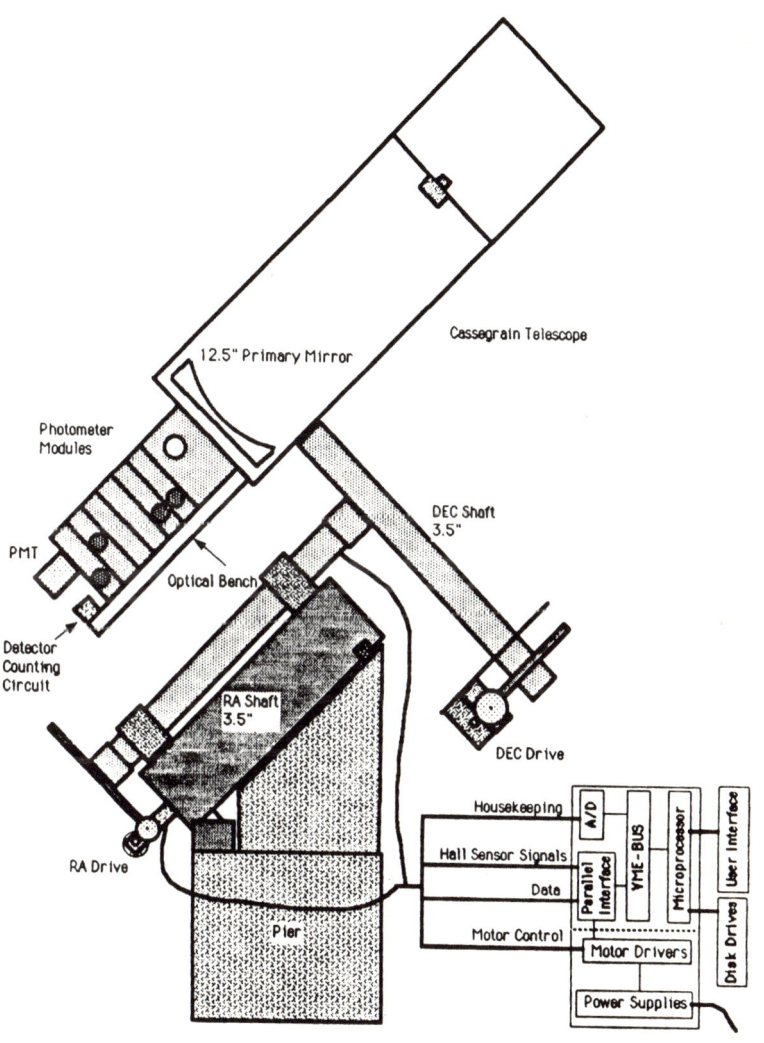

Figure 1. Telescope layout.

II. INSTRUMENT

The overall layout of the telescope/photopolarimeter is illustrated in a functional diagram in Figure 1. The system represents a combination of home−built and standard or purchased components. The drive, massive for a small telescope, was built at the University of Florida, based on a design which assures accurate tracking with minimum deflection. This was believed to be essential for an automated system that was to be used to test a variety of space hardware optical configurations. Two Responsyn 800−step four−phase motors, 1.5 amps per phase, were used to drive Thomas Mathis 15.2− and 10−inch worm/worm gears for the Right Ascension (RA) and Declination (DEC) drives respectively. No differentials or clutches were used in the system. The motor torques were sufficiently lower than the gear strength to assure that no damage would be incurred by the gears, and since the need for high speed slewing was not considered essential, no attempt was made to optimize the motor torque to gear strength. Motor system details, including slew rates and position resolutions are given in Table I. This Table provides interesting insight into this automatic telescope system.

Table I

Motor System Details

Function	Motor type	Steps/Rev	Gear Teeth Motor Spur	Worm Spur	Worm	Worm Gear	Total Gear Ratio	Arcseconds/Step	Tangent motion(μm)/Step	Max Speed Steps/sec
RA	Responsyn 165−800−4	800	32	160	1	360	1:1800	0.9	0.84	700
DEC	Responsyn 165−800−4	800	32	160	1	160	1:1575	1.0	0.20	700
FOV	Oriental PH264−02	200	30	90	1	360	1:1080	6.0	0.71	200
Filter	Oriental PH264−02	200	90	30	4	360	1:30	216.0	25.4	200
Shutter	Oriental PH264−02	200	90	30	4	360	1:30	216.0	25.4	200
Polaroid	Oriental PH264−02	200	90	30	4	60	1:6	1080.0	127	200

The 32 cm f/16 telescope was built at the University of Florida using Cassegrain optics from Star Instruments (Flagstaff, Az.). A motor−driven secondary mirror allowed automatic precision focus to be achieved. The telescope tube was oversized to allow an extensive dew cap while still maintaining an unvignetted field of view. Also the focusing motor, if left on, would generate enough heat to prevent dew formation under most conditions.

The photometer was made up of several discrete modules, each configured so that they could be combined in a variety of ways in order to test various optical designs. They were mounted on a U-channel optical bench which provided configuration flexibility. The configuration used for the comet observations was a standard photopolarimeter design for low-precision polarimetry, and involved the use of a rotating polaroid foil. The configuration is illustrated in Figure 2.

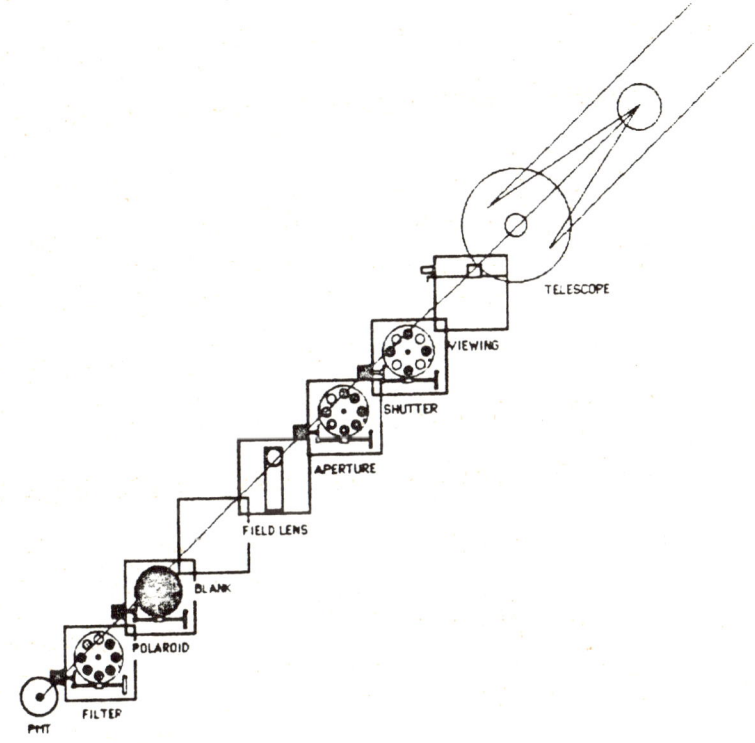

Figure 2. Photometer configuration.

A view finder was introduced to perform visual checks during the debugging of the system. This element was followed by a shutter wheel, a field-of-view selection wheel, a field lens which imaged the entrance aperture onto the detector, the rotating polaroid, and a filter wheel which included the IAU standard comet filters. An EMR 510 PMT detector package, originally developed for space application, was used in photon−counting mode to detect the incoming light. Its detection process was synchronized with the polaroid rotation. Four Oriental Motor 2−phase stepping motors, 0.4 amp per phase, were used to drive the photometer's polaroid foil, field of view wheel, shutter, and filter wheel. All four modules used the same basic worm gear drive arrangement, but, as seen in Table I, with different gear ratios.

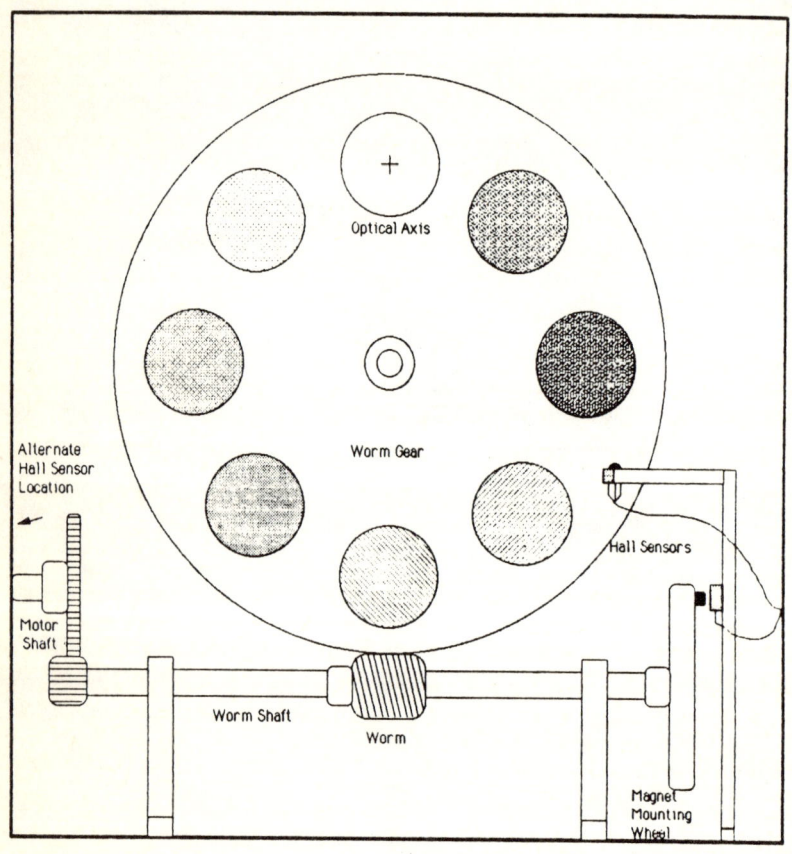

Figure 3. Module gearing arrangement.

The worm/worm gear arrangement is illustrated in detail for a nominal gear in Figure 3. In all cases the motor drives a spur gear which in turn drives the worm/worm gear. The accuracy with which each driven wheel had to be positioned is determined in practice by the ratio of the spur and worm/worm gear ratios. The speed with which the worm gear can be driven is established by the size of the motors and strength of the gears. In order to maintain a capability for interchange, the focus motor and all of the photometer motors were of the same type. The maximum speed of the system was determined empirically (*i.e.* by running the motors at as fast a speed as possible without frequent slippages). The simple two-phase stepping motors of 0.4 amps-per-phase were adequate for all photopolarimeter operations, but in order to maintain a reasonable speed for the RA and DEC motors, larger more powerful motors had to be used.

Early in the modification of the telescope/photopolarimeter system it was realized that it would not be possible to house the telescope in a shelter while operating on Mt. Haleakala, and therefore, if the electronic systems, power supplies, and microprocessor system were to be mounted with the telescope proper, environmental protection and security would have to be provided. This was not considered feasible. Further it was felt that locating the microprocessor system far from the operator might complicate the software programming and cause unacceptable delays in an already tight schedule. Consequently, 80 feet of cables were laid between the telescope and the power, electrical and computer systems. The result was that only a limited number of circuits, including detector, housekeeping, and power distribution were located at the telescope/photopolarimeter. Twisted pair wires were used to carry power and signals to and from the shelter where the electronics and computer systems were located.

The wires entering the shelter were connected into a distribution panel, from which they were distributed by ribbon cable to the computer system. It should be noted that AC power was not used at the telescope/photopolarimeter so as to avoid crosstalk over the extensive lengths of cable. DC power and stepping motor lines were bundled separately from housekeeping and data gathering lines, and all signal and control lines entering the microprocessor system were opto-isolated and went through the distribution box.

The detection system included its own high voltage and pulse amplitude detection circuits. The output was TTL level pulses which were counted by a 16 bit counting circuit, the output of which was brought down on parallel lines. The 5 volt power requirement for the EMR photometer system and counter was supplied by a dedicated power supply.

Measurements with this system were linked to the 5.52-second period of rotation of the polaroid wheel. The detector system counters were sampled every 4.6 ms, and refreshed. In all 100 of these samples were integrated by the computer to make a single measurement. An estimate of the measurements' uncertainty was also calculated. The duration of a single measurement thus corresponded precisely to 30 degrees of polaroid rotation, and in all 12 integrated measurements were made over a full rotation of the polaroid wheel. These measurements yielded the Stokes vectors I, Q, U, and their uncertainty, from which the intensity, the degree of polarization, the plane of polarization, and the estimate of uncertainty of each was determined. Each element of the raster was made up of at least one set of 12 measurements; consequently observations were in integrals of 5.52 seconds corresponding to the period of rotation of the polaroid.

A wedge field of view was provided to automatically locate and position stars. The length of time that the star remained in the wedge determined its position in declination, while its time of passage through the center of the wedge determined its right ascension. Corrections were then made to the telescope pointing to place the star on the telescope's optical axis, and an appropriate field of view chosen.

The secondary mirror was motorized so that the telescope could be automatically focused. In this process the secondary was positioned to minimize the duration of onset as a star was scanned by a wedge or circular field of view.

Non-sidereal tracking was achieved by providing precalculated pulses to the right ascension and declination motors. This was used to track the comet's motion against the background stars.

III. POSITIONING SYSTEM

In order to accurately position the various motor systems it was necessary to use a simple system that could be fitted to the existing photometer modules, and which was also low in cost. The positional feedback system and its implementation can be seen in Figure 2. A small magnet is located on a disk attached to the worm shaft and a Hall sensor is placed so as to sense this magnet. The Hall sensor generates a TTL-compatible 5-volt signal when it senses the proximity of the magnet. Since the number of motor steps necessary for a rotation of the worm was kept track of by the controlling microprocessor system, the loss or slippage of the stepping motor would register as a failure of the Hall sensor to activate at the proper step. In this way, with each rotation of the shaft, a small correction to the stepping motor count could be applied automatically, or in the case of large discrepancy, the operator alerted to a drive problem. The Hall sensor thus provided a mechanism for assuring that the motor system was functioning properly, and the correction of small slippages could be instituted automatically so that they would not accumulate. Another magnet and Hall sensor was located to assure the position of the worm gear. This sensor acted as a coarse zero or fiduciary, and when combined with the information from the Hall sensor associated with the worm shaft gave a precise positioning of the worm gear, generally to within one step of the motor. The potential accuracy of this placement is also given in Table I.

IV. OPERATIONS

The telescope/photopolarimeter system functioned without electronic problems of any sort. Surprisingly, no failures of the many cables laid between the instrument and the microprocessor system occurred. Provision had been made to backup the system power supplies as warnings abounded about the failure of these components on Haleakala, where high altitude conspires with high humidity and clouds to shorten power supply lives. However, by drastically under-rating all supplies and maintaining them indoors, problems were avoided.

The Hall Sensors did cause us some problems in the developmental stage. Chief among these was a hysteresis effect. The magnet approaching the sensor from a given direction would turn on the sensor almost always at the same step location and the sensor would stay on for several more steps and then go off very predictably. However, if the magnet approached the sensor from the other direction, the sensor would turn on over a different position range. That is, depending on which direction the magnet approached the sensor, the sensor would behave differently. The computer had to account for the direction of approach, and, as a consequence, the programming became more complex. This problem was further exacerbated by the need to take up backlash in the gears in approaching a target position by over shooting the position and coming back on it from "the standard direction."

The Hall Sensors needed to be small to allow the positional precision required. The handling, mechanical fixing, and precise adjusting of these sensors proved to be a real problem, and several sensors were destroyed during installation. Once properly placed, the sensor functioned reliably.

Mechanically the system suffered from its exposed location in which it was affected by strong winds. The slew rates on the RA and DEC axis were set slightly too fast, and frequent slippages of the motor resulted in this mode, however, the slippages were not too great to prevent resetting of the step counter using the Hall sensor position indicator. Also, during windy and gusty conditions some slippage was experienced during normal tracking, but again were correctable by the computer system.

Clearly one of the major problems in getting this telescope system on line was the shortness of time which forced us frequently to debug the mechanical, electrical, and software systems at the same time. Although some attempts to test the various functions independently were carried out, the need for speed required that the system be put together and operated before they were fully independently tested. Nevertheless a fully-operational system was assembled and fielded in time for the GIOTTO encounter.

V. CONCLUSION

The low-cost Hall-sensor solution to the problem of positioning the telescope/photopolarimeter proved to be accurate and, once the installation problems were overcome, reliable. But this was not achieved without a cost in complexity of software and in requirements for a fast microprocessor. However, if the RA and DEC positioning requirements were more modest than in this comet application, the complexities would be reduced. This approach can then provide a low-cost, highly reliable solution for an automated positioning system.

The use of commercially available (and commercially successful) microprocessor system interface boards, although possibly more expensive than what might be developed in the laboratory, assuming "free labor," are to be highly recommended as more reliable and, in the total systems sense, less costly.

The use of a fast versatile microprocessor system, such as the VME/68000, gives the total system a degree of flexibility that allows simpler correction of design errors and enhances the potential for reliable expansion in future modifications. A wide variety of support boards are available for the VME from a number of manufacturers. They are generally well engineered, built to appeal to a demanding commercial market. As a consequence it is felt to be worth the additional cost to use this microprocessor system in any application where reliability and flexibility are desired, including even those implementations that are less demanding than the comet project reported here.

VI. ACKNOWLEDGEMENTS

The systems reported here were built and deployed under NASA Contract NASW-3678. The authors are indebted to the student, faculty and staff members of the Space Astronomy Laboratory, Physics Machine Shop and Engineering Sciences Department of the University of Florida, and the staff of the University of Hawaii who contributed to the success of this project.

REFERENCES

Eichhorn, G. and Giovane, F. 1987 in *New Generation Small Telescopes*, ed. D. S. Hayes, R. M. Genet and D. R. Genet, (Mesa: Fairborn Press), p. 317.

A VME−BUS−BASED MICROPROCESSOR SYSTEM FOR AUTOMATIC TELESCOPE AND PHOTOMETER OPERATION

G. Eichhorn and F. Giovane

Space Astronomy Laboratory, University of Florida

I. INTRODUCTION

An existing 32−cm f/16 Cassegrain telescope system was automated during the winter of 1985/6 to allow multicolor photopolarimetric observations of the coma of comet Halley during the time of the ESA GIOTTO probe's encounter in mid−March of 1986. The anticipated mode of making these observations was to raster scan the comet's coma. Because of the complex nature of the motions needed to maintain tracking, positioning and timing control, a microprocessor system was needed to operate the telescope and photopolarimeter.

The microprocessor system had to be capable of controlling the sidereal and comet tracking of the telescope, as well as maintaining a knowledge of the position of the telescope during rastering so that the measurements could be made on the same areas of the comet in different colors. This system also had to operate the photopolarimeter system, which required maintaining synchronization of the polaroid analysis function and the collection of data from the detector. Other operations were required of the microprocessor system as well, including monitoring of housekeeping sensors, recording data, and focusing the telescope.

Because of delayed funding and subsequent reallocation of manpower, the time available for software development was short, and steps were taken within the modest budget to reduce hardware interface problems. Two basic decisions affecting the selection of the microprocessor system and implementation of the software were made. The first of these was the selection of a low—cost semi—open loop feedback system using Hall sensors to monitor the positions of the telescope and photopolarimeter functions, as the more conventional encoder systems were beyond the means of the budget. The other decision was to locate the microprocessor system remotely from the telescope. This was necessary because the observations were to be made from a field site with no shelter available at the telescope. It was decided that, for ease of operation and for the maintenance of security, it would be best to locate the electronics and microprocessor system together in a nearby shelter.

The use of a second, less—sophisticated microprocessor system at the telescope was rejected since it was expected that this would overly complicate the software development and require more time than was available. Consequently, few electronics circuits were located at the telescope, and many 80—foot wires were used to bridge the distance between the telescope and shelter.

The hardware configuration of the instrument is described in another paper in this Symposium (Giovane, et al. 1987). In this paper we describe the microprocessor hardware and software system that controls the telescope and photometer motors and records the data from the detector. In the concluding section some advantages and disadvantages of this system are pointed out as well as possibilities for further improvements.

II. SYSTEM DESCRIPTION

The selection of a suitable microprocessor system was determined mainly by three factors:

(a) The time available to develop both the hardware and software was rather short. This mandated a single−processor system for ease of programming.
(b) Time, money and reliability considerations mandated commercially−available electronics boards and limited the number of potentially suitable microprocessor−bus architectures.
(c) The speed of the required single−processor system (see (a)) limited the number of microprocessors suitable for the system.

Both (b) and (c) above suggested a VME−bus system based on a Motorola 68000 microprocessor. The number of commercially available boards for the VME−bus has risen dramatically in the last two years. As opposed to a PC−based system, a VME−bus system is virtually unlimited in modularity and expansion capabilities. This decision was aided by the fact that considerable expertise existed in our laboratory in developing VME−bus based microprocessor systems.

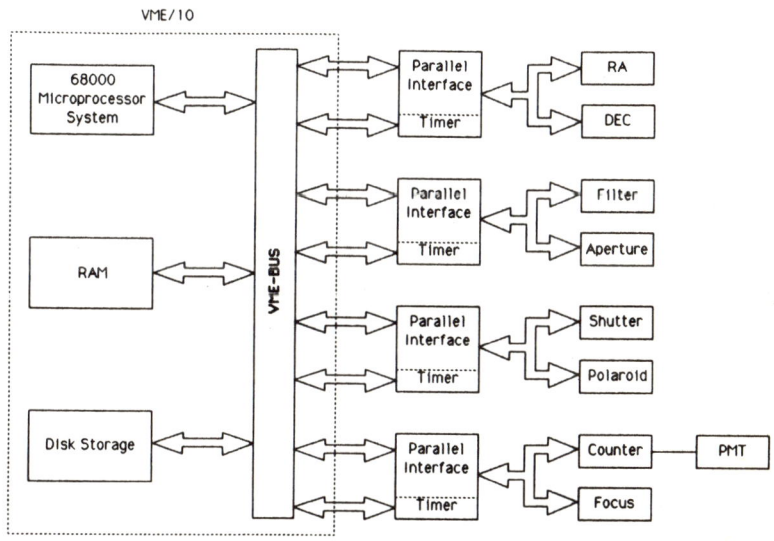

Figure 1. Hardware Configuration.

The speed of an MC68000 microprocessor is sufficient to operate a fairly complex system. If more capabilities are required it can easily be expanded with multiple processors on the same VME−bus or with a faster version of this processor family. The ease of implementing the multiprocessor capability of the VME–bus made this solution especially attractive. Figure 1 shows a schematic hardware diagram of the microprocessor system. As the main system in this application we used a Motorola VME/10 microprocessor development system since it was readily available at the time. The development system could be replaced by a much less expensive modular microprocessor system.

The configuration of the hardware modules and the software architecture was determined by the requirements of the different motor movements. Table I lists the different motors and drive speeds required for their different movements. It also shows the other time–critical parts of the system with their requirements. The timing is accomplished with separate hardware interrupts for each level. The interrupt levels called out in the table decrease in priority from left to right. The MC68000 microprocessor is ideally suited for these interrupt–driven prioritized operations and the interrupt system was designed to take advantage of these capabilities. Interrupts of higher priority will pre–empt execution of a lower level interrupt routine. This assures that, for instance, the clock and sidereal tracking (interrupt level 7) will always be executed at appropriate times regardless of movement of other motors. Similarly, the rotation of the polaroid and the frequency of sampling will be constant (level 4) even if the aperture wheel (level 3) is scanning. The interrupt routines provided the means for pulsing the stepping motors at constant rates which are determined by an interrupt timer. Every time an interrupt routine executes, the appropriate motors are advanced one step. The interrupt routine also checks that the internal software–maintained position counter is consistent with the readings from the Hall sensors and corrects small differences automatically. Correction may be necessary because the motors occasionally slip. This was especially true when slewing the right ascension and declination motors, but high winds could cause slips even during normal sidereal tracking.

The level 7 interrupt is the most time critical. It keeps track of the sidereal time and the associated telescope movement as well as the comet movement. This interrupt level is not maskable, which means it is executed even if another level 7 interrupt is currently executing. This assures that the sidereal tracking is always maintained. The level 6 interrupt is reserved for use by the operating system. The level 5 interrupt routine was used for slewing the right ascension and declination drives. Its timer was adjusted so that pulses were issued at the maximum rate that these motors could step without stalling. Since no data was taken during slewing, the slowing of the lower level interrupt routines and the command program execution was acceptable. The level 4 interrupt was used for rotating the polaroid and synchronizing data acquisition with this rotation. The photomultiplier counters were read out for every step of the polaroid and the readings accumulated into 12 sectors. The readings from these 12 sectors are then used to calculate the light intensity and polarization. The level 3 interrupt was used to move the filter wheel, aperture, focus and shutter. Since it was executed at a lower level than the data acquisition routine, the aperture could be scanned without disturbing the constant rotation of the polaroid. The lowest two interrupt routines were not utilized.

Electrical interfaces to the motors were commercially available parallel interface and commercially available stepper motor driver boards. By writing different bit combinations to the parallel port associated with a motor, the motor could be stepped in either direction. Some of the parallel lines were configured for input to the microprocessor. These lines were used to read the Hall sensor indications for the corresponding motor.

The software for the whole system was written in C. This language is well suited for an application like this because of its portability. Essentially every computer system supports a C–compiler. It is therefore fairly easy to port the software to a different microprocessor system. C also provides the capability of reading or writing directly to hardware addresses. Only the routine that sets up the interrupt routines with the operating system had to be written in assembler. Figure 2 shows the main parts of the software system.

TABLE I

System Timing Requirements

Interrupt timing [sec]	0.005	0.0005	0.005	0.005
- level	7	5	4	3
Peripheral				
RA	sidereal & comet tracking	slewing		
DEC	comet tracking	slewing		
POLAROID			counter & rotation	
COUNTER			data acquisition	
FOCUS				X
FILTER				X
APERTURE				X
SHUTTER				X
TIMEKEEPING	sidereal timekeeping			

The interactive part of the system reads and interprets commands, commands motor positioning, and reads, displays, and stores data.

Commands can be read from the terminal or from files. Command files contain predetermined sequences of commands that make up one or more observations, e.g. switching between the object to be measured and the background or changing filters and apertures. The command files can include loops over selected command sequences. This enables the user to preprogram complex observing sequences and have them executed without operator intervention. After reading a command, the command interpreter determines the necessary action. If a motor has to be moved, the new motor position is written into global memory and the timer for the responsible interrupt routine is started if necessary. The interrupt routines are executed as long as any motors they are servicing are not in the desired position. Temperature sensor signals that monitored the telescope, PMT, filter, and ambient temperatures were read along with the detector data.

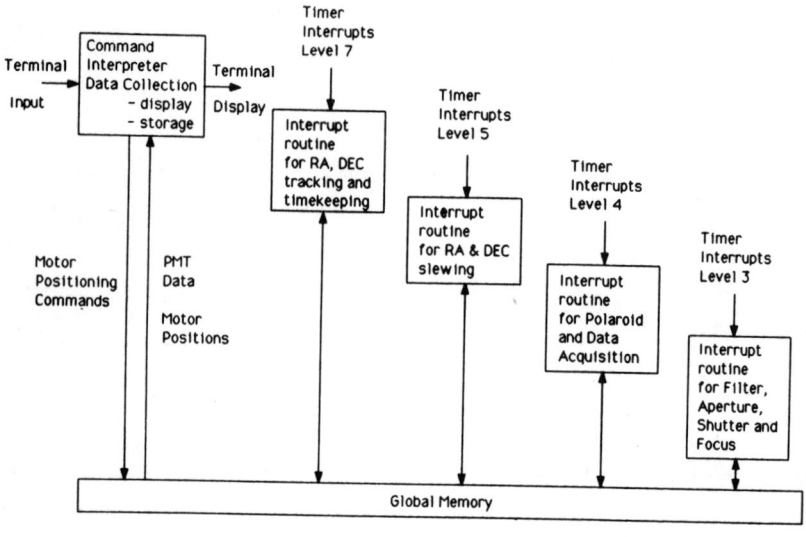

Figure 2. Software configuration.

Two Hall sensors were used for each device to check the real position of the wheel, one on the worm gear and one on the worm. The position with both Hall sensors active was called the zero position and was used to initialize the motors after power up. After this zeroing of the motors the internally kept position was updated whenever a Hall sensor signal was encountered. If a Hall sensor signal was not detected at a position where it should be, a warning was issued to the operator. Small discrepancies were automatically corrected. If the internally kept position differed by more than twenty steps from the Hall sensor reading, the affected unit was re-zeroed and then advanced to the last requested position.

III. EVALUATION

(a) Advantages: As a single processor system the telescope control system was implemented much faster than would be possible for a multiprocessor system. The MC68000 series of microprocessors with its prioritized interrupt structure is particularly well suited for this application. The VME-bus in the system insures that a wealth of different boards are available for such a system. This makes it possible to build a modular system from commercial boards, which is much faster and less expensive than building the necessary boards from scratch. It also improves the reliability of the final system and makes maintenance, repair and upgrades easier, less expensive and faster.

(b) Disadvantages: The position verification with the Hall sensors worked but was not optimal. The mechanical adjustment of the sensors was rather critical. They also showed a hysteresis effect between forward and backward movement of up to 10 steps. This was especially complicated when the direction of motor movement was reversed while a Hall sensor was active. Requirements for our precise positioning software handled all these corrections but because of their complicated nature, the software for position checks and updates had to be fairly elaborate. Since it executes in the interrupt routines, this was a performance-limiting factor. The use of any distinct fiducial, such as an optical switch, would have been preferred to the Hall sensor and its hysteresis problem. A shaft encoder, although significantly more expensive, would have made the programming task very much easier.

The microprocessor system as a whole can handle everything but one scenario: the rapid slewing of right ascension and declination simultaneously. Its fast interrupt repetition rate and the complicated position decoding used so much processor time that input from the terminal service routine was temporarily stopped. The operator had to wait until one of the two motors was in position before the system would react to his input again. Since data need not be recorded while slewing, this was acceptable in the present system. All other motors could be moved simultaneously without appreciably slowing down the system. It is possible that with a better position encoding system probably even the slewing could be handled with the present system.

IV. IMPROVEMENT AND FURTHER POSSIBILITIES

One improvement as mentioned above would be to use a better position reporting system. Another improvement might be the incorporation of a second processor in the system to handle the motor movement. This would free the main system for other tasks, such as real time data reduction and graphical displays. If long cables between the microprocessor system and the motors are a problem this second processor could become a stand—alone system located close to the telescope. This would require a new communications link between the two systems as well as environmental protection at the telescope. With a dial—up modem connected to the main system, operation of the telescope could be from any remote location. This expanded system would provide a completely automated telescope operating system.

ACKNOWLEDGMENTS

The systems reported here were built and deployed under NASA Contract NASW—3678. The authors are indebted to the students, faculty and staff members of the Space Astronomy Laboratory, Dr. Harold Doddington of the University of Florida, J. Rilum and B. Soderberg of the Royal Institute of Technology, Stockholm, and the staff of the University of Hawaii who contributed to the success of this project.

REFERENCES

Giovane, F., Ely, D., Weisenberger, A., Eichhorn, G. Rilum, J. and Soderberg, B. 1987 in *New Generation Small Telescopes*, ed. D. S. Hayes, R. M. Genet and D. R. Genet, (Mesa: Fairborn Press), p. 307.

A HIGH–SPEED PHOTOMETER INTERFACE FOR THE MICROVAX Q–BUS

Mark Trueblood

Winer Mobile Observatory

ABSTRACT: An Approach to interfacing a photon–counting high–speed photometer to the MicroVAX Q–Bus is described. All elements of the high–speed photometer system developed at the Winer Mobile Observatory are fully described, with emphasis on those located inside the MicroVAX chassis or in the same rack. The interface may be used with 16–, 18–, or 22–bit Q–Bus protocols, and with any LSI 11 or MicroVAX I or II processor. Direct memory access reduces processor loading. Custom circuitry not available off–the–shelf is minimized.

I. BACKGROUND

The Winer Mobile Observatory (WMO) is directing its efforts towards the construction of a mobile telescope facility optimized to observe minor planet occultations. Our plans for a 30-inch aperture trailer mounted telescope were described in Trueblood (1985). Our system concept is shown in Figure 1. A Ford E-150 van containing a computer and associated timing and navigation equipment, power generator, and observing equipment tows the telescope trailer to an observing site. The computer controls the alt-az telescope and gathers data from the high speed photometer.

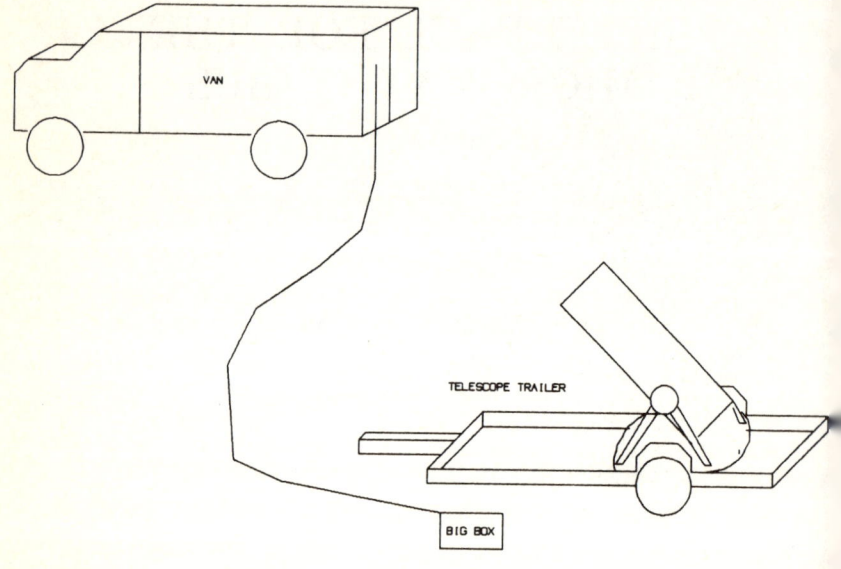

Figure 1. *WMO system concept.*

   Setup consists of mounting the photometer on the telescope, leveling the trailer, and connecting the signal and power cables between the van and the trailer. The observer then sights a few bright stars to tell the control computer where the telescope is pointing and how the trailer is aligned. From that point on, all operations could, in theory, be conducted from the heated van. However, the small diaphragm size (15 arcsec) often used in minor planet occultation observations makes it prudent for the observer to use the guiding eyepiece on the photometer head during critical periods.
   Development of the photometer has occupied our attention since 1985, when the machining of the photometer head, shown in Figure 2, was begun. The head was designed at Lowell Observatory, which performed the final installation and alignment of optics. Although the

head is designed for dual-channel operation, at present, only one channel is used with a bialkali response (visible) PMT.

Figure 2. High-speed photometer head.

The Mid-1985 Report (Trueblood 1985) indicated we were using an MDB Systems MLSI-11B direct memory access (DMA) bus foundation module as the primary interface to our LSI-11 Q-Bus computer. Although this board is well suited to many special purpose Q-Bus interface applications, it has two drawbacks for our application: (1) it is too general-purpose for our straightforward application, and needlessly complicates the interface to our custom circuitry, and (2) it only works with 16-bit Q-Bus protocols. We are now using an LSI-11/23 processor with 18-bit protocol, and will soon be upgrading to a 22-bit MicroVAX I. We therefore found it necessary to rework our overall design before continuing with construction. The redesign effort has been completed, and the results are reported here.

II. SYSTEM REQUIREMENTS

All development work at the WMO is performed in accordance with standard system—engineering practice. The first step is to determine the system requirements from an observational perspective. The end—to—end system, from photometer head to the computer, was designed to meet the following requirements:

*Single channel operation at up to 1,000 16—bit integrations per second, expandable to two or more channels

*Variable integration period, under software control, from one millisecond (mS) to at least 10 seconds

*Development and testing of the data acquisition software should be possible without the photometer head connected

*The same electronics should be usable with both low speed and high speed photometers, with any differences isolated to the photometer head

*The circuitry should be compatible with wide—bandwidth photon—counting preamplifiers/pulse height discriminators

*The primary data interface should work on all Q—Bus computers, including the MicroVAX

*For the time being, photometer operation, target centering, and guiding are all performed manually, though nothing in the design should preclude fully automatic operation

III. KEY ISSUES, OPTIONS, AND TRADE STUDIES

After the system's operational requirements are identified, the next step is to identify the key design issues and the options for their resolution. The various options are compared with each other, and the best one selected. The key issues in the design of the WMO high speed photometer system are listed below, along with the various options, and the resulting engineering recommendation.

Issue 1: Should separate computers be used for telescope control and data acquisition, or should both functions be combined in one computer?

Option 1a: Two computers
> Smaller, less powerful, and cheaper computers can be used

Option 1b: One computer
> Less total power consumption
> Greater reliability and availability, since larger computers tend to be better built and have fewer components than two smaller computers
> Only one rugged chassis able to tolerate portable operation is required
> Less volume is consumed inside the van
> Smaller load on van air conditioning
> Simpler communications between the telescope control software and the data acquisition software
> Less overall cost

Recommendation: Combine data acquisition and telescope control functions in one central computer

Issue 2: Should the computer interface to the photometer use DMA or interrupts?

Option 2a: DMA
> Imposes very light load on the processor
> Hardware more complex and expensive

Option 2b: Interrupts
> Imposes heavy load on CPU: assuming 1,000 interrupts per second, 100 instructions in the interrupt service routine and related operating system overhead (VMS), and 1 microsecond (uS) per instruction (MicroVAX II), a single channel photometer requires 100,000 uS per second, or 10% of the CPU per channel. This grows to 20% for two channels. The LSI-11/23 processor we have now requires about 25 instructions in the interrupt service routine and related operating system overhead (RT-11) and 4 uS per instruction, so the CPU load is about the same. This is quite heavy loading, considering that the telescope control applications are also running, and in real-time systems, total CPU loading should be kept below 50%.
> The MicroVAX operating system (VMS) can take up to 2 mS to begin interrupt servicing. At this rate, the interrupts would completely overload the computer.

Recommendation: Use DMA instead of interrupts

Issue 3: Should photon counting or current amplification be used?

Option 3a: Photon Counting

> Better signal-to-noise ratio
> Simpler, less expensive amplifier

> Easier to transport the signal from the telescope to the computer
> Easier to convert the signal into a number for the computer

Option 3b: DC Current Amplification
> Can handle large signals more accurately

Recommendation: Use photon counting, and stop down the telescope aperture for very bright stars

Issue 4: Should the photon pulses be integrated at the telescope and the counts sent to the computer, or should the pulses be sent to the computer and integrated there?

Option 4a: Integration at the Telescope
> Better correlation between integration timing and time of day used to time tag data blocks as they are written to disk
> No need to send timing pulses to the telescope, or to generate them there
> More circuitry can be placed inside the van, where it is warm and dry
> 120 nS dead time on discriminator output requires high bandwidth (10 MHz per channel) data path to the van

Recommendation: Send individual photon pulses to the computer for integration there

Issue 5: How should the PMT photon event pulses be sent to the computer? (Note: The preamp/discriminator must be located within six inches of the PMT output connector.)

Option 5a: Emitter coupled logic (ECL) pulses from preamp sent directly to van
> ECL pulses degrade in shape unless high grade coax is used; even this does not cure the problem, and adds another cable to be strung between the van and the telescope
> ECL has low voltage swings susceptible to noise from telescope motors, the power generator, and the computer

Option 5b: Optical fiber link
> Expensive
> Adds another cable to be strung between the van and the telescope

Option 5c: RS−422 Driver/Receiver Over Twisted Pair
> Designed for high speed data communications, with large voltage swings for good noise immunity

> Only two ICs needed to convert preamp output (ECL) to RS-422

Recommendation: Mount the preamp and ECL/RS-422 converter on the photometer head. Use 18 awg shielded twisted-pair cable to carry the pulses from the telescope to the van.

IV. MAKE/BUY TRADE STUDY

After identifying the key design issues and resolving them, the next step is to decide which system elements to make and which to buy. It is far less expensive (unless your labor is free) and easier to integrate existing off the shelf products into a working system than it is to design, fabricate, and test each unit before it is integrated into the system. Therefore, the goal is to use as many commercially available components as possible, while still meeting all system requirements.

The 9892B PMT, mounting barrel, socket assembly, and Model AD-100 preamp/discriminator were all obtained from Thorn EMI/Gencom. The Model 605 75-N HV power supply was obtained from Bertan Associates. The major make/buy challenges were the event counter and the Q-Bus interface.

Based on the decision to send individual photon event pulses to the computer, the Q-Bus market was surveyed to see if there existed a product that combined the functions of event counter and Q-Bus interface on the same board that met our requirements. Although several Q-Bus counter boards do exist, we know of none which use DMA. Furthermore, there were none with a high-speed pulse input receiver, such as RS-422 or fiber optic receiver.

In addition to these drawbacks, all of the boards we surveyed required that event counting cease while the count is transferred to the computer memory at the end of the integration period. Such a feature is of no concern to an observer using 10-second integrations, with several seconds between the end of one integration and the start of the next. However, it distorts the counts when one is gathering a continuous stream of 1 mS integration period counts. Therefore, two counters are needed: one for the arriving event pulses, and one that was filled previously whose count is being transferred to the computer. At the end of each integration period, the two counters reverse roles. This "ping-pong" effect is triggered by the integration timer.

No commercial Q-Bus board offers all these features. Consequently, we were forced to design the counting function into a custom circuit with an RS-422 input. We realized immediately the futility of attempting to design a Q-Bus DMA interface. Although Digital Equipment Corporation (DEC) does sell chip sets for interfacing to its buses, we decided to use a commercial general purpose Q-Bus DMA interface board. As mentioned previously, we initially chose the MDB Systems MLSI-11B, but decided later to use the DEC DRV11-WA.

This board performs all the chores of controlling DMA transfers to the Q—Bus, and presents a very simple interface to our custom logic. The DRV11—WA completes the list of commercial items.

The photometer head was custom built because it is designed specifically for the subject of our research program (minor planet occultations), and because it has proven itself in the field many times over to be a reliable and easy to use instrument that consistently produces high quality data. No commercially available photometer seemed as well suited to our observing program.

The only custom electronics in our system are the ECL/RS—422 converter board (using only two ICs and a power supply), and the custom counter board (consisting of 22 standard low—power Schottky TTL ICs and a power supply). The remaining "custom" element in the system is the signal cable, consisting of 15 individual 18 awg shielded twisted pairs, 50 feet long, terminated on each end with a 48—pin Amphenol MS connector. Past experience with field operations has shown that connectors tend to get walked on, run over, and banged into equipment, so it pays to use rugged military grade connectors at remote locations.

V. SYSTEM BLOCK DIAGRAM

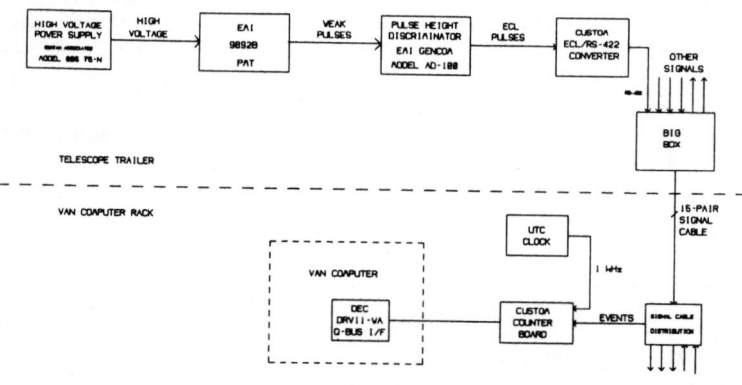

Figure 3. *HSP system block diagram.*

A block diagram of our system is shown in Figure 3. An equipment shelf is mounted on the PMT barrel on the photometer head, and holds the AD—100 preamp and what we call the "Small Box," which contains the ECL/RS—422 converter. Currently, the Bertan Associates Model 605 75—N HV supply is set next to the telescope and is connected to the PMT socket using eight feet of RG−59/U and an SHV high voltage connector. This power supply will soon be replaced with a Bertan Associates PMT−50A/N HV supply, which is more compact, has greater stability than the other supply, and has the added feature of being externally voltage programmable. This new supply will be mounted inside the Small Box, along with a pot for manually adjusting the HV output and a digital to analog (D/A) converter for eventual computer control of the HV output.

Also located at the telescope is what we call the "Big Box." The power cable from the generator inside the van, and the 15−pair signal cables from the computer rack both plug into this box. The Big Box serves as a distribution center for power and individual signal cables to points throughout the telescope. The power supplies for the AD−100 and the ECL/RS−422 converter reside inside the Big Box.

Inside the van, at the other end of the 50−foot signal cable, is a distribution panel with a 3−pin connector for each of the 15 shielded signal pairs. This panel and the chassis holding custom boards are mounted in the rear of the van computer rack. The custom counter board and its power supply are mounted in the custom board chassis. A pair of 40−pin ribbon cables carries the signals between the custom counter board and the DRVII−WA DMA Q−Bus interface.

VI. COMPUTER BLOCK DIAGRAM

The van computer rack is shown in Figure 4. Note that the Conrac Model 7111 high resolution color monitor positioned on top of the rack is not normally used in the van. From the bottom, the first item is a 3 kVA Topax line conditioning transformer with power monitoring meters. The next item is the Codar Technology Model 600 rugged Q−Bus chassis. Above the blank panel is the Qualogy DSD−880 disk subsystem, which will soon be replaced with 5.25−inch floppies and a hard disk which mounts inside the Codar chassis on a shock mounted equipment shelf. The space currently occupied by the blank panel and the disk subsystem will be used for mounting the terminal and keyboard in the rack.

A block diagram of our computer system is shown in Figure 5. The Data Translation image processing system is shown for completeness, and is the subject of future papers. Not shown in this diagram are the interface cards for the telescope drive motors, which are not yet installed. Our plans are to use a C−14 until field operations validate our concept for a mobile observatory.

Figure 4. Van computer rack.

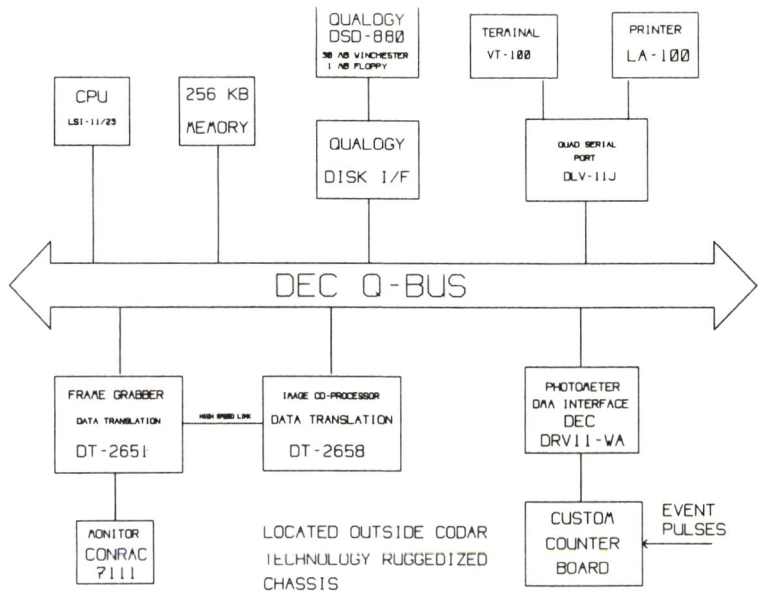

Figure 5. Van computer block diagram.

VII. PHOTOMETER INTERFACE DESIGN FEATURES

The functions allocated to the photometer interface are those of (1) counting the individual photon events during an adjustable integration period, and (2) providing a Q−Bus DMA interface. These two major functions are handled by two separate boards−−a DEC DRV11−WA provides the Q−Bus DMA interface, and exchanges data and control signals with the custom counter board.

The DRV11−WA works on any 16−, 18−, or 22−bit Q−Bus with any LSI−11 or MicroVAX processor. DEC sells the board for $850, and a third−party version is available for $582. It handles both 8− and 16−bit data transfers, and can use either DMA or individual interrupts on each data transfer. A standard VMS driver is available which can be customized as needed. The interface to the custom counter card is simple to use and well documented.

The custom counter board uses low power Schottky TTL integrated circuits (ICs), which have a typical toggle rate of 35 MHz. Most photometrists recommend using counters or prescalers with 100 MHz toggle rates for greatest accuracy, to overcome photon arrival rate statistics. This is recommended from the perspective of long integration period (10 seconds) high accuracy (0.5%) photometry. Bearing in mind the 120 nS dead time of the AD−100 discriminator and the relaxed accuracy requirements of high speed photometry, a 35 MHz bandwidth should be adequate. Low power ICs are used throughout WMO custom circuits as a general policy to help limit the total load on the electric power generator. If higher bandwidth is required in the future, there is adequate room on the board for a wide bandwidth prescaler, or high speed counters can be substituted.

The circuit is constructed on a standard 4.5 × 6.5−inch predrilled Vector #3677−2 board with a 22−pin card edge foil pattern. It is mounted in an 8.75−inch high rack mount box containing a backplane of 22−pin card edge connectors with a power bus strung through dedicated pins. A 5−volt DC power supply module mounted on another card supplies power to the backplane.

Aside from the interface to the DRV11−WA, the circuit requires two inputs: event pulses from the photometer, and a source of 1 kHz timing pulses. The latter are used to determine the integration period of the event counters. To meet the requirement of permitting software development and diagnostic testing without needing the photometer head connected, a 1 MHz crystal oscillator is used to generate a 1 MHz internal source of event pulses and a 1 kHz internal source of timing pulses. Each signal source can be selected independently of the other from software. Other features under software control are the choice of integration period in the range 1−65,535 mS, enabling or halting counting, and resetting all counters and control flip−flops to a known state.

VIII. HARDWARE AND SOFTWARE OVERALL DESIGN

The interaction between the hardware and the controlling software is shown in Figure 6. The parallels in the use of ping−pong dual buffering techniques in both hardware and software are obvious. The software that directly controls the hardware is a part of the operating system known as the "device driver." This software is typically written in assembler language, and must follow strict rules in its format and function, since it becomes part of the operating system after is installed. The driver has access to four registers on the DRV11−WA:

WCR (Word Count Register): This is a 16−bit register used to control the number of 16−bit data words to transfer. The device driver loads the 2's complement of the word transfer count into this register during setup. After each DMA transfer, the hardware in the DRV11–WA increments this register by one. When the register reaches zero, the DRV11–WA generates an interrupt to signal the driver that the N–word transfer is complete.

DAR (Data Address Register): This is a 22–bit register which holds the address in memory of where to read from or write to. At setup, the driver loads the address of the first word in memory. After each transfer, the DRV11–WA hardware increments this register by 2 (bytes), to point to the next word in memory.

DOR/DIR (Data Input/Output Register): This is a 16–bit register holding the data to be read or written.

CSR (Control and Status Register): This is a 16–bit register used to control the operation of the DRV11–WA and the custom board. The register has three function bits which are sent to the custom counter board, three status bits from the custom board, two extended address bits used with 18– and 22– bit Q–Bus protocols, and eight bits for control and status of the DRV11–WA itself.

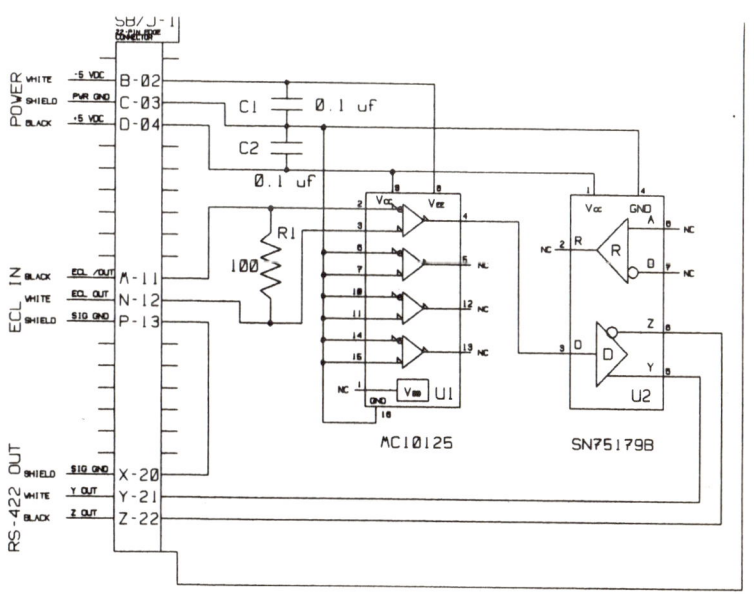

Figure 6. The relationship between the Counter Circuit and the Device Driver Software.

When the HSP application program starts, it first commands the driver to halt counting and to reset the custom board into a known state, in which all counters are cleared to zero, and counter switches point to Counter A to receive the next photon event pulse. The program then prompts the observer to enter a data file name, the name of the object being observed, the location, the weather conditions, the integration period as an integer number of milliseconds, and the filter being used. The program then opens the data file, and writes these entries into a header record.

Next, the program tells the driver the location and size of the data buffers. These are made the same size (256 words) as a disk block, to keep the disk interface clean and fast. The driver loads the integration period into the custom board using the DOR and CSR of the DRV11−WA, loads the address of Buffer A into the DAR, and loads the 2's complement of the length of Buffer A in words into the WCR. The observer is then prompted to indicate when to begin data acquisition. When the observer enters the command to proceed, the program tells the driver to set the GO bit in the CSR, and data transfers continue until the observer commands them to stop. At any time, the observer may change any of the setup parameters, including the integration period, and he may turn on and off both storage to disk and display of data on a screen.

After the GO bit is set, the DRV11−WA automatically transfers the next count after the end of each integration period to the next location in Buffer A, without executing a single processor instruction. During this time, the processor is free to perform other tasks, such as controlling the telescope tracking. When the last word of Buffer A is filled, the DRV11−WA issues an interrupt.

The interrupt service routine within the HSP device driver handles the interrupt. When using the DEC standard DRV11−WA, the highest priority and most time−critical task is to set up the DAR and WCR before the next DMA data transfer occurs, using the address of the first word in Buffer B instead of Buffer A. If 1 mS integration periods are being used, then no more than 1 mS can elapse from the beginning of the DMA transfer for the last word in Buffer A until the registers are set up and the CSR GO bit is set again. All this is done at the highest priority level, so that no other device can interrupt the CPU while this time−critical processing is occurring.

Simple operating systems, such as RT−11, have very little overhead (interrupt latency) from the time a device posts an interrupt until the first instruction of the interrupt service routine is executed. However, the VMS operating system used on the MicroVAX is quite complex, and its performance is a compromise among many competing design goals. One unfortunate result of this compromise is that it can take up to 2 mS from the time a device posts an interrupt until the interrupt service routine can start executing.

One can modify VMS to eliminate enough of the overhead to make 1 mS integration periods feasible, but a much simpler solution is provided by a third party vendor of a DRV11−WA emulator. Grant Technology

Division of Computer Products markets their Model 370, which is hardware and software compatible with DEC's DRV11−WA, except that it contains an optional first−in, first out (FIFO) buffer. Available options are 1, 2, 4, and 8 KB. We use the 1 KB option. There are different modes of operation, but in "flow−through" mode, when the custom board requests a DMA cycle, the 16−bit count goes right through the FIFO to the Q−Bus if a complete n−word transfer is not yet finished. After the last count is placed in the memory buffer, and the computer is busy servicing the interrupt, if another 1 mS count goes into the FIFO. Each succeeding one mS count goes into the FIFO until the interrupt service routine can set up the DRV11−WA (Grant Model 370) registers and initiates another DMA n−word transfer. After the GO bit is set (beginning another DMA n−word transfer), the Model 370 empties its FIFO by initiating a series of DMA cycles in rapid succession. With the FIFO now empty, the next 1 mS count flows through the FIFO directly onto the Q−Bus.

After the critical interrupt processing is complete, the device driver puts itself into "fork" mode, which permits its processing to be interrupted by another device, but not by a regular program. While in this mode, the driver gets the time from an external clock and puts it into a header area in Buffer A, then queues the buffer to the disk device driver for output to the disk. Optionally, the photometer device driver can return the buffer to the control program for display on a graphics terminal. The driver uses an internal flag to keep track of which buffer is receiving data, and which one is being written to disk. After all interrupt servicing is complete, control returns to whatever program was running at the time the DRV11−WA requested service.

IX. ECL/RS−422 CONVERTER SCHEMATIC DIAGRAM

The schematic diagram for the Small Box circuit that converts the AD−100 ECL pulses into RS−422 pulses is shown in Figure 7. There are only two active components on this board: U1, an MC10125 ECL−to−TTL converter, and U2, and SN75179B TTL−to−RS−422 converter and line driver/receiver. Only one of the four ECL−to−TTL converter circuits in U1 is used, and only the line driver portion of U2 is used. The power supply inputs are filtered using two 0.1 uf capacitors, and the balanced ECL input is terminated with 100 ohms. The driver output of U2 is fed to balanced 18 awg twisted shielded pair, which feeds to the Big Box signal cable distribution panel.

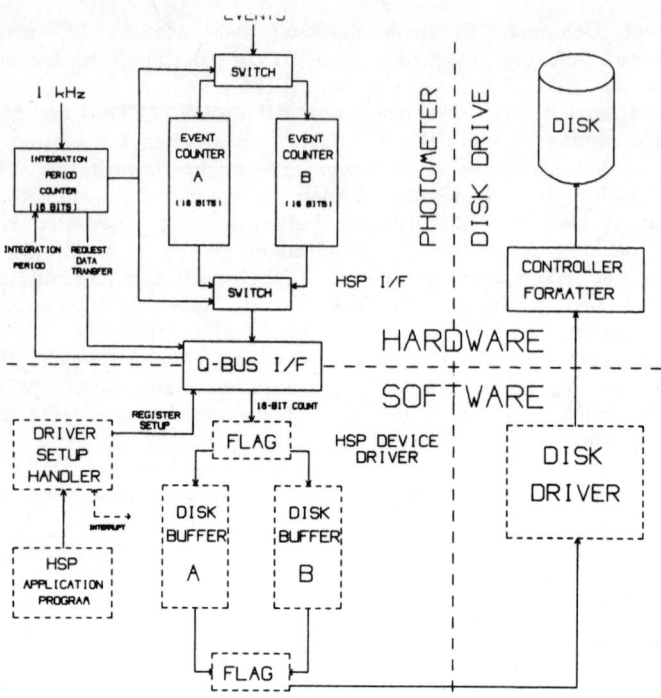

Figure 7. *ECL/RS-422 schematic diagram.*

X. CONCLUSION

The WMO HSP photometer system design is the result of an extensive system engineering analysis based on a well-defined set of operational requirements and an understanding of the operational environment based on observational experience.

(Editor's note. Details on Trueblood's Custom Counter Board are available by writing the author.)

REFERENCES

Trueblood, M. 1985 *I.A.P.P.P Comm.*, 22, 14.

SECTION 4–INTRODUCTION: PHOTOELECTRIC PHOTOMETRY

Donald S. Hayes

Fairborn Observatory

The preceding sections have emphasized the telescopes and auxiliary equipment which illustrate the concept of New Generation Small Telescopes or their precursors. The present section considers photoelectric photometry, including techniques of observation and reduction, and scientific programs which can be carried out with small telescopes. The emphasis here is on publishable science with small telescopes; the degree of automation of the telescope or the photometer is of secondary importance.

Photoelectric photometry can be carried out with small telescopes with great effectiveness. The reasons for this are clear, after a little thought. Photometry does not attempt to analyze the spectrum in great detail; the broad passbands (compared to spectroscopy) which characterize even narrow–band photometry allow more photons to be collected in a given time. Since relatively few colors are measured, the observations do not take much time. Additionally, photometry generally uses sensitive detectors, although the advent of CCDs has destroyed the advantage that photomultipliers had over other detectors previously used (such as film or plates). Finally, photometry commonly deals with large numbers of repeated observations, either in surveys of many stars or in the measurement of time–variability of single stars. The number of suitable

---

New Generation Small Telescopes, ed. D. S. Hayes, R. M. Genet, & D. R. Genet.
© 1987 Fairborn Observatory.

objects for surveys or for the study of variability is large, even among the bright stars, so the possibilities have not yet been exhausted. Having many suitable objects for study among the bright stars is the perfect justification for the continued use of small telescopes.

A further advantage of photometry, as a technique to be exploited with small telescopes, is that it is possible to construct a photometer with professional performance for modest expenditures of resources. It is practical for colleges and small universities, as well as advanced amateurs, to construct and use photometric telescopes. A paper in this section by Ken Ziegler, in fact, describes a photometric program which is being carried out in a high school! The results are of publishable quality. Publishable science done at the high-school level is, unfortunately, quite rare, but this paper illustrates what can be accomplished by a highly-motivated teacher, provided the program is chosen very carefully. One could argue that astronomical photometry offers one of the few instances where publishable science can be accomplished at the high-school level. In any case, the message is clear that photometry offers opportunities which are not limited to large institutions.

Photometry on small telescopes has generally involved differential photometry of variable stars in the *UBV* system. While this emphasis is entirely appropriate, there is nevertheless the opportunity for the use of other photometric systems and for surveys of non-variable bright stars. After the *UBV* system, the Stromgren *uvby* system is perhaps the photometric system most used by professional astronomers. Although the passbands are narrower than those of the *UBV* system, this fact does not preclude its effective use on small telescopes. The first paper in this section is a tutorial on the Stromgren system by David L. Crawford, of Kitt Peak National Observatory. Crawford was among those who established the Stromgren system, and he has lead the way in applying the system to astrophysical problems. The second opportunity mentioned above, that of surveys of non-variable stars, is not discussed in this section. Papers by Crawford and Hayes (1986) and Hayes and Crawford (1986), presented in last year's Fairborn IAPPP Symposium, discuss observing programs and the techniques of all-sky photometry with small telescopes. The papers by Upgren and Philip, in earlier sections of this Symposium, describe examples of such studies, but with somewhat larger telescopes in mind. The paper by Philip, while oriented toward spectrophotometry, concerns a study in which he has already made many observations in the Stromgren system, mostly with smaller telescopes.

The papers by Wood and Chen on the times of minima of eclipsing binaries, by Markworth and Rafert on observations of the W Serpentis stars, by Michaels on SZ Herculis, by Guinan on an interesting class of eclipsing binaries, by Ziegler on the photometry of asteroids, and by Eaton on IUE observations of chromospherically-active stars, illustrate some of the programs which can be tackled with small telescopes. The latter paper, while concerned with IUE observations, concerns classes of stars which are good candidates for visual-range photometry with small telescopes.

Given the best of equipment, a photometrist is not assured of good results; the proper observing techniques and reduction procedures are vital. Two of the papers in this section concern these topics: the paper by Olson and one of the two by Markworth and Rafert. More background information on these topics may be found in last year's Fairborn IAPPP Symposium (Automatic Photoelectric Telescopes, ed. D. S. Hall, R. M. Genet and B. L. Thurston, Mesa: Fairborn Press, 1986), in Photoelectric Photometry of Variable Stars (Second Edition, ed. D. S. Hall and R. M. Genet, Richmond: Willmann−Bell, 1988) and in Astronomical Photometry, by A. A. Henden and R. H. Kaitchuck, New York: Van Nostrand Reinhold, 1982.

REFERENCES

Crawford, D. L. and Hayes, D. S. 1986, in Automatic Photoelectric Telescopes, ed. D. S. Hall, R. M. Genet and B. L. Thurston, (Mesa: Fairborn Press), p. 81.

Hayes, D. S. and Crawford, D. L. 1986, in Automatic Photoelectric Telescopes, ed. D. S. Hall, R. M. Genet and B. L. Thurston, (Mesa: Fairborn Press), p. 87.

It was more difficult for some to travel to the symposium than for others. These astronomers came by raft. They claimed it was worth the effort.

A SHORT TUTORIAL ON STRÖMGEN FOUR–COLOR PHOTOMETRY

David L. Crawford

Kitt Peak National Observatory
National Optical Astronomy Observatories

It is quite likely that the Strömgren four–color photometric system has become the most widely used system. I would like to describe the system briefly in this paper, and compare it to the *UBV* system. I will not describe the uses, as many papers using it exist in the literature. A few references will be given at the end, for further reading. A future paper will discuss how one might use the system on small telescopes, as a tutorial for those who are not yet using the system or for those not familiar with it.

Perhaps two things distinguish the system from the *UBV* system, in the main. One, the system is an intermediate–band system rather than a wide–band system. By this I mean that the filters are about 200 to 300 A wide rather than 800 to 1000 A wide as are the three in the *UBV* system. There are a number of advantages because of this, some of which I will mention below. One disadvantage is that the number of

---

New Generation Small Telescopes, ed. D. S. Hayes, R. M. Genet, & D. R. Genet.
© 1987 Fairborn Observatory.

photons passing through the filters is lower than with the wider filters. If the difference is a factor of four (ratio of the bandwidths), then one would have a 1.5 mag brighter limiting magnitude for the same precision. However, the added "scientific resolution" (see below) more than makes up for this for most research programs.

Second, the bandpasses that define the system are filter defined, rather than defined by a combination of the filter, the atmospheric transmission, and the photomultiplier response. Hence, the filters can be used at any site and with almost any photomultiplier (1P21 or GaAs for example) and still produce the same photometric system to high accuracy.

The filters are approximately as follows:

Filter:	Material:	Wavelength:	Bandpass:
u	Glass	3500 A	300 A
v	Interference	4100	200
b	Interference	4700	200
y	Interference	5500	200

(See the Standard Star paper listed below for details.)

As such, the system is most commonly called the "uvby system." The letters stand for "ultraviolet, violet, blue, and yellow."

The band passes can be defined by filters, by slots in a spectrograph, or by a combination of both.

Why did Stromgren choose these specific bandpasses and widths? If one considers the spectral distribution of typical main sequence stars, the dominant features are:

1. Color (due mainly to temperature differences). One can and does define different color indices, depending on the wavelength of interest. Examples are $B-V$ and $b-y$, or $U-B$ and $u-b$.

2. The Balmer discontinuity.

3. Blanketing absorption due to heavy elements.

4. Strong individual absorption lines.

Stromgren designed the uvby system to measure the first three of these, and in a way in which the parameters are well separated. Note that the Hβ photometric system is designed to measure the strength of the Hβ absorption line. Other systems have been designed to measure Hα, the CN band, and other features.

The u-band is located entirely below the Balmer discontinuity, and above the region of the atmospheric cutoff. It is wide enough to fill most of the region between these two wavelengths.

The v–band is located above the Balmer discontinuity, in the region where blanketing is strong. It does have the Hδ line near the center of the band.

The b and y bands are located above the point (about 4500 A) where blanketing becomes important (for stars hotter than the Sun, anyway). As such, the $b-y$ color index is rather free of blanketing effects, considerably more so than is $B-V$ of the UBV system.

We can use $u-b$ as a color index too, rather like $U-B$. The latter index is not a clean measure of the Balmer discontinuity, as both U and B slop over the region of the discontinuity. Note that $U-B$ is also adversely affected by blanketing. For some stars, it is used as a discontinuity measure, ignoring blanketing effects, while for other stars it is used as a blanketing measure, ignoring discontinuity effects.

Strömgren defined a new index, actually a "color difference," called $c_1 = (u-v) - (v-b)$ as a measure of the discontinuity (the continuous hydrogen absorption). In essence, $(v-b)$ defines a color gradient, and c_1 measures how much u differs from that gradient. As the blanketing in u is about twice what it is in v, the c_1 index is free of both color and of blanketing effects, and therefore is a rather clean measure of the Balmer discontinuity.

In a similar way, Strömgren defined another color difference to measure the blanketing, $m_1 = (v-b) - (b-y)$. Here, $(b-y)$ defines the color gradient, and m_1 measures how much v differs from that gradient.

The y band is located at 5500 A, the same central wavelength as the V of the UBV system. As there are generally no strong features in the V bandpass, the intermediate–bandpass y–filter magnitude can be accurately compared to the wide–bandpass V–magnitude. So a magnitude measured with y can be accurately transformed to the V–magnitude system of the UBV system. Hence, no new V–magnitude system was defined for the uvby system. One just transforms directly to the V system.

The parameters of the system therefore are:

1. A V magnitude, measured with the y filter.

2. A $(b-y)$ color index, measuring color, and hence used to determine temperature, interstellar reddening, etc., rather like $(B-V)$ is used in the UBV system. The $(b-y)$ index is freer of blanketing effects than $(B-V)$ is, however. The wavelength separation of the b and y filters is about 70 percent less than that of the B and V filters, so the scale of the $(b-y)$ index is about 70 percent of the scale of the $(B-V)$ system. No problem.

3. The m_1 index, measuring blanketing, used particularly for determining abundance differences from star to star. There are calibrations between the measured parameter and [Fe/H], for example. In practice, one uses an index δm_1, which is the difference between a m_1 relation for the Hyades cluster stars and the measured m_1 index, to measure blanketing differences. It is this δm_1 that is calibrated to [Fe/H].

4. The c_1 index, measuring the Balmer discontinuity. This index is used to measure temperature (hence intrinsic color) for stars hotter than the those where the discontinuity is at a maximum (about spectral type A2), and to measure absolute magnitude (surface gravity) for those stars cooler than the maximum. [The Hβ parameter measures temperature (intrinsic color) for the cooler stars and absolute magnitude (surface gravity) for the hotter ones.]

So a combination of these four parameters plus the Hβ one offers us the major things we want to know about the star, for an initial look—see, whether for astrophysics or for use in galactic structure studies. We get a measure of apparent brightness and of absolute brightness, hence of distance. We get a temperature measure, well calibrated with spectral type. We get a measure of the star's [Fe/H] abundance. From a combination of these, we also get a measure of the star's age, through stellar evolution model calibrations.

We also can estimate the interstellar reddening and absorption of the starlight, as we have a good measure of intrinsic color and we measure the $(b-y)$ color. So the uvby, Hβ systems are excellent for studies of the interstellar absorbing matter as well as for studies of the stars themselves.

Note that transformation to the standard system (the system being defined as it is in the UBV system by reference to standard stars) is easier and more accurate than with the UBV system, due to the narrower bandwidths of the filters. No second order color terms are needed for extinction, so extinction transformations are easier also.

A typical transformation equation is:

$$c_1 = C + D \times c_1(\text{obs}) + F \times (b-y)$$

where: C is the zero point correction,
D is the scale term,
c_1(obs) is the observed index, corrected for extinction,
c_1 is the value on the standard system, and
F is a color term, which allows for Hδ effects on v.

One could use the Hβ index in place of $(b-y)$ in the last term, and this has been occasionally done when the index was available before the uvby transformations where done.

When done with care, accuracies of 0.003 mag. have been achieved in routine work.

One should go to the astronomical literature to see applications; there are many. The system is excellent for all sky photometry and for work on variable stars as well. I will note here only a few more items, to assist in understanding the literature:

1. As is done with m_1, a standard c_1 relation is defined. Rather than use the Hyades relation, a relation that defines the Zero Age Main

Sequence is used. It is essentially the lower envelop of a c_1 vs. $(b-y)$ or Hβ diagram. So a parameter called Δc_1, which is the difference between the measured value and the standard value, for the same $(b-y)$ or Hβ, is used. This has been calibrated in terms of absolute magnitude for A- and F-type stars.

2. $E(b-y)$ is the color excess due to interstellar absorption. It is the difference between the measured color and the intrinsic color as estimated by either c_1 or Hβ. We call the intrinsic c_1 and m_1 parameter c_0 and m_0. The apparent magnitude corrected for interstellar absorption is called V_0.

3. The two plots mostly used are c_0 vs. $(b-y)_0$ and m_0 vs. $(b-y)_0$. Occasionally Hβ is used instead of $(b-y)$ in these diagrams. (Remember that Hβ is a measure of absolute magnitude for B-type stars and of temperature for A- and F-type stars, while c_0 is a measure of temperature for B-type stars and of absolute magnitude for A- and F-type stars.) The axes in these plots are chosen so that cooler temperatures are always to the right and the brightest stars are to the top of the diagrams.

REFERENCES

"A Catalog of Bright uvby Hβ Standard Stars", by Charles Perry, Erik Olsen, and David Crawford 1987 *Publ. Astron. Soc.Pacific*, in press. (This paper includes many references to published data by many authors, including the earlier standard star papers.)

"Four—color uvby Photometry for Bright O— to G0—Type Stars South of Declination +10 degrees", by Gronbech and Olsen 1976, *Astron. Astrophys. Suppl.*, 25, 213.

"Standard Stars for uvby Photoelectric Photometry South of Declination +10 degrees", by Gronbech, Olsen and Stromgren 1976, *Astron. Astrophys. Suppl.*, 26, 155.

"Standard Stars for uvby Photometry", by Crawford D. L. and Barnes J. V. 1970, *Astron. J.*, 75, 978.

"The Relation Between the MK System and Photometric Classification", by D. Crawford 1984, in *The MK Process and Stellar Classification*, ed. R. Garrison (Toronto: David Dunlap Obs.), p. 191.

"Empirical Calibrations of the uvby System, II: the B—type Stars", by D. Crawford 1978, *Astron. J.*, 83, 48.

"Interstellar Reddening in the UBV, uvby, and Geneva Systems", by D. Crawford and N. Mandwewala 1976, *Publ. Astron. Soc. Pacific*, 88, 917.

"Quantitative Classification Methods", by B. Stromgren 1963, in *Basic Astronomical Data*, ed. K. A. Strand (Chicago: U. of Chicago Press), p.123.

"Spectral Classification Through Photoelectric Narrow—Band Photometry", by B. Stromgren 1966, in *Ann. Rev. Astron. and Astrophys*, 4, 433.

Crawford 1978, in Astronomical Papers Dedicated to B. Stromgren.

A QUALITY-CONTROL CHECK ON SCALE-FACTOR TRANSFORMATION COEFFICIENTS

Edward C. Olson

Department of Astronomy
University of Illinois, Champaign-Urbana

In differential variable star photometry, only the scale-factor transformation coefficents and differential extinctions enter the determination of standardized differential magnitudes. A typical observing procedure consists in observing (1) comparison, variable, and check stars, and (2) standard stars, to determine extinctions and transformations. Usually, comparison, check, and variable are close enough together so that mean extinctions can be used. An exception could arise if the variable needs to be followed to large air mass, where air-mass differences could become significant. However, in such cases, extinctions can be found from the comparison observations themselves. So, if accurate standardized differential magnitudes are desired, one mainly needs accurate scale-factor transformations.

I have done four years photometry at the Mount Laguna Observatory, an excellent site, using the Illinois 1-meter reflector. I do $uvbyI$ intermediate-band photometry of long-period interacting binaries. In a given run of 1 1/2 to 2 weeks, I will do all-sky photometry of Crawford-Barnes (1970) standards on a few of the best nights, often partly in twilight. To keep count rates to reasonable values, I use 2.5 or 5.0 mag masks attached to the upper end of the telescope. These masks contain many small apertures distributed to sample uniformly the incoming beam, thus minimizing problems with sensitivity variations across the photocathode. Some 30 to 60 standards will be observed in a typical run. I have results from 13 runs over the past four years. Standard deviations from averaging coefficients from these runs vary from 0.006 to 0.015 mag. for the uvby filters, with somewhat higher values for infrared transformations (these arise from small photocathode temperature variations due to the use of dry-ice cooling). While these precisions may seem reasonable, at least part of the variations may be inevitable errors in standard star solutions, because of the moderate numbers of standards observed. Uncertainties of 0.01 mag. are large enough to inject spurious scatter into high-precision differential observations.

When starting a new binary, I choose two check stars, and initially observe these fairly often. When a reliable check-comparison pair is found, the check is subsequently observed once (two passes through the five-filter sequence) per night. The reduction program calculates differential natural magnitudes, corrected for differential extinction. Usually, these stars differ significantly in color (color matching with intermediate band systems is unnecessary). These check/comparison observations then provide an excellent independent check on the stability of scale-factor transformation. If these coefficents change, then so will the check/comparison color differences.

For each run, I calculate mean check-comparison color differences. Standard deviations from all observing runs are shown in the statistics at the quoted levels. For the other two binaries, this is no longer true, at least in the violet and ultraviolet, and particularly for KU Cyg. From these data, we can conclude that color differences remained stable over this four-year period to better than 0.004 mag, which implies that transformation coefficients were about two or three times more stable than standard star solutions suggested. Thus, for the highest precision, at the end of a four- or five-year observing program, I determine the best mean transformations from more than 400 standard star observations, and re-reduce all the observations. The resulting light curves show smaller scatter than those reduced with transformation coefficients from individual nights' standard solutions.

This conclusion is not a by-product of using masks for standards, as similar conclusions had already been reached for an earlier set of observations of S Cnc by Etzel and Olson, in which no magnitude masks were used for standards. Moreover, to the precision that individual solutions give, there is no difference in coefficients with and without masks. Similar checks should be made with broad-band observations. Choosing check and comparison stars of significantly different colors should then become the standard procedure, no pun intended. If experience shows, in general, that scale-factor transformations remain very stable, then it is a waste of valuable observing time to observe large numbers of standards each night, in the hope of finding "the best" coefficients. These comments do not, of course, apply to zero points and extinctions. But even here, it is best to apply the long-term average scale factor transformations in fixing those coefficients.

Standard Deviations of Check-Comparison Star Differences

Binary	y	(V-I)	(b-y)	(v-y)	(u-b)
RW Per	0.004	0.004	0.002	0.003	0.004
RZ Oph	0.003	0.005	0.002	0.009	0.005
KU Cyg	0.004	0.007	0.008	0.007	0.007

REFERENCES

Crawford, D. L. and Barnes, J. V. 1970 *Astron. J.*, 75, 978

PHOTOELECTRIC DATA REDUCTION FOR EVERYONE

Norman L. Markworth

Department of Physics and Astronomy
Stephen F. Austin State University

James B. Rafert

Department of Physics and Space Sciences
Florida Institute of Technology

I. INTRODUCTION

The many man—months of effort described in other papers in this Symposium by the authors have but one goal—the acquisition of usable photoelectric data. The night was perfectly clear, the equipment operated flawlessly (quite contrary to the astronomical interpretation of Murphy's

Law), and you are basking in the warm glow of a job well done. Now, what do you do with these stacks of numbers so skillfully collected? Several years ago it became clear to us that the proliferation of automated photometric telescopes and fully or partially automated data acquisition systems can easily inundate even the most well−intentioned of us in mounds of unreduced data. This problem, however, is precisely the one we want to have. The sheer quantity of data that an automated telescope can produce argues strongly for nightly reductions. Nightly reduction keeps data from piling up or being stockpiled for "reduction soon," and serves as a useful incentive to actually using the data rather than losing it. It also provides immediate feedback concerning your observational technique, the status of your equipment, and the quality of the previous night's sky. This is more important than is usually imagined, since the details of the previous night's events can only be remembered clearly for a short time.

Our nightly reduction procedure has been briefly described in Rafert, *et al.* (1985), but since that time it has been enhanced. We will restrict the following discussion to the initial data reduction. The observation and use of standard stars and the transformation to the standard system are not considered here, but useful discussions may be found in Henden and Kaitchuck (1982) and Hall and Genet (1982). We consider the initial data reduction to be an integral part of the telescope operating system described in the previous chapters. There are many different approaches to photoelectric photometry. Each may require different photomultipliers, filters, apertures, and observing sequences. These differences in instrumentation and data collection usually require different approaches in data reduction.

Most initial data reductions, however, are still performed on large mainframe computers, which may not be accessible to the growing community of photometrists. We have found that the initial reductions can be accomplished on microcomputers without the feared loss of precision that has retarded their use on such problems in the past. We present here a series of programs written in BASIC for the Commodore 64 microcomputer. These programs carry the data from the raw form gathered at the telescope to a usable light curve.

II. A FEW ASSUMPTIONS

The programs described in this chapter are intended for the reduction of differential photometry. They may be used for either pulse−counting or analog photometry. The observer is, of course, responsible for using some sort of sensible standard observing sequence. Specifically, for our use this means that the filters and tube should closely approximate the broadband *UBV* response, the comparison star should be of about the same color as the variable star, and the comparison star should be near the variable star. In addition all star readings should be bracketed by sky readings, all variable star readings by comparison star readings, and all check star readings by comparison star readings. The programs will, however, accommodate slight departures from these guidelines. The input data are those produced by our TAU program, which include Universal Time, intensity, filter code, object code, and integration length. The data records are assumed to be in chronological order.

If pulse−counting photometry is done, the photometer must be calibrated to determine the dead−time coefficient. Since we use the Starlight−1R photometer, the red−leak correction is also applied to the ultraviolet and blue filters.

III. PRELIMINARY WORK

Two programs are used prior to the actual data reduction. These programs perform a wide range of editing functions on the file, should they be necessary. The first program (named PRINT/PLOT) produces a printed listing of the data, ordered by object and filter. This listing is very useful in checking for trends in the data. PRINT/PLOT then produces high resolution (320 × 200 pixels) plots of the data by color. Figure 1 shows the red data for RW TAU obtained on 1 January 1987. Notice that each object is displayed in a different symbol for quick identification. Using the plot and the printed listing allows the user to identify sections of the data of questionable usefulness.

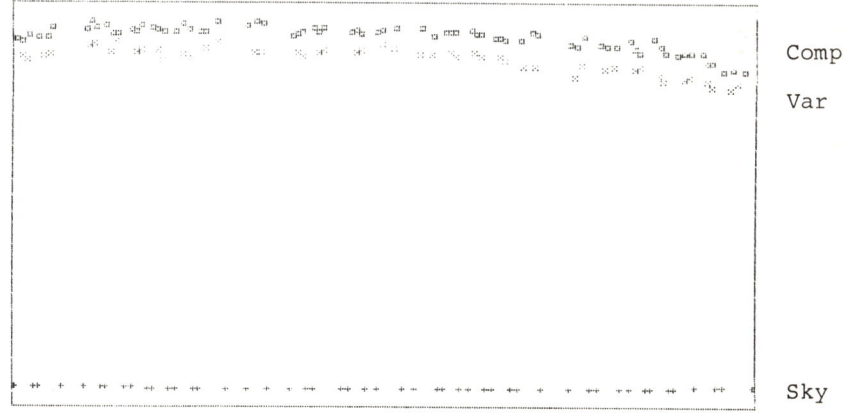

Figure 1. "Red" data obtained for RW TAU.

The second program in this set is called EDIT. This program is really a sequential file editor, written specifically with photoelectric photometry in mind. Among its many options are inserting, deleting, and correcting records. These alterations of the original raw data should not be done without good reason. Ideally, the observer should be the person responsible for the editing of the data file, but in any case, good observing notes are essential. We do not encourage selective deletion of records to "improve" otherwise faulty data, but neither should several hours of useful results be discarded for twenty minutes of passing clouds. Even if the full initial reduction cannot be performed on a daily basis, we strongly urge that reductions be carried through this stage of editing. EDIT also can be used to average repeated observations and split large data files into segments, which is useful when the memory capacity of the computer is limited. The edited file is renumbered in chronological order and saved onto diskette upon exiting EDIT.

IV. THE INITIAL REDUCTION

Our approach is modeled on a more extensive FORTRAN program by Markworth (1977). The program REDUCE requests the edited file as well as an information file on the stars observed. The information file contains the names, coordinates, epochs, and color indices of the variable, comparison, and check stars used, as well as the ephemeris for the variable star.

REDUCE begins by correcting the readings to the center of the integration interval and converting to counts per second. If pulse counting photometry was done, REDUCE next applies the deadtime correction. We use the iterative approach described by Henden and Kaitchuck (1982.) It bears repeating that this approach imposes a limit of brightness on the stars observed. The largest number of counts per second n that can be successfully corrected must be less than $1/(te)$, where t is the deadtime coefficient, and e is the base of the natural logarithms. For the 104−cm telescope, this limit is about 1,758,000 counts/sec, or about a fourth−magnitude star.

The bulk of REDUCE performs a series of linear interpolations in time. These interpolations are required in all single−channel photometry. We assume that the chronologically ordered data vary smoothly in time. All interpolations use the intensities rather than the magnitudes, in order to avoid loss of precision. First the sky readings in the same color as the star readings are interpolated to the time of the star reading and subtracted. This procedure does not specify a sky reading as "belonging" to either the variable or comparison stars. Our procedure has been to select a region of sky free of stars and roughly midway between the variable and comparison stars.

Since we use a Starlight−1R photometer, we must next correct for the measured red leak in the ultraviolet and blue channels. During our tests on the photometer, we uncovered a previously unpublished correction which should be applied. Since this photometer has a rather wide circulation, we present it here. Unlike the standard Starlight−1, the Starlight−1R uses the red sensitive EMI 9798A (S−20) photomultiplier. Along with the usual U, B, and V filters, we have a red (R) filter and a red leak (I) filter. When properly corrected, the Starlight−1R closely approximates the extended $UBVR$ system. Because of the red sensitivity of the PMT and the spectral response of the ultraviolet and blue filters, a significant amount of long wavelength radiation is measured in these channels. The extent of this red leak was sufficient that EMI chose to measure it rather than filter it, as is usually done. The I channel is intended to measure the long wavelength component of the ultraviolet and blue filter measurements. Once the I channel is measured, the U and B channel readings can be corrected in the reduction stage. In fact the I channel slightly underestimates the red leak and the effect is color dependent. The empirical relation

$$I' = 1.094 - 0.37\ (B-V),$$

where $B-V$ is the color index of the star, should be multiplied by the I intensities to obtain a better red leak correction. This correction should be applied after deadtime and sky subtraction have been done.

Next REDUCE searches for comparison star readings that bracket variable star readings of the same color. The bracketing comparison star readings are interpolated to the time of the variable star readings and subtracted in the sense (Variable − Comparison). When no bracketing comparison star readings can be found, REDUCE finds the comparison star reading closest in time and performs a simple subtraction.

Now the color indices are formed, using all normal combinations (*i.e.*, $U-B$, $B-V$, $V-R$, *etc.*) available. Since we assume a single-channel system is being used, some decision is now necessary to assign times to the color indices. Linear interpolations are performed to tie the $U-B$ color index to the time of the blue differential intensity that helps to form it. Similarly, the $B-V$ index is assigned the time of the visual differential intensity and the $V-R$ index is given the time of the red differential intensity.

Finally, the ephemeris of the variable star is used to compute both heliocentric Julian date and orbital phase. Before output to the printer and disk files, the differential intensities and color indices are converted into magnitudes.

An input option of REDUCE allows the comparison star readings to be used to compute extinction coefficients for the night. These coefficients are printed but not applied to the data. A separate file is created for the extinction data.

V. CONCLUSION

The accuracy of the REDUCE program has been checked by comparison of its output with the FORTRAN program mentioned earlier. Naturally, this does not remove any code—dependent errors. It does, however, reveal the ability of the C—64 to retain enough numerical precision that use of this program is not limited by the machine itself. Although the program is written for the C—64, very little machine—specific BASIC is used. After minor modifications of print and disk drive commands, the program should run equally well on any microcomputer.

The authors are anxious to make this program available to all those who are interested in it. Please contact one of us to arrange a mechanism whereby the program can be transmitted.

REFERENCES

Hall, D. S. and Genet, R. M. 1982 *Photoelectric Photometry of Variable Stars* (Fairborn: I.A.P.P.P.).

Henden, A. A. and Kaitchuck, R. H. 1982 *Astronomical Photometry* (New York: Van Nostrand Reinhold).

Markworth, N. L. 1977 PhD Dissertation, University of Florida.

Rafert, J. B., Michaels, E. J. and Markworth, N. L. 1985 *I.A.P.P.P. Comm.* No. 21, 28.

OBSERVATIONS OF THE W SERPENTIS STARS: A PROGRESS REPORT

Norman L. Markworth

Department of Physics and Astronomy
Stephen F. Austin State University

James B. Rafert

Department of Physics and Space Sciences
Florida Institute of Technology

I. INTRODUCTION

The class of binary stars which Plavec (1980 a, b) called the W Serpentis stars can almost certainly be identified with the later stages of rapid mass transfer. Beta Lyrae is the best known case, in which Wilson and Lapasset (1978) and Wilson (1982) found it necessary to include an optically and geometrically thick disk enshrouding one of the components of the system. The loss of mass and angular momentum shown by previous observations must be directly associated with the formation of the thick disk, but neither the observational base nor the present theory is adequate to further develop these ideas.

The observational problems are enormous. While many of the W Ser stars have spectroscopic coverage, photometric coverage is rather spotty. Three reasons contribute to the poor photometric data base.

1. The orbital periods are long (several days to several hundred days), meaning that few observatories can schedule the time necessary to complete the phase coverage.

2. The light curves of W Ser stars are rarely well-behaved. Cycle-to-cycle variations are the rule rather than the exception. In several cases the variations are erratic enough that the binary nature is difficult to discern, even if eclipses are present.

3. Partial phase coverage of the light curve is usually far less useful than isolated spectroscopic observations. Important basic data can be obtained spectroscopically in a night or two, whereas the equivalent amount of information may take months or years to gather photometrically. It should be noted, however, that the derivation of orbital elements from the radial velocity curves of active binaries is far more uncertain than from the light curves.

We began a photometric program on the W Ser stars at the observatories of Stephen F. Austin State University and Florida Institute of Technology two years ago (Wilson, *et al.* 1984). This chapter describes the progress in that program to date.

II. INSTRUMENTATION AND DATA COLLECTION

We have used both the 46-cm and 104-cm telescopes of the SFASU Observatory (described in other papers in this Symposium) as well as the 36-cm telescope at FIT for this program. All three instruments are equipped with Starlight-1R pulse-counting photometers from Thorn EMI. All three use the TAU operating system described elsewhere in this Symposium (the FIT telescope uses only the data logging features of TAU, since it has not been put under computer control as yet). During the summer of 1985, one of us (JBR) used the 76-cm telescope of the McDonald Observatory on the W Ser program, using a Texas photometer modified to give approximately the same spectral response as the

Starlight—1R. All of the telescopes used in the program operate in an all—sky, differential photometry mode.

The observing sequence consists of 56 observations of the variable and comparison stars and a representative sky position in each group. We use the BVR standard broadband filters, with a separate "red—leak" filter to correct the B bandpass. Comparison stars are carefully selected for nonvariability.

Standard stars are observed each night and are used to compute the first order extinction coefficients and zero points. These nightly standards are generally red—blue pairs selected from the $UBVRI$ extinction star network of Barnes and Moffett (1979). In addition, cluster standards are observed seasonally to determine the second order extinction terms and the transformation coefficients. The cluster standards we use are selected from the lists compiled in Henden and Kaitchuck (1982).

An important component of our W Ser Program is the inclusion of student observers. Major funding for the program comes from a Research at Undergraduate Institutions grant from the US National Science Foundation. The structure of the telescope operating system allows us to train student observers during actual observing time. While this naturally limits the number of star groups that can be observed in any one night, it richly enhances the students' understanding of observational astronomy. Our most active season is the summer, when academic schedules are more flexible. During the first two years of operation, we averaged 6 to 10 star groups per night. These first two years have also been used to test and improve the operating system, install new electronics at the 46—cm telescope, and learn the idiosyncrasies of each telescope used in the all—sky mode. With most of the testing complete, our present average is 10—15 star groups per night.

The logistics of the observational program virtually demand timely initial reduction of the data. These are stored on floppy disk for later transformation to the standard system and data analysis. Of the 22 W Ser stars or related objects on our present catalog, we estimate that the observational component of the program is 40% complete. For the majority of the program objects, one set of 56 measures per night is sufficient, and the resulting differential magnitudes are averaged into nightly means. Thus each point on our light curves represents one night. For the shorter period systems, however, several sets of observations are obtained throughout the night.

Figures 1 and 2 show representative light curves for two of our program objects. The data shown are from the SFASU data base only. The plotting symbols differentiate between telescope (104— or 46—cm) and observing year (1985 or 1986). Transformations to the standard system have not been incorporated into these plots. The cases shown indicate the problems of observing the W Ser stars. V356 Sgr shows evidence of a thick disk both spectroscopically and photometrically (Wilson and Caldwell 1978), but shows a relatively clean light curve, with deep eclipses and fairly repeatable maxima. W Ser, on the other hand, shows a very complex light curve. Although the SFASU data base does not contain reliable eclipse data as yet, it indicates that a general system brightening occurred in 1986, with a commensurate increase in photometric activity.

III. DATA ANALYSIS

Several of the program objects have sufficient phase coverage to permit some initial analysis. Ultimately, our aim is to use a modified version of the Wilson—Devinney differential corrections program (Wilson 1982) that incorporates a thick disk about one of the component stars, in an attempt to more completely model the W Ser stars and related objects. Even at this stage, however, some preliminary work can be done to improve the ephemerides of the program stars.

None of our program objects has an orbital period short enough to permit a direct computation of the time of minimum on a single night. Since photometric observations of the W Ser stars are sparse, however, we expect that the ephemerides now available are insufficiently precise for the detailed analysis we would like to do on these stars. This is particularly true since our observations will likely span three to four years before detailed analysis begins. Inaccuracies in the light elements will generate spurious phase dispersion in the light curves.

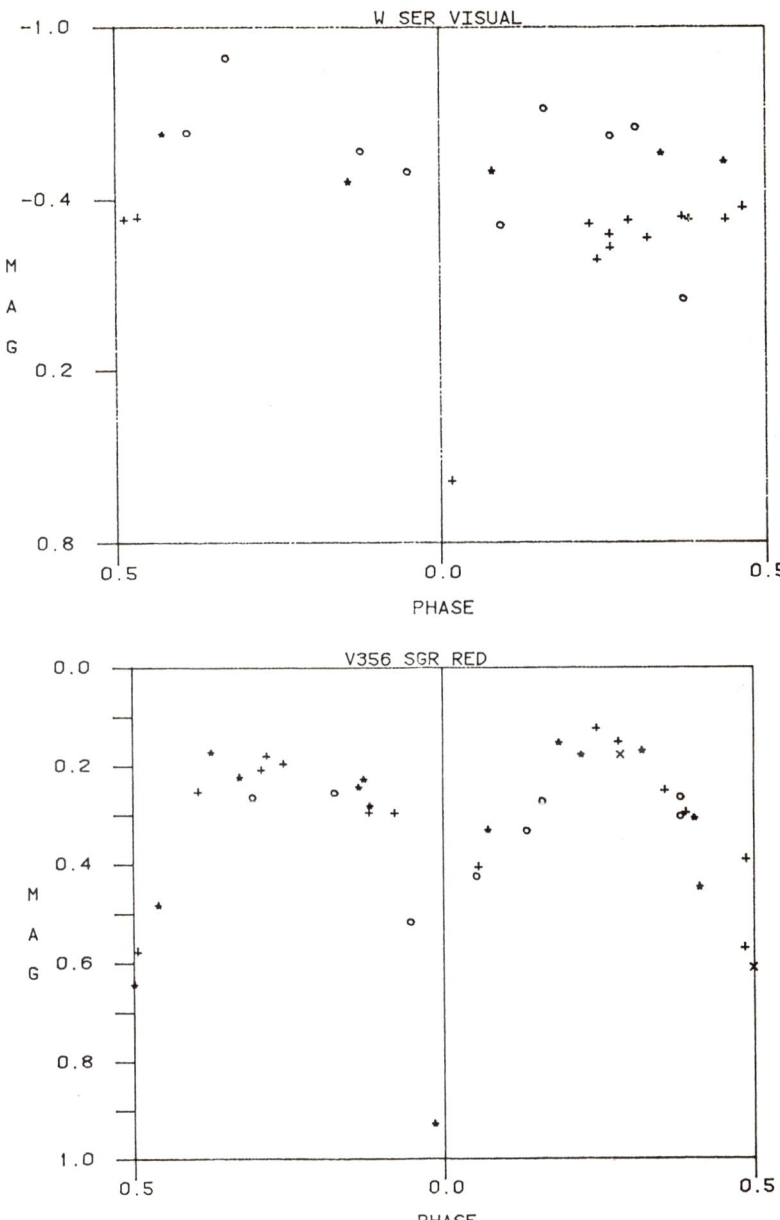

Figures 1 and 2. Representative light curves for two of our program objects.

If no ephemeris is available or if the ephemeris is doubtful, we use the Jurkevich method (see DuPuy and Hoffman 1985.) Usually, however, the spectroscopic light elements are already accurate enough so that the Jurkevich method cannot improve them. A refined period is determined by selecting a range of periods about the suspected value. Phases are recomputed for the observed points for each of the chosen periods. We then assume that the light curve has the form:

$$I = A_0 + A_1 \cos(2*\text{phase}) + A_2 \cos(4*\text{phase}),$$

where I is the measured intensity, A_0, A_1, and A_2 are constants. A least squares fit is used to compute the As and the sum of residuals squared is found for each test period. A plot of the residuals versus period will be parabolic in the vicinity of the correct period. We take the minimum of this parabola to be the correct period. The functional form of the assumed light curve can, of course, be varied from star to star to more closely agree with the observed light variation. We feel, however, that simple harmonic series (rather than more complete and complex eclipse functions) will be sufficient to improve the light elements.

IV. CONCLUSIONS

The observational aspect of the W Ser program is approximately 40% complete. Initial period studies are in progress. By the end of the coming summer, several of the program objects will be ready for the detailed analysis of the thick disk Wilson–Devinney program.

REFERENCES

Barnes, T. G. and Moffett, T. J. 1979 *Publ. Astron. Soc. Pacific* 91, 289.
DuPuy, D. L. and Hoffman, G. A. 1985 *I.A.P.P.P. Comm.* No. 20., 1.
Henden, A. A. and Kaitchuck, R. H. 1982 *Astronomical Photoelectric Photometry* (New York: Van Nostrand Reinhold).
Plavec, M. 1980a "The Impact of IUE on Binary Star Studies," *U.C.L.A. Astronomy and Astrophysics* Preprint No. 95.
Plavec, M. 1980b "Mass Loss from Interacting Close Binary Systems," *U.C.L.A. Astronomy and Astrophysics* Preprint No. 112.
Wilson, R. E. 1982 *I.A.U. Colloquium* No. 69.
Wilson, R. E. and Caldwell, C. N. 1978 *Astrophys. J.* 221, 917.
Wilson, R. E. and Lapasset, E. 1981 *Astron. Astrophys.* 95, 328.
Wilson, R. E., Rafert, J. B., and Markworth, N. L. 1984 *I.A.P.P.P. Comm.* No. 16, 1.

SZ HERCULIS: NEW PHOTOELECTRIC OBSERVATIONS AND PERIOD STUDY

E. J. Michaels

Dept. of Physics and Astronomy
Stephen F. Austin State University

I. INTRODUCTION

SZ Herculis is an Algol–type eclipsing binary star that is well known for its changing period. A complete discussion of the early period changes and the computation of photometric elements was published by Broglia, Masani, and Pestarino (1955). Since this study, SZ Herculis has had only marginal photoelectric coverage. Amateur astronomers, on the other hand, have accumulated a wealth of new observations in the form of visual times of minima for the deep primary eclipse. In view of the scarcity of recent photometry and the large number of times of minima observed since 1954, it was decided that a new photoelectric study of SZ Herculis would increase and update current knowledge of this eclipsing binary. New photoelectric observations resulted in eight times of minima and a new orbital solution using the light—curve synthesis program of Wilson and Devinney (1971). A literature search was undertaken to locate the rather large number of minima times, thus allowing a complete period study to be made.

II. THE OBSERVATIONS

SZ Herculis was observed on seven nights in 1985 (May 23/24, 24/25, 26/27 and June 7/8, 8/9, 14/15, 20/21) using the 46—cm Ritchey—Chretien Telescope of Stephen F. Austin State University Observatory. The observing site is located in East Texas 10 miles north of the town of Nacogdoches. The telescope at this time was equipped with Kearfott Singer stepper motors controlled by a dual AIM—65/Commodore—64 computer system (Gann 1986). Commands for telescope positioning were issued by the Commodore—64 to the AIM—65 via a communication interface. The AIM—65 computed the correct number of tracking or slew pulses and then issued the pulses to the motors. This system provides for automatic cycling between comparison, variable, check and sky positions at the completion of each filter sequence. On the average, one complete cycle (comparison V, R; variable V, R; comparison V, R) was repeated every 4 minutes. The Commodore—64 also controlled the photon counter by issuing start and stop integration commands, reading the counter via the user port, and finally storing the observation (UT time, integration length, filter, object, and counts) on floppy disk. This configuration has recently been replaced (spring 1986) with a new system consisting of Superior Electric stepper motors, Superior Electric drive electronics, and a single Commodore—64 computer.

The photometer was a "Starlight—1" photon counting system equipped with a red sensitive (S—20) EMI 9798A PMT. A 51—arcsec diaphragm and standard V and R filters were used to make a total of 615 observations. Integration times ranged from 10—30 seconds depending on sky conditions and variable—star brightness.

The comparison star and check stars are identified on two excellent charts in earlier publications. On a chart by Broglia (1955), the check star is labeled "f" and has a visual magnitude of 10.77. The comparison star is identified on a second chart in *Sky and Telescope* (Vol. 25, p. 227, 1963). This star is designated with the letter "C" and has a visual magnitude of 10.4. The comparison star was chosen because of its almost equal brightness and proximity to SZ Her (2 arcmin south and 8 arcmin east). This provided faster cycling time of the rather slow automatic moves and also eliminated the need for an extinction correction. Since this comparison star was not used by past observers of SZ Herculis (Broglia 1955), a special point was made to observe the check star several times each night. This resulted in a total of 70 differential check star magnitudes which were then formed into nightly means by color. The standard deviation of the nightly means was 0.02 mag. in both the visual and red bandpasses. Brightness variations in a single night were typically only a few hundredths higher than the average value given above. The differential check star magnitudes confirmed the constant brightness of the comparison star over the observing period.

Differential magnitudes and phases where computed for the observations using the reduction program described by Rafert, Michaels, and Markworth (1985). The resulting visual light curve is presented in Figure 1. The observational error was determined by making a linear least−squares fit to the comparison magnitudes vs. air mass for each individual night. The resulting average error of an observation (over the seven nights) was 0.03 mag. in the visual and 0.02 mag. in the red. This represents a rather typical error for the observing site and sky conditions of late spring. These observations resulted in eight new times of minima (Michaels 1985).

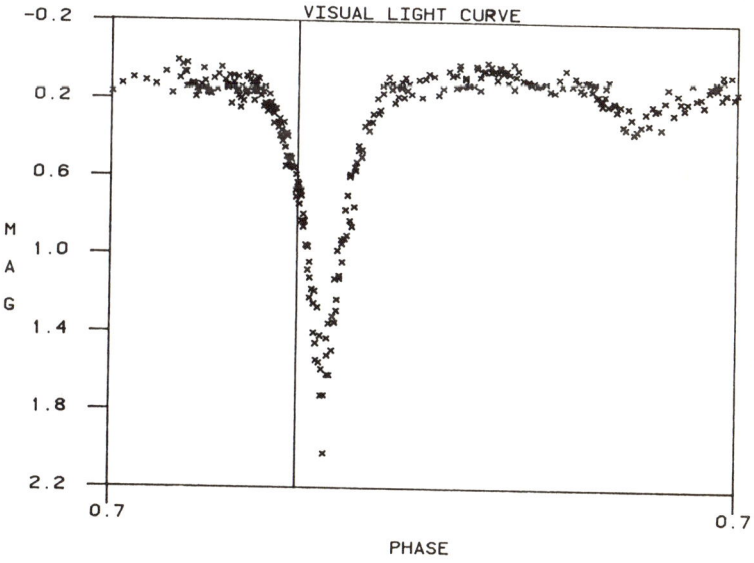

Figure 1. The individual differential magnitudes of SZ Herculis in the sense (variable−comparison).

III. THE PERIOD STUDY

SZ Herculis is one of the most observed eclipsing binary stars in this century. A vast majority of the observations involved the determination of times of minimum light by amateur astronomers using visual techniques. The literature search for these minima was a large project in itself. A total of 466 times of minima where located in 82 different references! The minima covered a time interval from 1902 to 1986 which corresponds to a total of 37,497 cycles (complete orbits)!

The period study was accomplished using the Honeywell 6000 mainframe computer at SFASU. The FORTRAN software required inputs consisting of the best available ephemeris, the times of minimum, and the corresponding weights of each minimum. The program computed the cycle count of each observation, the time of minimum from the given ephemeris for each observation, and then formed residuals (Observed Minimum − Calculated Minimum or O−C). A linear least−squares fit was then made to the residuals which resulted in an accurate orbital period. The initial ephemeris was taken from the "Finding List for Observers of Interacting Binary Stars" (Wood *et al.* 1980) and is given by:

$$HJD(MinI) = 2434987.3852 + 0.81809378 \, E.$$

This ephemeris predicted primary minimum 51.5 minutes earlier than the observed times of this work (see Figure 1). The weights applied to each time of minimum are those used by Duerbeck (1975) who finds the mean error of a visual observation to be six times that of a photoelectrically determined observation and a photographic minimum to be twice that of a photoelectric one. The weights given to the reported minima of Broglia differ from the ones indicated above (Broglia 1955). Those weights were the result of a normalization procedure applied by Broglia and therefore were retained in this period study.

The very large number of visually−determined minima resulted in an initial O−C diagram of very high density and considerable apparent scatter. This observational scatter does not necessarily mean the observations are of poor quality. The actual visually−determined time of minimum light for a star with deep eclipses such as SZ Herculis is very good, given reasonable seeing conditions and a fairly experienced observer. From a series of visual observations in the early 1960s (the period of SZ Herculis appeared to be well behaved at that time) it was found that errors in determining minimum light for SZ Herculis were typically less that ten minutes (*Sky and Telescope* Vol. 25, p. 227, 1963). This corresponds to a residual error of less than .008 days! Contained within the many O−C residuals is information on apparently real short term period variations. It was obvious that a more accurate period history would result if the cycle counts and residuals computed from the times of minima were normalized. Any minor period changes that may have occurred would then become apparent.

To normalize the residuals required the total range of cycles (−23159 to 14338) were separated into bins containing 100 cycles each. This bin−width was determined by a trial and error inspection of the O−C diagrams. The effort was directed towards maximizing information yet minimizing scatter. Cycle counts and residuals falling within a given bin were used to form a weighted average for each normal point. The resulting O−C diagram is presented in Figure 2.

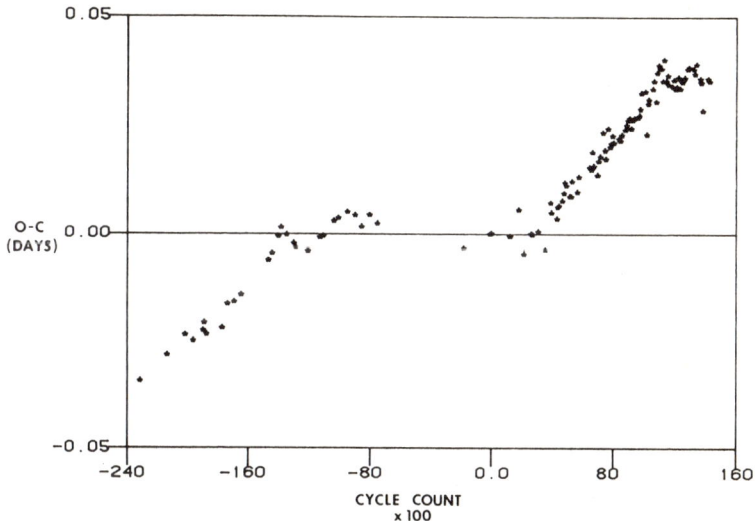

Figure 2. The (O−C) diagram for the normalized minima of SZ Herculis.

From Figure 2, it is immediately apparent that three clearly−defined period changes have occurred over the duration of the observations. Following each of these rather sudden changes of orbital period are epochs of constant period with, perhaps, occasional small, but temporary, period variations. The last major period change has only recently occurred with many smaller changes apparently still in progress.

Of historic interest is the rather large gap in times of minima from cycle counts −9016 to 0000. This time period corresponds to the years around World War II, when observers were obviously in short supply (like everything else).

For a complete analysis of the period changes and the computation of a new ephemeris it was decided to segment the normal points into the four time epochs with each epoch ending with a major period change. Using the same input ephemeris a linear least−squares solution for T_o (the initial epoch) and P (the period) was made for each segment. This analysis is illustrated in Figure 3. The resulting residuals for the linear fit to each segment are plotted together in Figure 4. As can be seen from this figure, the straight line provides a very reasonable fit to each segment. The period and initial epoch for segment 4 was computed from a total of 3311 cycles. These values represent the best current ephemeris for SZ Herculis.

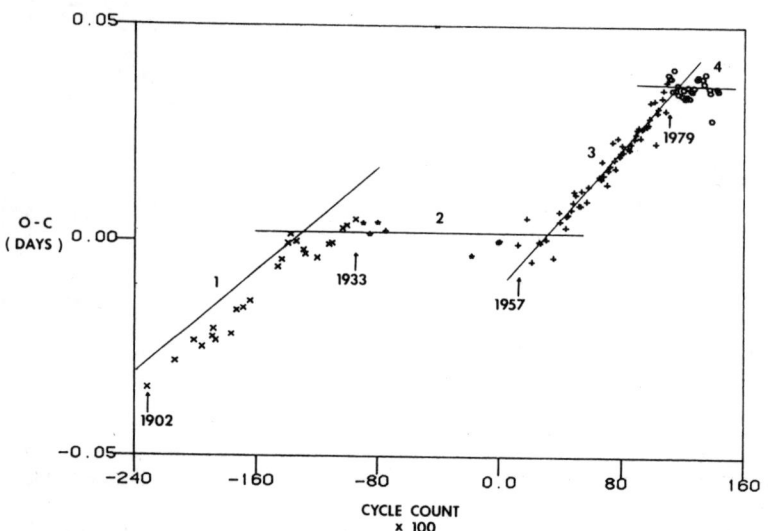

Figure 3. The (O−C) diagram of the normalized minima illustrates how the residuals were divided into four segments. Each segment begins and ends at a major period change. The different plotting symbols and calendar dates indicate the break points for each segment.

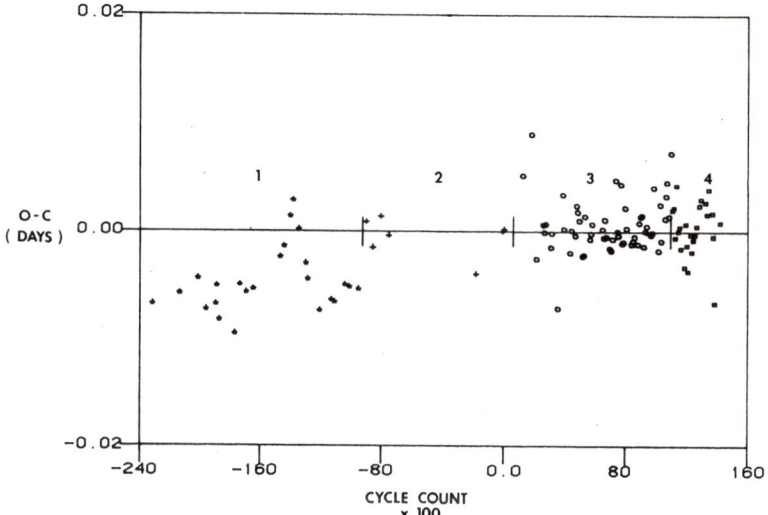

Figure 4. The residuals for the linear fit to each period segment.

The major period changes of SZ Herculis are of an alternating character as indicated by the change in slope of the residuals at the beginning of each period segment. These abrupt period changes are then followed by epochs of reasonably constant orbital periods. From the O−C diagrams, it can be seen that many short−term minor changes have also occurred over the observational history of SZ Herculis. Early observers suggested these small−scale period changes were possibly periodic (Broglia 1955), thus leading to explanations involving light time changes or apsidal motion. In Figure 2, there are perhaps hints of periodicity visible in segments 1 and 4 but none are visible in the well observed time interval of segment 3. Apsidal or light time effects would be revealed by a continuous periodic change in the residuals throughout the entire data base. This is not observed in the residuals of SZ Herculis; therefore, it is unlikely that apsidal motion or light time changes contributes to the observed period changes of this star.

V. CONCLUSIONS

The nature of the period changes strongly suggests that a mass transfer process is active between the component stars. Alternating period changes of the type observed for SZ Herculis, have been well studied in other binaries (Sahade and Wood 1978). In those cases Biermann and Hall (1973) suggested mass transfer was the primary cause of the period changes. In a semi−detached system such as SZ Herculis (the semi−detached classification was determined in a recently completed light curve analysis by the author) mass transfer occurs from the less massive secondary component, which fills its Roche surface, to the more massive and hotter primary star whose radius is less than the critical surface. When a sudden burst of mass from the secondary is transferred to the primary, a sudden increase in period will normally occur. If the transferred matter and angular momentum is temporarily stored as rotation in the surface layers of the primary star or onto an orbiting disk, then the period will decrease (Biermann and Hall 1973). A rapid collapse of the disk onto the primary would return the angular momentum to the orbit with a resulting period increase.

The sudden alternating period changes of SZ Herculis are very likely caused by a sudden and brief mass exchange between the component stars. Spectroscopic observations are certainly needed to confirm this and to provide a more accurate determination of the orbital and physical parameters of this eclipsing binary.

ACKNOWLEDGEMENTS

This period study and others like it would not have been possible if it were not for the many dedicated amateur astronomers from around the world who keep watch on these stars. The value of past and future observations of this sort can only increase with time. The combined efforts of both professional and amateur astronomers will most certainly lead to an increased knowledge and understanding of binary stars such as SZ Herculis.

Also, I wish to express my thanks to Norman Markworth for many helpful suggestions in the analysis of the observations and to my wife, Cathy, for her support and encouragement.

REFERENCES

Biermann, P. and Hall, D. S. 1973 *Astron. Astrophys.* 27, 249.
Broglia, P. 1961 *Mem. Soc. Astron. Ital.* 32, 43.
Broglia, P., Masani, A., and Pestarino, E. 1955 *Mem. Soc. Astron. Ital.* 26, 321.
Czerlunczakiewicz, B. and Flinn, P. 1968 *Acta Astron.* 18, 3.
Duerbeck, H. W. 1975 *Astron. Astrophys. Suppl.*, 22, 19.
Flinn, P. 1969 *Acta Astron.* 19, 2.
Flinn, P. and Slowik, A. 1967 *Acta Astron.* 17, 59.
Gann, R. L. 1986 Masters Thesis, Stephen F. Austin State Univ.
Michaels, E. J. 1985 *Infor. Bull. Variable Stars*, No. 2792.
Pohl, V. E. and Kizilirmak, A. 1964 *Astron. Nach.* 288, 69.
Rafert, J. B. and Canton, D. B. 1981 *Publ. Obs. Appalachian State Univ.*, 1, 257.
Rafert, J. B., Michaels, E. J. and Markworth, N. L. 1985, *I.A.P.P.P. Comm.* No. 21
Sahade, J. and Wood, F. B. 1978 in *Interacting Binary Stars* (New York: Pergamon Press), p. 65.
Wood, F. B., Oliver, J. P., Florkowski, D. R. and Koch, R. H. 1980 in *A Finding List for Observers of Interacting Binary Stars*, (Philadelphia: U. of Pennsylvania Press).

TIMES OF ECLIPSING MINIMA OF BINARY STARS

Frank Bradshaw Wood and Kwan-Yu Chen

Rosemary Hill Observatory, University of Florida

Many amateur astronomers wish to go beyond enjoying the beauty of the skies to make observations of permanent value. Most of them only have access to relatively small telescopes. However, aided by proper instrumentation—such as a photoelectric photometer—much useful work can be done. The term *useful* implies that the work will be published in a reputable astronomical publication. Much of the research now produced is being done with "small" telescopes. (For example see the paper on DH Cephei by H. C. Lines, et al. 1986, in which amateur and professional astronomers collaborated). There are many such opportunities and the present paper emphasizes only one of these, the observation of the times of light minima of eclipsing variable stars.

There are several things which make this a useful field. There is little need for extremely precise color systems. While at one time it was suggested that the precise time of minimum depended on the color (Tikhoff–Nordman effect), this suggestion has long been proven nontenable. Further, observation of a time of minimum cannot be replaced by later observations, but will be increasingly valuable. Before proceeding

---

New Generation Small Telescopes, ed. D. S. Hayes, R. M. Genet, & D. R. Genet.
© 1987 Fairborn Observatory.

further, we should deal with the concepts of heliocentric time and Julian day.

In calculations which involve long periods of time, it is convenient and advantageous to use a system of chronological reckoning by days. J. Scaligen, in 1582, proposed the system based on the Julian year (365.25 days) with the starting point at January 1, 4713 B.C. Then the time recorded for any astronomical phenomenon is simply expressed as the numbers of days and fractions of a day elapsed since the original Epoch; and to be more specific, at 0^h GMAT (Greenwich mean astronomical time = universal time + 12θ) of that particular day. This is the Julian day number, or simply Julian day (JD). Then it is a simple matter to find the number of days between any two events by taking the difference between their Julian day numbers.

However, there is a slight complication. The time of an observation recorded on earth (geocentric) contains the factor of light traveling from the source, say a star, to the observer. Because of the earth's orbital motion around the sun, the earth−star distance varies depending upon the time of the year and the position of the star. In order to have a time of the observation of a given stellar event with a fixed stellar distance, the heliocentric time is used corresponding to the sun−star distance. The heliocentric (Hel.) JD can be obtained by adding to the geocentric (Geo.) JD a light time correction, also known as the heliocentric correction. One reference on the calculation of heliocentric corrections is given by Landolt and Blondeau (1972), in which the values of these corrections were computed and tabulated for every hour in right ascension, every ten−degree interval in declination, and in ten−day intervals.

As an example, the eclipsing binary KR Persei was observed to have a light minimum on January 15, 1983, at 13:24:05:U. T. (Geo. JD 2445350.0585). The 1900.0 coordinates of this star are R.A. = 04 30 01 hms, dec = +44° 00′.4. From the Landolt−Blondeau Tables, the heliocentric correction is 0.0039 d. Thus this observed time of minimum is Hel JD 2445350.0624.

Note that observations must be given in decimals of a day. While tables can easily be constructed to permit conversion of hours and minutes to decimal days, some observers may wish to note that the construction of a clock showing time in decimals of a day is not a difficult task (Blitzstein, Thorpe and Wood 1951). Indeed, by computing the correction in advance, the clock can be set so that the recorded time is ready for publication.

Once minima have been obtained there is then the necessity of making them a permanent part of the astronomical literature. This means publication in a journal which will be widely distributed. If the observations themselves cannot be published, at the very least the times of minima derived from them and the observational procedure should be. Many research journals usually prefer papers requiring theoretical discussion as well, but some do exist which confine themselves to variable star work. As examples only, in addition to the *I.A.P.P.P. Communications* themselves, there exist others (such as the *Mitteilungen*

Uber Veranderlichen Sterne, published by the Sonneberg Sternwarte, or the American Association of Variable Star Observers) which confine themselves to variable stars. In general variable star work, there is also the *Information Bulletin on Variable Stars* published at the Konkoly Observatory, Budapest, under the auspices of IAU Commission 27. However, we are aware of only one group which deals almost exclusively with observed times of minima of eclipsing stars and the light elements derived from them. This has the rather formidable title of "Bedeckungsveranderlichen Beobachter Der Schweizerischen Astronomischen Gesellschaft" which translates as the "Eclipsing Binary Observers of the Swiss Astronomical Society," more familiarly known as the BBSAG. Despite the title, the active membership includes many observers outside of Switzerland. All the data collected are now stored on computers and serious observers of eclipsing star minima should by all means make contact with this group.

Any astronomer wishing to do serious work in this field should become familiar with the publication *Rocznik Astronomiczny Obserwatorium Krakonskiego* which is published annually. Information as to its availability can be requested from the editorial office, Obserwatorium Astronomiczne Uniwersytetu Jagiellonskiego ul Orla 171, 30-244 Krakow, Polska. Among other items, this extremely useful pamphlet contains predicted minima for many eclipsing variables and, for others, the most recent light elements which can be used for making individual predictions.

An astronomer wishing to contribute to this useful work must first select one or more systems which seem to him of interest. Then using the light elements (or consulting the Krakow predictions in the *Roznik*), he determines when the next favorable times of observation are i.e., when the star reaches light minimum; the times and observations are recorded and times are then further reduced by calculating intervals before or after primary minima—*i.e.* the phase.

A free-hand curve through these will usually give a satisfactory time of minimum. However, if more than one minimum is observed, they may all be plotted as a function of brightness versus phase, and a better light curve obtained. If this is put on tracing paper, it can then be fitted to future observations and a time of minimum determined, even when a complete run through primary is not obtained. This is described in some detail in Chapter 6 of *Photoelectric Astronomy for Amateurs* (Wood 1963). For a more objective method, that of Kwee and van Woerden (1956) can be used. As one example there is the paper on YZ Cas by R. Diethelm and R. D. Lines (1986) which determines times of minimum and gives an interesting discussion of this system.

The exact choice of a program is, of course, a matter of individual choice. The latitude of the observer versus the declination of the star is an important matter, as is the magnitude limit of the equipment. If a given system is being extensively observed elsewhere, there is, perhaps, little need for duplication. A good deal of valuable work could be done by selecting long period systems, which are generally neglected. These, however, offer difficulties; observations may have to be made in several minima before a satisfactory light curve can be drawn, and minima are far apart. An astronomer working solely on these would need to have several on his program in order to have a candidate available on any clear night.

An observer with access to astronomical literature may wish to go further and complete a period study. Conventionally, this is done by plotting the (O−C)−the difference between the observed time of minima and that computed from a given set of light elements−as a function of the period. Sometimes these show periodic changes which can be interpreted as the rotation of the major axis of an elliptical orbit or changes in light time as the system moves around the center of mass of itself and a third star in the system. This is discussed in more detail in the chapter by Wood (1963). Many more systems show sudden, and so far unpredictable, period changes. Many years ago (Wood 1950) it was noted that systems showing such changes almost always had at least one component near the stability surfaces (zero velocity surfaces), beyond which the star would not be stable and now it seems generally accepted that for most cases mass loss either out of the system or by transfer to the other component is the cause of these fluctuations. Thus careful and continuous monitoring of such systems is of prime importance. When an observer has tentatively selected his program, he would be well advised to check with the BBSAG to see whether or not it is one of the systems they are monitoring closely. All in all, it is clear that contributions of permanent value can be made by the observation and publication of times of minima. All future studies will use the results so produced.

We hope the preceding has given at least some feeling for the contribution which can be made by a sustained program of observation of minima. However, even the determination of a very few times of minima can be of considerable value provided they are published. Just possibly they may fill in part of one of the time gaps of several years which do occur in the study of many of these systems.

REFERENCES

Blitzstein, W., Thorpe, J. K. and Wood, F. B. 1951 *Sky and Telescope*, July 1951.
Diethelm, R. and Lines R. D. 1986 *Information Bull. Variable Stars* No., 2924.
Lines, H. C., *et al.* 1986 *Information. Bull. Variable Stars*, No. 2932.
Kwee, K. K. and van Woerden, H. 1956 *Bull. Astron. Instit. Netherlands*, 13, 327.
Landolt, A. U. and Blondeau, K. L. 1972 *Pub. Astron. Soc. Pacific*, 84, 784.
Wood, F. B. 1950 *Astrophys. J.* 112, 196.
Wood, F. B. 1963 *Photoelectric Astronomy for Amateurs* ed. F.B .Wood (New York: Macmillan Company), p. 141, ff.

ECLIPSING BINARY STAR SYSTEMS AS TESTS OF GENERAL RELATIVITY

Edward F. Guinan and Frank P. Maloney

Department of Astronomy and Astrophysics
Villanova University

I. INTRODUCTION

Eclipsing binary stars are rich sources of important astrophysical information, not only about the physical properties of the component stars, but also about the most fundamental of our physical laws. The fortuitous alignment of their stellar orbits in, or nearly in, our line of sight produces mutual eclipses of the stars. These periodically—recurring eclipses result in comparatively short but measurable decreases in the total apparent brightness of the binary system, ranging from a few hundredths of a magnitude up to a few magnitudes. But even when the eclipses are not taking place, the light continues to change in many of the systems because of the tidally and rotationally distorted shapes of the components, and also because of the heating of each star by the other. And so the light that comes to us from eclipsing binaries is coded with a wealth of information, both about the stars themselves, and the physical laws which govern their behavior.

About 3000 eclipsing binaries have been discovered so far. Nearly every type of star is represented as a member of an eclipsing binary. These include main sequence stars, giants, and supergiants—with spectral types from O to M, as well as subdwarfs, white dwarfs, and neutron stars. The decoding or analysis of their light curves yields information on their orbital properties, and on the relative sizes and shapes of the components, as well as the fractional brightnesses, temperature ratios, and limb darkening of the stars. In addition, for those eclipsing binary systems with spectroscopically—determined radial velocity curves, the absolute sizes of the stars, their masses, and densities can be calculated. These data are **fundamental** to our understanding the structure and evolution of stars.

However, the astrophysically—important data which can be obtained from observations of eclipsing binary star systems transcend the information on the physical properties of the star or the characteristics of their orbits. In eclipsing systems with eccentric orbits, an application of Newtonian physics makes it possible to "see" inside the stars and to determine their internal structure by studying the rotation or precession of the binary system's orbit. In addition, there is a contribution to this orbital rotation due to General Relativity. This orbital rotation is known as apsidal motion or periastron advance. The line of apsides is the long or major axis of the orbital ellipse. The orientation of the orbit in space is given by the longitude of the periastron ω, and is the angle measured in the plane of the true orbit, in the direction of the companion's motion, between the line of nodes (where the orbital plane intersects the plane of the sky) and the periastron point. The longitude of periastron ω can have any value between 0 and 360 degrees. In eclipsing binaries with eccentric orbit, the secondary minimum (the eclipse of the cooler component by the hotter star) is displaced from the half—period point in proportion to the product of the eccentricity and the cosine of the longitude of periastron, $e \cos \omega$. Thus the timings of the mutual eclipses are a sensitive measure of ω, and these timings, taken over many years, can reveal changes in ω, yielding the rate of apsidal motion, denoted by $\dot{\omega}$. The apsidal motion of a binary system is shown in Figure 1.

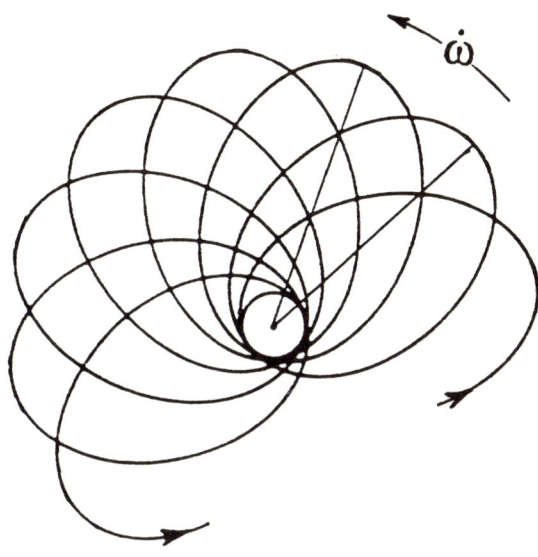

Figure 1. The apsidal motion in a binary system, showing the longitude of periastron and the advance of the line of apsides.

II. APSIDAL MOTION

Classically, in a gravitationally–isolated binary star system, one component, $\dot{\omega}_{CL}$, of the apsidal motion arises from the deviation of the figures of the stars from spherical symmetry. These deviations arise from tidal and rotational deformation of the binary components, which depend on the stars' fractional radii, their axial rotation rates, and their internal mass distributions. The fractional radius $r = R/a$ is the ratio of the star's radius R to the semi–major axis a of its orbit. The classical apsidal motion is a strong function of the stars' fractional radii—varying as r^5! Fortunately, the values for r can be determined quite precisely from the analysis of the light curves of eclipsing binaries. The rotational velocities of the stars can be found from the projected rotational velocities vsini, which are measured from the profiles of the absorption lines in the spectrum. The internal mass distributions of the stars are computed from stellar interior models (see *e.g.* Jeffrey 1984), and from the observation of apsidal motion in eclipsing systems with distorted components. But there is an additional component, $\dot{\omega}_{GR}$, to this apsidal motion due to General Relativity, and it depends on the mass of the binary system and the size and shape of the orbit. In GR, the stars must move thorough the space and time distorted by the presence of their own mass, and these distortions produce the additional apsidal motion contribution. The expressions for calculating classical and relativistic apsidal motions are given in most texts on Mechanics and can also be found in a previous paper by us (Guinan and Maloney 1985).

Careful observations of the planet Mercury have revealed that the perihelion of its orbit is advancing more rapidly than can be accounted for by the presence of the other planets. One of the great triumphs of GR is the apparent resolution of this 43"/100-year discrepancy between the observed and the classically-predicted apsidal motion. In close binary systems with massive components in eccentric orbits, the relativistic apsidal motion is frequently hundreds or even thousands of times greater than that of the Mercury-Sun system. This results from the greater masses of these binaries which produce greater curvature in space-time than the Sun does. In the vast majority of eclipsing binary systems, however, the relativistic component $\dot{\omega}_{GR}$ is much smaller than the classical component $\dot{\omega}_{CL}$ so that small uncertainties in the determination of the physical properties of the stars or in the nature of their orbit can mask the relativistic contribution. Because of this we cannot use these binary systems to test the theory of General Relativity.

Rudkjobing (1959) called attention to the 8th magnitude, eccentric eclipsing binary DI Her as an important test case for studying relativistic apsidal motion, since the relativistic component is greater than the classical component. DI Her (= HD175227) consists of two main-sequence B5 stars moving in an eccentric orbit (the orbital eccentricity e = 0.49) with an orbital period of 10.55 days. The binary system's orbit is viewed nearly edge on (i = 89 degrees) and is composed of components having small fractional radii relative to the semi-major axis a of their orbit: r_1 = 0.062 and r_2 = 0.057. Because of this, the system produces deep, narrow eclipses that permit a very accurate determination of the apsidal motion from the measurement of the displacement of the secondary minimum from the primary minimum. The orbital and stellar properties of DI Her are given in Table I, and its relative orbit with the components drawn to scale is shown in Figure 2. Excellent determinations of the orbital and stellar parameters of the system have been made from the combined analysis of the system's radial velocity data and UBV light curves (see Figure 3), which permit the theoretical classical $\dot{\omega}_{CL}$ and relativistic $\dot{\omega}_{GR}$ components of the apsidal motion to be determined with reasonable accuracy:

$$\dot{\omega}_{CL} = 1°.93 +- 0°.26/100 \text{ yr., and}$$
$$\dot{\omega}_{GR} = 2°.34 +- 0°.15/100 \text{ yr.}$$

TESTS OF GENERAL RELATIVITY 387

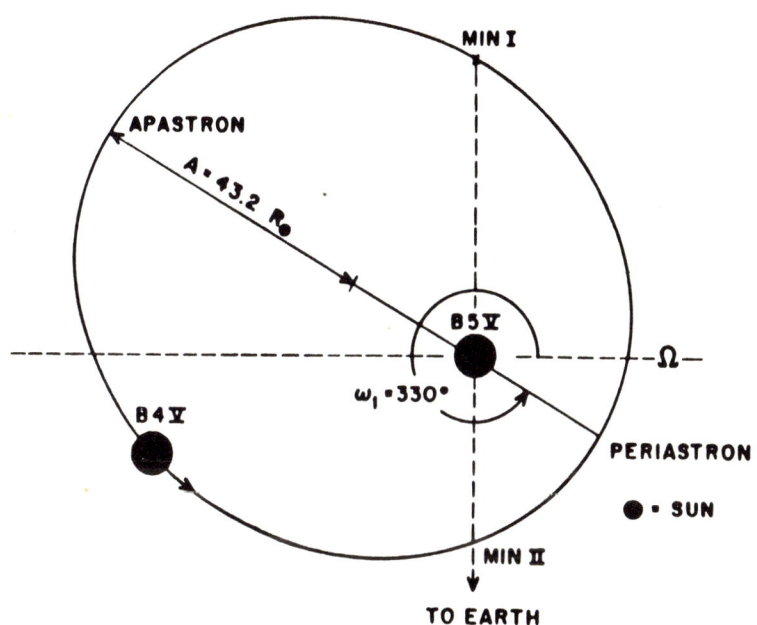

Figure 2. The relative orbit of DI Her, drawn to scale, from the quantities given in Table I. The points of primary (Min I) and secondary (Min II) eclipses are indicated, as well as periastron and apastron.

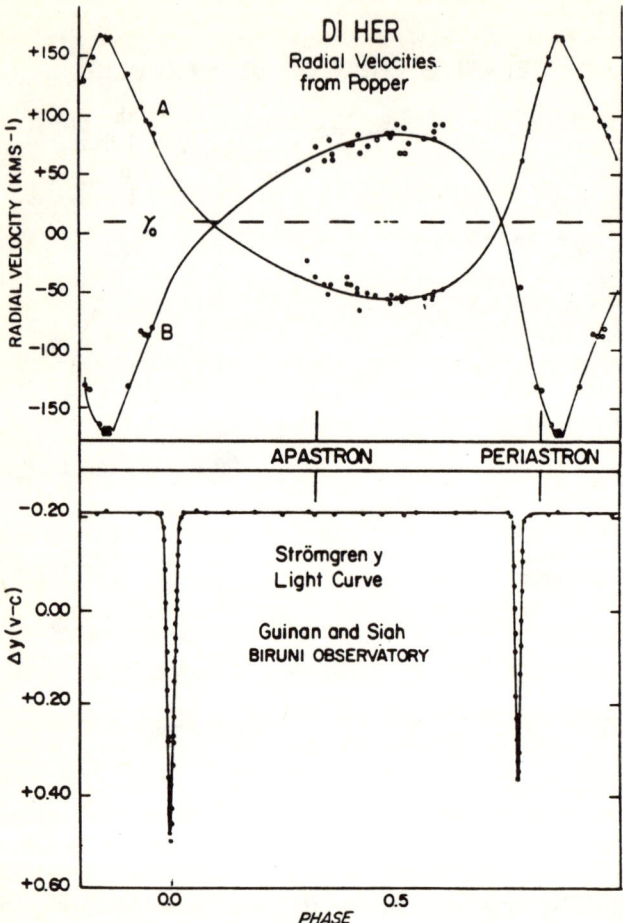

Figure 3. (Upper panel) radial velocity observations and adopted spectrographic orbits of DI Her, where A and B refer to the more massive and less massive components, respectively. Apastron and periastron are indicated in the figure. (Lower panel) Strömgren intermediate–band y curves for DI Her, showing primary and secondary minima, obtained by Guinan and Siah during 1977–1978.

We note that the theoretical apsidal motion expected from GR for DI Her is nearly 200 times greater than the 43"/100−year advance predicted for the orbit of Mercury. Least—squares solutions of the timings of primary and secondary eclipse minima, extending over an 84—year interval, yield a small apsidal motion:

$$\dot{\omega}_{OBS} = 0°.65 +- 0°.18/100 \text{ yr}.$$

This observed apsidal motion is about one—seventh the theoretically predicted value $\dot{\omega}_{CL+GR} = 4°.27/100$ yr. that is expected from the combined classical and relativistic effects, and results in a discrepancy of $-3°.62/100$ yr., a deviation of 20σ. This discrepancy is indeed large, and we are faced with the puzzling situation of either an undiscovered or misunderstood classical effect present in this system, or a problem with General Relativity.

In addition to DI Her, there is another 8th magnitude eccentric binary system, AS Cam, with well—determined orbital and physical properties (see Table I), that has an observed apsidal motion significantly less than that predicted from the combined classical and relativistic effects. AS Cam consists of a pair of B9V stars with an orbital period of 3.43 days. Accurate determinations of the orbital and stellar properties of the system have been made, which again permit the contributions to its apsidal motion to be reasonably well determined:

$$\dot{\omega}_{CL} = 35°.7 +- 3°.1/100 \text{ yr}., \text{ and}$$
$$\dot{\omega}_{GR} = 7°.9 +- 1°.6/100 \text{ yr}.$$

We have gathered timings of primary and secondary minima from published sources and have supplemented these with historic eclipse timings from 1899 to 1920 obtained from the Harvard Observatory plate collection. Least—squares solutions of the eclipse timings reveal a smaller than expected apsidal motion of:

$$\dot{\omega}_{OBS} = 13°.6 +- 1°.5/100 \text{yr}.$$

This work is being carried out with Sallie Baliunas, Robert Donahue, and Jenny Loeser of CfA, and Patricia Boyd of Villanova University. This observed apsidal motion is about one—third of that expected form the combined classical and relativistic effects $\dot{\omega}_{CL+GR} = 43°.6 +- 3°.5/100$ yr. However, AS Cam is not so good a dynamic test of General Relativity as DI Her, because the larger fractional radii of the components of AS Cam produce a larger contribution to the apsidal motion from classical tidal distortion than from GR. Nonetheless, the properties of the system are well—determined, as is the observed apsidal motion. Thus, AS Cam joins DI Her in having an observed apsidal motion significantly smaller than that expected from the combined classical and relativistic effects.

III. DISCUSSION

The large apparent discrepancy between the observed and the theoretically-expected apsidal motions in DI Her and AS Cam is interesting in view of the confidence we have in the value for the observed apsidal motions, and the overwhelming acceptance and vindication of the classical and relativistic components of apsidal motion. This discrepancy between theory and observation is well determined and is too large to be accounted for by observation uncertainties or errors in the physical properties of the two systems. We have identified five possible causes for the discrepancy. They are; a) a rapid circularization of the orbit; b) the presence of gas clouds in which the systems reside; c) the existence of a third member of the systems in a highly inclined orbit; d) the misalignment of the orbital plane with the equators of the stars; and e) a problem with General Relativity.

a) Rapid Circularization of the Orbit

We have previously mentioned that the observed apsidal motion is determined by measuring the change in the displacement of the time of secondary eclipse form the time of primary eclipse. This displacement is proportional to the product $e\cos\omega$. Thus, a change in the displacement could be caused by changes in e, the eccentricity of the orbit, rather than in ω. For DI Her, we calculate that in order to resolve the $-3°.62/100$ yr. discrepancy between the observed and the expected apsidal motion, the change in e, denoted $\dot{e}$ dot, must be $\dot{e} = -0.0136/100$ yr. Using a characteristic timescale $T = e/\dot{e}$, we calculate that $T = 3600$ yr. This corresponds to an extremely rapid decrease in orbital eccentricity, which could circularize the orbit of DI Her in only 3600 years. This rate is at least 1000 times that theoretically expected from tidal dissipation for stars with radiative envelopes, again because of the stars' small fractional radii. In addition, this timescale implies that 3600 years ago, the DI Her system had an eccentricity of unity, implying some type of capture interaction between the two stars.

It is possible that the theoretical timescale could be shortened to 3600 years by the following mechanisms:

1) If there were increased tidal dissipation of DI Her's angular momentum, is could be occurring due to induced oscillations in the component stars produced by the strongly-varying gravitational field arising from the eccentric orbit. This mechanism is not so important for DI Her, however, because of the small fractional radii of the stars. Further, some types of oscillations or instabilities produce small luminosity variations which may be detectable. Photometric observations made at Villanova University Observatory in three separate three-hour runs near the time of periastron passage have revealed no light variability greater than 0.015 mag.

2) Another mechanism that could be important in the dissipation of the angular momentum is through large mass loss, possibly manifesting itself in the form of stellar winds. Ultraviolet observations of DI Her taken with the International Ultraviolet Explorer satellite near periastron, when the tidal forces are at their greatest, again do not show any evidence of significant mass loss.

In AS Cam the observed value of the longitude of periastron is ω = 227° so that the secondary minimum occurs before the half−period point of its orbit. Thus, if the orbital eccentricity is rapidly decreasing then the apparent apsidal motion would be **longer** than predicted by theory rather than being smaller.

We are forced to conclude that a rapid circularization of the orbits of DI Her and S Cam seems most unlikely.

b) Circumbinary Clouds

An eccentric binary star system which is imbedded in a resisting medium will undergo a slow recession of the longitude of periastron, by an amount proportional to the density and extent of the medium. This recession is produced by the modification of the gravitational field of the stars by the presence of the material between them, and not by the resistance of the medium on the stars' motion. If we assume for simplicity that a hypothetical cloud is uniformly distributed in the DI Her system, then in order to account for the $-3°.62/100$ yr. apsidal motion discrepancy, we calculate that the cloud must possess a density $\rho = 10^{-10}$ g/cm^3. Further, if this cloud is composed mainly of hydrogen, then we obtain a density of 10^{14} H atoms/cm^3. This value for the gas density is very high, from 10^2 to 10^4 more dense than that found for circumstellar shells observed around other binary systems. Moreover, ground−based optical and satellite UV spectroscopic observations of DI Her do not reveal the presence of any significant amount of circumbinary material. We therefore conclude that the apsidal motion discrepancy in DI Her cannot be explained by the presence of a circumbinary cloud.

For AS Cam the discrepancy between the observed and theoretical apsidal motion of $\Delta\omega = 30°/100$ yr. is large and would require an unrealistically large gas density of $\rho \simeq 10^{-9}$ gm/cm^3. The optical and ultraviolet spectra of AS Cam, however, reveal no telltale spectrographic signatures of circumbinary gas. Thus, as in the case of DI Her, we conclude that the discrepancies in apsidal motion of AS Cam cannot be attributed to the system being embedded in gas.

c) The Presence of a Companion to the Eclipsing Pair

A possible explanation for the apsidal motion discrepancy is the presence of a third star, revolving in a highly inclined ($i'>45°$) orbit in the DI Her system. This third star can produce a recession of the longitude of periastron, by an amount dependent upon the mass of the stars, the period P′ of the third member and the period P of the binary, the eccentricity e' of the third member and the eccentricity e of the binary, and the inclination i' of the third member relative to the plane off the binary. The existence off this hypothetical third star in the DI Her system also produces a *light–time* effect, which arises from the variation of the orbit of the close pair around the barycenter of the triple system. This will produce an apparent periodic variation in the times of the primary and secondary eclipses of the close pair. The amplitude Δt of this variation depends upon the masses of the components, their periods P and P', and inclinations i and i'. Further, the presence of a third member in the DI Her system may reveal itself simply because of its stellar luminosity.

From observations of the DI Her system, we can provide some constraints on the properties of the hypothetical companion.

1) Since the visual and photographic times of minimum are less precise, we examined only the photoelectric times of minimum, which reveal no oscillatory light–time effect, at least not to the precision of the timings, about 3 minutes, or 0.002 day.

2) An examination of the *UBV* light curves of the system reveals that the luminosity L_3 of the companion must be less than ~ 0.03 ($L_1 + L_2$), where L_1 and L_2 are the luminosities of the components of the close pair. This luminosity upper limit corresponds to a stellar mass less than 2.5 $M_\odot$, which in turn translates to a star of main–sequence spectral class B9 to A0 or cooler.

It is possible then to create mathematically a variety of theoretical companions for the DI Her system, whose effect on the apsidal motion of the close pair is sufficient to resolve the $-3°.62/100$ yr. discrepancy, but whose light time effect and third light contribution is too small to be observed. For example, if we create a companion $M_3 = 1$ $M_\odot$, in a circular orbit whose period $P' = 5$ years, at a relative inclination $i' = 59°$, the apsidal motion discrepancy will be resolved, the light time variation will be only 0.0018 day, and this main–sequence G2 companion will be effectively invisible–masked by the light of the close pair, even during their eclipses.

However easy it is to make a hypothetical companion for DI Her or AS Cam, its existence remains improbable on both statistical and dynamical grounds. Statistically, only about 20% to 30% of close pairs are attended by companions. Dynamically, the presence of a third member sufficient to produce the required perturbation in ω will also produce perturbations in all the orbital elements of the close pair. Notably, we expect a change in the orbital inclination i with time, which will manifest itself observationally in a change in the depths of the eclipses. This change in eclipse depths is a very sensitive measure of the change in the orbital inclination of the close pair. For the DI Her system, the rate of change in the magnitude of the eclipse depths with inclination is about 0.2 magnitude per degree change in i. And there will be no eclipses at all if the inclination drops below 81°.7. An examination of the depths of the eclipses from photoelectric data reveals no variations greater than 0.005 magnitude over a 15 year interval, which translates to a change in the inclination of less than 0°.03. Moreover, an examination of the photographic eclipse data from as early as 1898 reveals no variation in eclipse depth greater than about 0.05 magnitude. Thus it appears unlikely that the DI Her system can contain a third member, a star which can produce just the right amount of apsidal motion perturbation, but otherwise, in all other observables, can remain invisible. Moreover, it would be doubly strange if AS Cam too possessed an invisible close companion that would perturb its orbit in the same manner as DI Her.

d) Misalignment of the Orbital Plane with the Stars' Equators

From dynamical considerations, it is expected that the equatorial planes of stars in close binaries should be coplanar, or nearly so, with the plane of the orbit (see Kopal 1978). And, as discussed by Plavec (1960), there is no convincing observational evidence to refute these theoretical predictions concerning coplanarity. However, if the stars' equators are not coincident with their orbital plane, then the contribution to the classical apsidal motion from rotational deformation decreases, even reversing sign, as the angle between the equator and the orbit increases. At inclination angles greater than 45°, $\dot{\omega}_{CL}$ from tidal distortion is small and the stars are spinning rapidly. This mechanism, however, appears unlikely to explain the apsidal motion discrepancy in DI Her and AS Cam for the following reasons:

1) It requires the stars in DI Her and AS Cam to spin rapidly, $v_{rot} > 300$ to 400 km/s, with their equators tipped 60° to 90° relative to the orbit plane. This rapid rotation should produce effects similar to those observed in a class of rapidly rotating B stars—the Be stars. The components of DI Her and AS Cam do not exhibit the photometric or spectroscopic signatures associated with Be stars: line emission, light variability, or infrared excesses. Further, the observed projected rotational velocities vsini of the components are in accordance with those expected if their equators are coplanar with their orbit.

2) If binary stars form from fission mechanisms, as most close binaries appear to do, then it is more likely that stars' equators and their orbit planes will be coincident. However, if a binary system forms by a capture process, then initially both stars would have their equators randomly oriented with respect to the orbit plane. But over timescales short compared to the lifetimes of the stars, torques exerted on the stars by each other tend to force their equators into coplanarity with their orbit.

In the near future, we plan to determine directly the orientations of the spin axes of the components of DI Her and AS Cam from high-resolution radial velocity observations obtained during the eclipses. The rotational velocity of each star can be determined from the analysis of the line profiles as the eclipses progress.

e) Problems with General Relativity

In the absence of a suitable classical explanation for the discrepancy in the apsidal motions observed in DI Her and AS Cam, there exists the possibility that General Relativity, in its present formulation, is incomplete. This possibility is most disturbing in view of the success of GR in its experimental verifications. Nonetheless, Moffat (1979, 1984) has proposed a non-symmetric gravity theory (NGT) which is consistent with the apsidal motions of DI Her and AS Cam, as well as with those of several other eccentric binary systems. In his formulation of General Relativity, Einstein employed Riemannian geometry, in which the basic mathematical tools are symmetric—the equations have certain parts that can be interchanged without affecting the results of the calculations. But a true mathematical description of nature may require both symmetric and anti-symmetric elements in the equations. Moffat employs non-symmetric geometry to reflect this. The addition of anti-symmetric elements produces an additional term in the expression for calculating the relativistic apsidal motion. This term becomes important only in binary star systems which contain massive stars in eccentric orbits. Also, in GR only advances in the line of apsides are possible, whereas in NGT, recessions are permitted in these more massive binary systems. However, NGT requires the evaluation of a constant of integration l, or "free parameter", which appears to be a function of the masses and chemical compositions of the stars. Thus, NGT is not able to predict *a piori* the apsidal motions observed in binary stars systems, but must itself be calibrated by them. However, once the calibration is done, NGT is then consistent with the solar system relativity experiments, as well as being consistent with the relativistic effects observed in the binary pulsar PSR 1913+16. This distinguishes NGT from other rival theories of gravity, which have not successfully fit all of the observational tests.

An application of most recent calibration of NGT to the DI Her and AS Cam systems results in the following (see also Table I):

DI Her - $\dot{\omega}_{OBS} = 0°.65 +- 0°.18/100$ yr., $\dot{\omega}_{CL+NGT} = 15°.0/100$ yr.

AS Cam - $\dot{\omega}_{OBS} = 13°.6 +- 1°.30/100$ yr., $\dot{\omega}_{CL+NGT} = 15°.0/100$ yr.

Thus the application of NGT to both systems appears to show apsidal motions close to the observed values. However, we must be reminded that this expected apsidal motion depends upon the value of (n) extracted empirically from the apsidal motions of several eccentric binary systems, including both DI Her and AS Cam.

IV. FUTURE WORK

The large discrepancies between the observed and the theoretical apsidal motions off DI Her and AS Cam are both interesting and perplexing. For both systems, the rates of apsidal motion are well determined and based on over 80 years of observations. Moreover, the physical properties of these systems are known quite precisely, as are the theoretical classical and relativistic terms of apsidal motion. Furthermore, for DI Her, the relativistic component of its apsidal motion is larger than the corresponding classical component. And for AS Cam, the relativistic component makes a significant contribution to that system's apsidal motion. Thus, in both systems there exists either an undiscovered or misunderstood classical effect, or an inconsistency with General Relativity.

Progress towards the resolution of this problem can be made by continued observations of DI Her and AS Cam. Photoelectrically-determined eclipse timings over the next five to ten years are crucial for more accurately determining the apsidal motions of each system, and for possibly uncovering the presence of close companions. As discussed previously, a nearby companion can be detected by the influence it should have on the motion of the barycenter of the eclipsing pair, and by the change in eclipse depths caused by the possible perturbation in the orbital inclination.

Other eccentric eclipsing binaries can provide additional tests of our understanding of apsidal motion in close binaries. However, suitable candidates whose physical properties are well-determined, whose systemic masses and orbital eccentricities are large, and whose fractional stellar radii are small, are very rare. Moreover, to distinguish between the predictions of GR and NGT, it is necessary to study **massive**, main-sequence stars in close, eccentric orbits. Because massive stars are rare to begin with, and have relatively short lifetimes, it is not surprising that there are few close binary systems consisting of massive main-sequence components. Stars which have evolved off the main sequence are not so useful because their large fractional radii can produce large classical apsidal motions which can overwhelm the relativistic contribution. Moreover, the requirement that the massive binary system also undergoes eclipses and has an eccentric orbit diminishes the number of suitable candidate systems. The need for the system to be an *eclipsing* binary is obvious because the masses and radii of the components must be known in order to compute the theoretical apsidal motion, while the eclipses provide an accurate means of measuring the apsidal motion.

Table II is a list of binary systems, including DI Her and AS Cam that may be suitable for studying relativistic apsidal motion. The HD numbers of the stars and their 1950-epoch celestial coordinates are given, along with spectral types of the components, their apparent visual magnitudes, stellar masses, orbital periods, and eccentricities. Also given are the observed values of apsidal motion and the theoretical apsidal motions expected form the combined GR and classical effects. The apsidal motion rates are expressed in units of degrees/100 years.

1) As shown in the table, the observed apsidal motions of AG Per, ι Ori, HR 1952, α Vir, DI Her, and AS Cam are less than the theoretically expected apsidal motions from GR and classical effects. For these systems, NGT appears to give a better fit to their observed apsidal motions than does general relativity. For AG Per, however, the classical contribution to apsidal motion dominates the GR or NGT contributions, so that small errors in determining the radii and internal mass distributions of the components of AG per produce uncertainties in $\dot{\omega}_{CL}$ which can exceed the smaller $\dot{\omega}_{GR}$.

2) On the other hand, EK Cep and CW Cep have observed apsidal motions that agree better with GR than with NGT. However, both systems have theoretical apsidal motions that are dominated by classical effects rather than by GR or NGT.

3) For V889 Aql and V541 Cyg the observed and theoretical apsidal motions are relatively small and it is not possible at present to distinguish between GR and NGT. Both of these systems are important test cases for studying relativistic apsidal motion because in each case $\dot{\omega}_{GR}$ is significantly larger than $\dot{\omega}_{CL}$. We are currently refining the observed apsidal motions for these systems by determining eclipse timings back to 1900, using the Harvard College Observatory plate collection.

4) Photometric studies of NY Cep, QX Car, and AR Cas are currently underway to determine their observed and theoretical values of apsidal motion. Recent photometry of NY Cep by Helen and Richard Lines reveals a shallow primary eclipse, but no secondary eclipse.

5) ι Ori, HR 1952, and α Vir are probably **not** eclipsing binaries. Their apsidal motions are determined directly from their spectroscopic orbits. The spectroscopically-determined apsidal motion for ι Ori and α Vir are well determined, but the apsidal motion for HR 1952 is poorly known, and a modern radial velocity study should be undertaken. ι Ori and α Vir show small light variations due to interaction effects and possibly stellar pulsations. The HR 1952 system, however, has not been studied photometrically, but could have small brightness changes from possible eclipses or from binary star interaction effects—*i.e.* tidal and reflection effects. Photometric observations of these three systems would still be useful to determine the nature of their light variability and to constrain their orbital properties.

6) A recent study of the system V1143 Cyg by Andersen *et al.* (1987) reveals that its observed apsidal motion falls about midway between the theoretical predictions of GR and NGT. More eclipse timings and accurate, complete light curves are important for this system.

V. CONCLUSION

The study of eclipsing binary star systems has revealed a wealth of scientific information; not only about the physical parameters of the stars and their orbits, but also about the most fundamental laws of nature which govern their behavior. Historically, eclipse phenomena have revealed the size of the Earth, Moon, and Sun, the value of the speed of light, the masses of the planets, the length of the astronomical unit, the existence of planetary rings, and the precise positions of many radio sources.

That which binds two stars in a binary system is gravity. It binds all objects in the universe to one another, and is the most important force in the macroscopic universe. It was Newton who first formulated a mathematical description of gravity, and he used that description to explain the motion of the planets and Halley's famous comet. But his description was incomplete, in that it failed to explain the detailed motion of the planet Mercury. A more complete description of gravity was made by Einstein in his theory of General Relativity, which was able to explain the anomalous motion of Mercury, and predict the previously unseen effects of gravitational redshift, gravitational radiation, and gravitational bending of electromagnetic radiation—all of which have been observed, reobserved, and confirmed. Presently GR is the standard-bearer for all scientific inquiry into the nature of the physical universe. But that should not mean that it represents the absolute truth. GR must be tested in all possible arenas. One such arena is the rare class of eccentric, massive eclipsing binary star systems. And in that arena, the two known systems—DI Her and AS Cam—seem to defy GR. If their apsidal motions are not consistent with GR, then scientists are faced with the situation of having an undiscovered new phenomenon, or a problem with Einstein's theory of General Relativity.

This arena is one in which the non-professional astronomer, working carefully and conscientiously but with only modest equipment, can make a significant and valuable contribution, not only to the study of apsidal motions in binary systems, but also to our most fundamental laws of nature. The binary systems listed in Table II are an ideal set of candidate stars for an Automatic Photometric Telescope (APT) system. Complete light curves of these systems in a standard photometric system such as UBV or Strömgren $ubvy$ would be highly desirable so that their physical and orbital properties could be better determined from an analysis of their light curves. Also, it would be valuable to secure accurate timings of the eclipses so that their apsidal motions can be better defined and possible anomalies in their orbital periods could be uncovered. Since all but one of these stars are brighter than 10th magnitude, the observations can be carried out effectively with a small telescope equipped with a photoelectric photometer. We would be happy to supply anyone interested in observing these stars with ephemerides, finding charts, and corresponding comparison and check stars.

TABLE I

PROPERTIES OF DI HER AND AS CAM

	DI Her	AS Cam[1]
Period :	10.550164 days	3.4309687 days
Eccentricity :	0.489	0.170
Longitude of periastron :	330°	227°
Orbital inclination :	89°.3	88°.4
Fractional radii r_1	0.062	0.150
r_2	0.057	0.110
Spectral type :	B4V + B5V	B8V + B9V
Mass :	$5.15 + 4.52\ M_\odot$	$3.3 + 2.5\ M_\odot$
Apsidal Motion :		
Classical	1.93 ± 0.26 °/100 yr.	35.7 ± 3.1 °/100 yr.
General Relativistic	2.34 ± 0.15	7.9 ± 1.6
Total CL+GR	4.27 ± 0.30	43.6 ± 3.5
Observed	0.65 ± 0.18	13.6 ± 1.3
Total CL+NGT	0.53 ± 0.20	15.0 ± 2.2

1) Khaliullin and Kozyreva (1983)

TABLE II

PROPERTIES OF THE CANDIDATE BINARY STAR SYSTEMS

	HD	RA h m s	DEC ° ' ''	SPECTRAL CLASS
AG Per	25833	4 3 43.5	33 18 54	B4V + B5V
V889 Aql	181166	19 16 34	19 0 30	B9V + B9V
EK Cep	206821	21 40 30	69 27 48	A0V + G0V
NY Cep	217312	22 56 41.1	62 48 32.6	B0IV + B1IV
ι Ori	37043	5 32 59.1	-5 56 34	O9III + B1III
V541 Cyg	BD +30.3704	19 40 31.9	31 12 14	B9V + B9V
HR 1952	37756	5 38 18.3	-1 9 13.1	B1V + B5V
α Vir	116658	13 22 33.3	-10 54 4	B1IV + B3V
V1143 Cyg	185912	19 37 34	54 51 24	F5V + F5V
QX Car	86118	9 52 57	-58 11 0	B5V + B5V
CW Cep	218066	23 2 1	63 7 36	B1IV + B1IV
DI Her	175227	18 51 21.8	24 12 54	B4V + B5V
AS Cam	35311	5 24 16	69 27 23	B8V + B9V
AR Cas	221253	23 27 43	58 16 24	B3V + B5V

TABLE II (continued)

FOR THE STUDY OF APSIDAL MOTION

m_V	$\dot{\omega}_{obs}$ °/100 yr	$\dot{\omega}_{theo}$ °/100 yr	Mass suns	Period days	Eccentricity
6.7	544.	719.	4.4 + 4.0	2.03	0.07
8.52	1.5	1.5	2.5 + 2.5	11.12	0.38
7.99	8.82	7.91	2.0 + 1.12	4.43	0.13
7.4	-	-	28 + 13.5	15.28	0.49
2.77	14.4	54.5	31.9 + 18.9	29.10	0.76
10.5	0.96	1.10	2.5 + 2.5	15.34	0.49
4.93	-6.7 (?)	12.6	8.3 + 5.3	27.16	0.76
0.97	296.	393.	7.16 + 4.55	4.01	0.18
5.85	3.40	4.20	1.3 + 1.3	7.64	0.54
6.6	-	-	5 + 5 (?)	4.48	0.28
7.6	925.	650.	11.7 + 11.0	2.73	0.028
8.3	0.65	4.27	5.15 + 4.12	10.55	0.49
8.7	13.6	43.	3.3 + 2.5	3.43	0.17
4.88	36.	-	8 + 5 (?)	6.07	0.25

REFERENCES

Andersen, J., Garcia, J. M., Gimenez, A. and Nordstrom, B. 1987 *Astron. Astrophys.* 174, 107.
Guinan, E. F. and Maloney, F. P. 1985 *Astron. J.* 90, 1519.
Jeffery, C. S. 1984 *Mon. Not. Roy. Astron. Soc.* 207, 323.
Khaliullin, Kh. F. and Kozyreva, V. S. 1983 *Astrophys. Space Sci.* 94, 115.
Kopal, Z. 1978 *Dynamics of Close Binary Systems* (Dordrecht: D. Reidel).
Moffat, J. W. 1979 *Phys. Rev.* D 19, 3554.
Moffat, J. W. 1984 *Astrophys. J. Lett.* 287, L77.
Plavec, M. 1960 *Bull. Astron. Czech.* 11, 197.
Rudkjobing, M. 1959 *Ann. d'Astrophys.* 22, 111.

PHOTOELECTRIC PHOTOMETRY RESEARCH AT THE HIGH-SCHOOL LEVEL

Kenneth W. Zeigler

Gila Astronomical Research Institute

ABSTRACT: A program of astronomical research involving high school students and utilizing the techniques of photoelectric photometry and astronomy is described. By introducing students to the concepts and techniques of scientific research at the high-school level, the student is better prepared for studies in science and engineering in college.

In recent years there has been an ever-growing concern as to the quality of education within the American public school system. After all, the scientists, engineers, and businessmen of tomorrow are being educated in our schools today. If America is to remain a technically and industrially competitive nation in the future, the education of the youth who are our future must be a primary goal.

Recent data from the National Science Foundation indicates that producing professional scientists, or those who will play a significant professional role in science and engineering, could be a very serious problem in the future if not addressed now.

For example, it is projected that by the year 1995 there will be a 30% decline in the number of college-age American students. If we only maintain the present proportion of students entering science, math, and engineering from among this group, there will be almost 700,000 fewer graduates receiving bachelors degrees in scientific and technical fields, which means that the pool from which we draw professionals would be significantly smaller than currently. Gifted scientists and engineers are a vital national resource in a world of increasing technological complexity. For this reason many educators have proposed new and different techniques of science and mathematics education.

In order to increase the number of future scientists and engineers in the United States to meet the increasing demand, it is necessary to encourage more high school students to become interested in pursuing careers in these areas of study. In order to accomplish this, and assist the student in selecting a career, it is important to introduce the student to the methods and instrumentation of original scientific research. Although many high school science courses incorporate laboratory experience into the curriculum, this laboratory experience is most often of the cookbook variety. This is to say that the student follows a precisely prescribed laboratory procedure and makes observations and physical measurements as instructed. Although such a laboratory activity is quite useful in introducing the basic laboratory techniques and principles of the natural sciences, it does not teach the student how to think like a scientist. Such a laboratory experience does not give the student first hand information as to what scientific research is all about. The majority of students have no contact with actual programs of scientific research until their second or third year of college. By this time the student has probably made a career selection. If students were given the opportunity to participate in a program of original scientific research at the high school level or at an earlier point in their college education, they would then have additional experience upon with which to base a career selection. The experience of true scientific discovery can motivate many students to achieve more and decrease the present science phobia that plagues many students.

There are many things to be considered when selecting a research program for high school students. The program must be one that will arouse student curiosity. If upon initial inspection the program seems too dry or complicated to students, they will hesitate to become involved in it. The selected program must offer the potential of discovery as well as being capable of capturing the imagination of the student from the start. The student must feel that there is a reasonable possibility that the objectives off the proposed research program can be achieved.

Any research program that is initiated at the high school must be able to operate on a shoe-string budget. Programs that have very large start up costs will only be able to be initiated at the largest and wealthiest school districts. Even the most interesting and scientifically worthwhile student research programs will fail if they are prohibitively expensive to equip and sustain.

Regardless of the amount of equipment, money, and interest present, a research program will fail if there is not a knowledgeable and qualified teacher to direct it. If there are no qualified instructors within an educational institution to direct a specific research program then there must be a means made available by which an instructor may gain a knowledge of the use of the instrumentation and techniques utilized in that research program.

Research in astronomy is well-suited to high school students for several reasons. The starry universe holds a certain fascination for young people. This fascination is reflected in the immense popularity of motion pictures such as the Star Wars trilogy, E.T., and the Star Trek series. Students sense a certain challenge in the exploration of the depths and secrets of space using a telescope. The start-up costs of a simple program in research astronomy are quite modest with a very low annual maintenance cost. Astronomy is also a science in which a significant number of science teachers have at least some knowledge of the basic concepts. It also presents an area of study in which significant original research may be done with very modest instrumentation. Currently, there are so many relatively bright objects of astronomical interest that there are just not enough professional astronomers or telescopes to study all of them in sufficient detail. For approximately $2,500 an educational institution, such as a high school or community college, could purchase an 8-inch Schmidt-Cassegrain telescope and a solid state photometer. With such a system a variety of research programs ranging from photoelectric photometry of bright variable stars to photometric studies of bright asteroids may be accomplished. Such a system also offers the possibility of a large variety of new and different evening laboratory activities for interested students.

At Globe High School in Globe, Arizona an ambitious program of astronomical education has already been initiated. Many people would consider high school students between the age of 14 and 18 too young and inexperienced to become involved in a program of serious astronomical research. The experience of the author has shown that this is not the case.

The astronomy research program developed at Globe High School by the author is based around an intensive two semester course in astronomy in which participating students receive both high school and college credit. First semester course topics include the design of astronomical instruments, the nature of light, stars, stellar evolution, and galactic structure. The second semester course is primarily a study of planetary science. Second semester course topics include the history of ancient astronomy, the structure of the earth, comparative planetology, and the theories of solar system formation. Both the first and second semester courses include laboratory experience. Laboratory activities include the use of the telescope, astronomical coordinate systems, and a number of experiments involving different techniques in photoelectric photometry, spectroscopy, and celestial mechanics. For students who wish to continue their studies in astronomy for a second year, a research astronomy course is offered. Students in the research astronomy course gain direct experience in the area of astronomical research. Students involved in this course develop an original research program of their own or may participate in one of the ongoing research projects at Gila Observatory.

Gila Observatory is an astronomical observatory dedicated to student research at the high school and community college level. Gila Observatory is located on a 3700 foot elevation ridge two miles west of Globe, Arizona. Construction of this unique observatory began during the spring of 1983. During the construction phases of this project much of the labor was provided by students. The funding to construct this original observatory was provided by the author. By July of 1983 the initial observatory was completed. The original telescope was an 11-inch Schmidt-Cassegrain telescope housed in a 2.5-meter diameter dome built by Observe-Dome Laboratories. The original instrumentation included a 35 mm camera and an Optec SSP-3 solid state photometer interfaced with a Radio Shack Model 4 microcomputer. This computer-interfaced system allowed rapid computer-data-logging of photometric information gathered using the photometer. Photometric data was transmitted directly from the photometer to the computer where it could be reduced, plotted, or stored on disk for later analysis. Early student research projects included VRI photometry of the supergiant star Betelgeuse, photoelectric photometry of RS CVn variable stars, and photometry of asteroids directed at determining the rotational periods of asteroids for which no published rotational period existed.

During the time since 1983 a number of additions have been made at Gila Observatory. These additions include an EMI Starlight 1 photon counting photometer on loan from Fairborn Observatory, a Hopkins HPO photoelectric photometer, a 14-inch Celestron Schmidt-Cassegrain telescope, and a 12-inch Cassegrain telescope. Several of the new instruments will be housed in a roll-off roof observatory adjacent to the current structure. When completed, this roll-off roof observatory will include a warm control room, a workshop, and a photographic darkroom. Funding to purchase the HPO photometer and build the new observatory were provided largely by Globe School District and through a grant from GTE Corporation.

Since March of 1986 Lowell Observatory has provided telescope time for the author and his students to conduct photometric research using the 31- and 24-inch telescopes. An average of four nights of telescope time per month has been provided for this research program. The opportunity to have direct contact with professional astronomers and conduct research at a major observatory using a larger telescope has been a very exciting event for the students involved.

The scientific results of this program over the past several years have been impressive. Students involved in the observational aspects of this program have been involved in the data analysis and publication aspects as well. To date, this program has generated seven papers that have appeared in a variety of scientific journals and other publications. Since 1983 this program has produced photometric data on 50 asteroids, a number of variable stars, and has resulted in the discovery of one bright variable star. In addition to this, students have won numerous awards through the science fair by presenting their research results there as well.

Thus, this program has provided a means by which students in the Globe area may participate in the process of scientific discovery. This experience has shown that, given the opportunity, high school students can successfully reach for the stars. If the high school students in a small mining town in Arizona can do this, then students from other high schools and small colleges can as well.

Currently the scope and financial status of the program has not allowed it to include other schools beyond the Globe area. A program such as this has the potential to benefit many instructors and their students at the high school and undergraduate college level, to say nothing of the potential benefit to the scientific community as a whole. Future plans for this educational endeavor include a program that would assist other high school and college astronomy instructors in the development of their own observationally−based astronomy program as well as assist in the coordination of research efforts with other amateur and professional astronomers. The proposed educational program would be coordinated through a non−profit educational corporation known as Gila Astronomical Research Institute Incorporated(GARI). The program will by subdivided into the three separate educational projects listed below:

1) Astronomy Teaching Material Program

This program would make available to teachers laboratory materials for the implementation of observationally−oriented programs in astronomy. A variety of field laboratory activities, covering topics such as the proper use of the more popular varieties of astronomical telescopes, doing astronomical photoelectric photometry, and doing astrophotography, would be available. Classroom activities in astronomy such as spectroscopic identification of elements, photoelectric photometry and the inverse square law of light, and the trigonometric determination of the heights and depths of lunar and planetary surface features would also be available. In addition to written laboratory materials, computer software and hardware as well as laboratory instrumentation useful in doing astronomy laboratory experiments would be made available to the teacher. Gila Astronomical Research Institute would also correspond with teachers in order to assist them in the designing of an astronomy course that would meet the specific learning objectives of their program.

2) Cooperative Astronomical Research Program

Through the cooperative astronomical research program high school and undergraduate college teachers and their students would be assisted in the selection and implementation of research projects at their individual schools. Such aid would include technical advice in the design of an observing area, supplying computer software to aid in the data reduction process, and assisting in the selection of the appropriate astronomical instrumentation. In addition to these services, GARI would assist schools in coordinating their efforts with other schools, and amateur and professional astronomers.

3) Teacher Training Program

To conduct an effective research program in astronomy requires that the teacher have a working knowledge of the techniques and instrumentation of astronomical research. Although a significant number of science teachers have a relatively strong background in basic astronomical concepts few have been involved in actual astronomical research and are, thus, not familiar with the instrumentation used. For this reason a summer workshop is proposed in order to acquaint high school and college science instructors with many of the techniques of research astronomy. Such topics as photoelectric photometry, astrometry, and astrophotography would be covered. This workshop would be conducted using commercially available telescopes and related instrumentation of the type that could be utilized in an economical astronomy research program. This course would be offered for college credit and would prepare teachers to pass the research skills learned on to their students. The workshop would be conducted at Gila Observatory in Arizona. June is the clearest month of the year in Arizona and Gila County offers some of the darkest and clearest skies in the world.

Carol Spring Mountain Educational Observatory

The most ambitious program proposed is an educational observatory to be placed on Carol Spring Mountain, a 6620−foot peak in the Tonto National Forest approximately 100 miles east of Phoenix. This an excellent site far from the lights of the city and with excellent seeing. The mountain top is a long ridge which runs from northwest to southeast and is currently the site of a radio transmission facility. An excellent and well−maintained road of 1 1/4−mile length runs from this site to US route 60. The mountain is on forest−service land and the initial reaction of the National Forest Service has been favorable.

The observatory would house a 32−inch Cassegrain reflector which utilizes a new thin−mirror technology. This telescope should provide the clearest image of any telescope of its size. This telescope would be housed in a dome 18 feet in diameter. Like advanced professional telescopes this telescope would be operated most of the time from a warm control room. The telescope would be equipped with a photoelectric photometer, CCD imaging system, and perhaps a small spectrometer.

High school and college teachers and their students would be able to take advantage of the size and quality of this instrument to do advanced research. In order to be awarded telescope time the teacher and students would submit an observing program proposal to GARI. With the assistance of professional astronomers the proposal would be reviewed. If appropriate, revisions that might make the program more effective would be suggested to the teacher applying for telescope time.

Teachers would also have the opportunity to cooperate with many professional astronomers in areas of research in which they are involved using this telescope. At most professional observatories the prime objective of a research program is the gathering of scientific data directed to the development of a scientific model of specific astronomical phenomena. The prime objective at this observatory is the educational value of the research experience itself while the scientific value of the research conducted is considered a bonus. For this reason proposed observing programs whose aims are to produce educational materials for the classroom such as CCD images of stars and nebulae will be considered worthwhile observing programs. The educational value of such a program is great in that it develops a very positive attitude on the part of the student toward science which will persist throughout the educational process. It is hoped that such students will be more likely to choose science or an engineering-related profession as a career.

    Currently this program is in its infancy. The articles of incorporation of Gila Astronomical Research Institute have been accepted by the Arizona Corporation Commission, and the non-profit status of this institute will soon be decided upon by the Internal Revenue Service. The initial grant proposals for this program are being prepared and will be submitted to the appropriate agencies during 1987. In the event that sufficient funding were available the Astronomy Teaching Materials Program would be instituted during the fall semester of 1988 and the Cooperative Astronomical Research Program would begin during the spring semester of 1989. Depending upon funding, the Teacher Training Program would be offered first during the summer of 1989. The Carol Spring Mountain project would be developed if it could be shown that sufficient interest existed within the educational community. At the present time there are a number of gifted persons involved in the development of this project with the proper skills to make this ambitious venture work. The one remaining ingredient still needed is funding. The potential educational and scientific benefits that could potentially be derived from this program far exceed the costs.

ULTRAVIOLET PHOTOMETRY WITH THE IUE SATELLITE

Joel A. Eaton*

Astronomy Department, Indiana University

I. INTRODUCTION

The ultraviolet is a potentially promising spectral region in which to study the properties of cool stars. Atomic physics combines with the prevailing physical conditions in just the right way to make the ultraviolet sensitive to conditions in both the chromosphere and upper photosphere. Whereas in the optical, the atmospheric opacity is primarily absorption by H− and H−I; in the ultraviolet much of the opacity comes from a thicket of lines of iron−peak elements which principally scatter photons produced at deeper levels. Thus the ultraviolet continuum is sensitive not only to atmospheric temperature but also to the concentration of iron−peak elements in the atmosphere. This latter factor

---

* Guest observer with the International Ultraviolet Explorer Satellite.

New Generation Small Telescopes, ed. D. S. Hayes, R. M. Genet, & D. R. Genet.
© 1987 Fairborn Observatory.

varies both because of differences in abundances in the atmosphere and from changes in ionization of the atoms that are there.

Likewise, the ultraviolet is an effective region in which to study the chromospheres of cool stars. With the low densities and moderately high temperatures (10^4K) they obtain, most of the common chemical elements are singly ionized. The strongest emission lines of these, which provide most of the cooling of the chromospheric gas, are located in the ultraviolet. Examples (Figure 1) are the strong magnesium feature at 2800 A, the carbon feature at 2325 A, and a complex grouping of iron lines in the range 2550–2900 A. Because of the way such emissions are excited in cool giants, a significant fraction of the cooling line emission comes out in the magnesium doublet at 2800 A. Thus it is a key feature for studying the variation of chromospheric heating and extent among the whole group of cool stars.

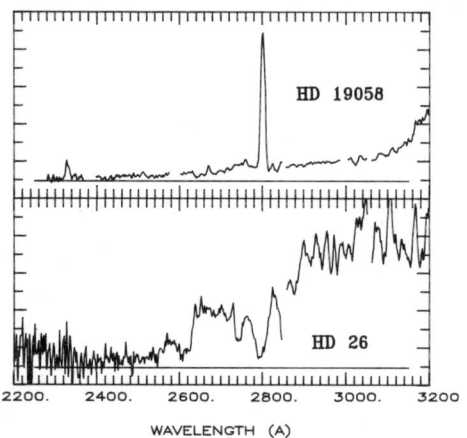

Figure 1. Low dispersion IUE spectra of two characteristic cool stars. HD 26 is a moderately metal poor CH giant (roughly G−K spectral type with elevated carbon abundance) and shows well the features seen in ultraviolet spectra of solar−type stars. HD 19058 is the M4 giant ρ Per and shows the typically strong Mg II chromospheric emission of such objects and a relatively smooth photospheric continuum that grades out into line emission shortward of about 2700 A. The emission feature at 2325 A is primarily a group of lines from singly ionized carbon that may be used to estimate the electron density in the chromosphere.

We know from past studies that the brightnesses of cool stars and the shapes of their ultraviolet spectra change with spectral type (Wu et al. 1980, 1983). However, we are now in a much better position to use the collection of spectra obtained with the IUE satellite over ten years of its operation to study trends with spectral type in a systematic way.

There now exist on the order of 50,000 images made with IUE, many of them for stars cooler than A0, and these constitute the data base of the present study. We will be using them to start to consider several questions pertaining to chromospheric emission.

Specifically: (1) Does the chromospheric heating seen in strong ultraviolet emission lines also manifest itself in the ultraviolet continuum? IUE has shown a range of roughly 10^4 in surface flux of lines such as $Mg\ II\ \lambda\ 2800$ and $C\ IV\ \lambda\ 1550$ (Hartmann, Dupree, and Raymond 1982; Eaton et al. 1985). If differences in the UV continuum surface brightness are the result of non-radiative heating, we would expect that there would be a good correlation between chromospheric flux and ultraviolet continuum brightness. Likewise, if the ultraviolet surface brightness at a given effective temperature depends primarily on metallicity, there should be differences in ultraviolet flux due to the increased ionization expected from irradiation of the photosphere by ultraviolet chromospheric emission. The ultraviolet excesses and peculiar iron abundances of RS CVn binaries may be a result of this process. Similarly, are the ultraviolet surface brightnesses of W UMa stars, which seem to have so much surface activity it is saturated (Vilhu and Rucinski 1983), dominated by surface activity?

(2) How does the chromospheric emission differ among stars in the same classes and between stars of different classes? It is now clear that stars of some evolutionary classes have either very high or quite low chromospheric emission. This effect can be illustrated with ultraviolet colors, which can be used further to determine the scatter of chromospheric properties among the members of a given group of stars.

(3) Does membership in a binary system really lead to elevated chromospheric emission in some circumstances? It already seems clear that it does, but we as yet cannot say why a binary component should have stronger chromospheric emission than a comparable single star.

II. OBSERVATIONS

The observations consisted of about 400 low-resolution LWR and LWP spectra of about 200 stars later than A0 in spectral type. Spectra for most of the ordinary field stars were taken from the IUE archives. Particularly useful were the group presented as an atlas by Wu et al. (1983) and the set of K and M giants studied by Steiman-Cameron et al. (1985). These were augmented by archival spectra of other G, K, and M giants where possible. Spectra of various other groups of stars were collected. The spectra of three FK Com stars (Bopp and Stencel 1981) were obtained. Spectra of many carbon stars (R, N, and CH stars) came primarily from programs of Hollis R. Johnson and collaborators, as did spectra of a few S stars. Spectra of W UMa binaries were obtained by Eaton (1983) and from the archives.

TABLE I

Definition of Photometric Bands

Band	Wavelengths Covered	Feature
IUE	2585–2785 + 2815–3200 A	Ultraviolet continuum
Mg II	2785–2815 A	*Mg II* UV1 doublet
B2	2900–3025 A	Region of strong Fe I
B3	3030–3200 A	Continuum
B4	2530–2630 A	Region of strong Fe II
B5	2640–2740 A	Comparison for B4

Colors: $Fe\ I = B2 - B3$
$Fe\ II = B4 - B5$

Because of the nature of opacity in the 2400–3200 A region, we were able to choose several photometric bands and indices that characterize a cool star's energy distribution. In Figure 1, one can see several apparent absorption edges. At 2630 A there is a sharp feature in F–K stars due to the resonance multiplet of singly ionized iron (*Fe II* UV1). Longward of this, indeed extending to about 2900 A, is a shelf which is depressed by the resonance lines of singly–ionized and neutral magnesium (*Mg II* UV1 and *Mg I* UV1). Between 2900 and 3030 A the continuum is depressed by two strong ground–state multiplets of neutral iron (*Fe I* 9 + 10 = UV1). In addition, stars that are in any way chromospherically active tend to have *Mg II* UV1 at 2800 A in emission. This feature is prominent for a variety of reasons: It is intrinsically strong since magnesium is a common element, which is mostly ionized in chromospheres, is easily excited by collisions with electrons, and emits all of its energy over a very narrow range of wavelength. Furthermore, the chromospheric emission does not have to compete with much photospheric light. In contrast, the strength of *Fe II* emission should be comparable to that of *Mg II*, but it is spread over many more spectral emission lines and does not appear as prominently in ultraviolet spectra.

The photometric bands chosen to measure these various features are identified in Table I. Combinations of bands define colors at the bottom of the table. Magnitudes for all of the bands were determined from each observed spectrum, values for the spectra of each star were averaged, and the results are presented in Table II The magnitudes were derived and are related to observed flux through the equation:

$$M = -2.5\log(f) - 21.087 \text{ mag.}, \qquad (1)$$

where the flux, f, is the mean flux at the earth in the band [ergs cm^{-2}s^{-1}A^{-1}] and the constant provides a normalization to M = 0.0 mag. at 3.64 x 10^{-9} ergs cm^{-2}s^{-1}A^{-1}, as in the *V* band of the *UBV* system.

The (*IUE−V*) color measures the ratio of brightness in an *IUE* band, extending from 2585 A to 3200 A but excluding the 2785−2815 A range containing *Mg II* UV1, to brightness in the *V* band of the *UBV* system. It is sensitive to such things as the temperature of the star, how deep in the atmosphere the ultraviolet continuum is formed, whether the star is deficient in heavy elements, and possibly whether there is appreciable mechanical heating in the upper photosphere. The (*Mg II−V*) color measures the mean brightness in the 2785−2815 A *Mg II* band to brightness at *V*. For the warmer stars the *Mg II* band detects mostly photospheric radiation in the wings of the broad absorption line. For stars of spectral type middle K and later, however, most of the flux is chromospheric line emission, and for these stars (*Mg II−V*) is an indicator of the fraction of the star's luminosity emitted in chromospheric lines. The *Fe II* index is a color measuring the ratio of brightness on the two sides of the 2630 A break, in analogy to the way (*U−B*) roughly measures the strength of the Balmer jump in *B−F* stars. The *Fe I* index gives the same information for the *Fe I* break near 3040 A.

As a check of our procedures, for about 30 stars with spectral types between A0 and G2, we calculated the color (2200−*V*) and compared it with the standard relations of Wu *et al.* (1980) in a (2200−*V*) vs. (*B−V*) diagram. Agreement was excellent, especially for the numerous dwarfs for which the scatter from the standard relation was less than 0.1 mag.

Some of our stars, especially the supergiants, are appreciably reddened, and their ultraviolet colors should be corrected as well as possible. Reddening relations for the various colors were determined by artificially reddening spectra of standard stars by various amounts, calculating the photometric indices, and then finding how each index changed with E(*B−V*). This was done for the reddening curve of Savage and Mathis (1979) and led to the results:

TABLE II

Visual and Ultraviolet Photometry for Late-type Stars

Star		Spectrum	V	B-V	V-R	IUE-V	MgII-V	FeI	FeII	E(B-V)
HD 26		K0 III-CH	8.22	1.09		2.37	4.03	0.13	1.17	
HD 352	5 Cet	K2 III+cF	6.17	1.36	1.08	3.61	3.12	0.32	0.52	0.20
HD 2151	Bet Hyi	G2 IV	2.80	0.62		1.51	3.00	0.19	0.91	
HD 3712	Alp Cas	K0 III	2.23	1.17	0.79	4.47	5.60	0.58	0.93	
HD 4128	Bet Cet	K1 III	2.04	1.02	0.72	3.33	4.24	0.54	0.78	
HD 4408	57 Psc	M4 III	5.38	1.65		5.83	5.49:	0.62		0.05
HD 4502	Zet And	K1 II+cool	4.06	1.12	0.84	3.53	3.64:	0.58	0.54	
HD 4614	Eta Cas	G0 V+M0 V	3.44	0.57	0.50	1.20	2.62	0.04	0.92	
HD 4656	Del Psc	K5 III	4.43	1.50	1.17	6.11	5.95	1.14	0.67:	
HD 5223		R0-CH	8.33	1.43		3.24	4.62	0.01	1.17	
HD 5516	Eta And	G8 III	4.42	0.94	0.73	3.20	4.07	0.79:	0.88	
HD 5820	WW Psc	M2 III	6.11	1.67		6.48	5.86	0.69	0.90:	0.07
HD 6268		G0p	8.09	0.84		2.41	3.65	0.27	0.77	0.20
HD 6833		K2 III	6.74	1.17		3.88	5.34	0.69	1.04	
HD 6860	Bet And	M0 IIIa	2.06	1.58	1.24	6.33	5.69	0.89	0.55:	
HD 6903	Phi3 Psc	G0 III	5.55	0.69		1.89	2.78	0.28	0.70	0.05
HD 10307		G2 V	4.95	0.62	0.53	1.53	2.97	0.17	0.96	
HD 10476	107 Psc	K1 V	5.24	0.84	0.68	2.59	3.83	0.52	0.95	
HD 10588		G8 III-IV	6.31	0.89	0.68	2.52	3.65	0.64:	0.94	
HD 10700	Tau Cet	G8 V	3.50	0.72	0.62	1.94	3.47	0.22	1.05	
HD 12311	Alp Hyi	F0 V	2.86	0.28		0.91	1.72	0.14	0.49	
HD 12479		M2 III	5.87	1.56		6.39	5.77:	0.69	0.88	
HD 16115		R3	8.16	1.23	0.88	2.98	4.74	0.34	1.04	
HD 16458		G8p	5.78	1.30		4.05	5.72	0.57	1.04:	
HD 17584	16 Per	F2 III	4.23	0.34	0.30	1.03	1.85	0.16	0.53	
HD 17709	17 Per	K7 III	4.53	1.56	1.19	6.25	5.90	0.95	0.66	
HD 18191	RZ Ari	M6 III	5.91	1.47		6.50	4.70	0.66	0.17:	
HD 19058	Rho Per	M4 II-III	3.39	1.65	1.78	6.67	5.51	0.71	0.44	0.05
HD 19373	Iot Per	G0 V	4.07	0.59	0.54	1.46	2.93	0.23	0.99	
HD 19557		R5	7.35	2.24		6.73	7.98	0.86	0.66	
HD 20234	TW Hor	N0	5.74	2.47	2.00	7.27	7.22	0.83		
HD 20630	Kap1 Cet	G5 V	4.83	0.68	0.56	1.80	2.93	0.26	0.90	
HD 20902	Alp Per	F5 Ib	1.79	0.48	0.45	1.95	3.44	0.27	0.83	0.22
HD 22049	Eps Eri	K2 V	3.73	0.88	0.73	2.75	3.44	0.59	0.80	
HD 22649	BD Cam	S3.5/2	5.10	1.63		6.10	5.31	0.70	0.22	
HD 25408	UV Cam	R8	7.90			6.44	7.01	1.18		
HD 26609	YY Eri	G2 V	8.23	0.66		1.65	2.38	0.26	0.73	
HD 26630	Mu Per	G0 Ib	4.14	0.95	0.77	2.65	3.39	0.22	0.36	0.19
HD 26736		G3 V	8.09	0.66		0.86	1.93	0.27	0.81	
HD 26756		G5 V	8.46	0.70		2.07	3.06	0.26	0.83	0.04
HD 27176	51 Tau	F0 V	5.65	0.28		0.69	1.36	0.12	0.42	
HD 27383		F9 V+G3 V	6.88	0.56		1.19	2.29	0.20	0.87	
HD 27524		F5 V	6.80	0.44		0.89	1.88	0.15	0.73	
HD 27808		F8 V	7.14	0.52		1.07	2.26	0.13:	0.91:	
HD 27836		G1 V	7.62	0.60		1.47	2.57	0.13	0.90	
HD 28527		A6 IV	4.78	0.17	0.17	0.59	1.02	0.11	0.34	
HD 29051		K5 III	7.10	1.40		6.83	6.22			
HD 29139	Alp Tau	K5 III	0.85	1.54	1.22	6.12	5.83	0.90	0.77	0.03

TABLE II - Continued

Star		Spectrum	V	B-V	V-R	IUE-V	MgII-V	FeI	FeII	E(B-V)
HD 29712	R Dor	M8 III	5.40	1.58		7.74	7.12	1.53	0.48	
HD 31398	Iot Aur	K3 II	2.69	1.53	1.06	4.31	4.43	0.97	0.49	0.17
HD 35155		S3/2+disk	6.77	1.80	1.32	5.33	4.18	0.34	0.08	
HD 36673	Alp Lep	F0 Ib	2.58	0.21	0.22	1.47	2.44	0.20	0.51	0.06
HD 37212		R8	7.84	2.27	1.71					
HD 37824		K1 III	6.70			3.43	2.93	0.54	0.10:	
HD 39801	Alp Ori	M2 Iab	0.50	1.81	1.59	6.30	5.34	0.64	0.53	0.21
HD 40136	Eta Lep	F0 IV	3.71	0.33	0.31	0.65	1.32	0.10	0.48	
HD 42995	Eta Gem	M3 III+?	3.28	1.60	1.50	5.42	5.31:	0.59	0.46	
HD 44478	Mu Gem	M3 III	2.88	1.64	1.54	6.36	5.54	0.69	0.72	0.04
HD 44537	Phi1 Aur	K5 Iab	4.91	1.97		5.74	5.21	0.50	0.65:	
HD 44984	BL Ori	C6,3	6.19	2.36	1.78					
HD 46687	UU Aur	N3	5.29	2.60	1.98	10.07	9.24	0.62	0.37:	
HD 48329	Eps Gem	G8 Ib	2.98	1.40	0.95	4.64	4.55	0.70	0.69	0.36
HD 49368		S3/2	7.65	1.78	1.35	6.66	5.87	0.89	0.74	
HD 51208	NP Pup	N0 I	6.42	2.30	1.87	7.56	8.00	1.18		
HD 52432	V614 Mon	R5	7.29	1.83	1.73	5.60	6.73	1.10		
HD 52877	Sig CMa	K7 Ib	3.46	1.74	1.32	5.86	4.97	0.67	0.62	0.12
HD 54605	Del CMa	F8 Ia	1.86	0.65	0.52	2.56	3.96	0.35	0.86	0.10
HD 56167		R0	8.00			3.80	5.74:	0.51:	1.21	
HD 58881		S5-/6	8.60	2.12	1.55	7.43:	6.92			
HD 59612		A5 Ib	4.85	0.23	0.26	1.49	2.10	0.14	0.39	0.13
HD 59643		R9p+disk?	7.99	2.24	1.52	4.53	3.78	0.08	-0.01	
HD 61110	Omi Gem	F3 III	4.90	0.40	0.38	1.07	1.98	0.17	0.62	
HD 61338	74 Gem	K5 III	5.05	1.56	1.20	6.14	5.68	0.76	0.74	
HD 61421	Alp CMi	F5 IV-V	0.38	0.42	0.41	0.07	1.97	0.17	0.71	
HD 62509	Bet Gem	K0 III	1.14	1.00	0.75	3.33	4.70	0.65	0.91	
HD 69267	Bet Cnc	K4 III	3.52	1.48	1.13	5.97	6.03	1.10	0.60	0.05
HD 72324	Nu2 Cnc	G9 III	6.36	1.02		3.46	4.99	0.61	0.89:	0.04
HD 73665	39 Cnc	K0 III	6.39	0.98		3.33	4.65	0.53	0.78:	
HD 75021	UZ Pyx	R8	7.10	1.85	1.55	6.24	6.97	1.06	1.16:	
HD 75156		M3 II-III	6.61	1.67		6.52	5.59	0.78		0.08
HD 76151		G3 V	5.99	0.67	0.55	1.07	2.27	0.11:	0.87	
HD 76294	Zet Hya	G8 III	3.11	1.00	0.72	3.30	4.78	0.54	1.00	0.05
HD 76644	Iot UMa	A7 IV	3.14	0.19	0.21	0.50	1.08	0.08	0.32	
HD 78647	Lam Vel	K5 Ib	2.21	1.66	1.24	5.88	5.76:	0.74	0.55	0.04
HD 78712		M7 IIIMS	5.95	1.67	2.00	7.34	6.46	1.09:		
HD 82328	The UMa	F6 IV	3.17	0.46	0.44	1.00	2.22	0.18	0.79	
HD 83950	W UMa	G0 V	7.87	0.64		1.36	2.12	0.20	0.73	
HD 84441	Eps Leo	G0 II	2.98	0.80	0.66	2.43	3.71	0.36	0.91	0.07
HD 85503	Mu Leo	K2 III	3.88	1.22	0.91	4.76	5.69	0.99	0.89	0.06
HD 86663	Tau Leo	M2 III	4.70	1.60	1.39	6.21	5.56	0.95	0.18:	
HD 87696	21 LMi	A7 V	4.48	0.18	0.18	0.43	0.92	0.06	0.29	
HD 89025	Zet Leo	F0 III	3.44	0.31	0.32	1.46	2.37	0.14	0.53	
HD 89758	Mu UMa	M0 III	3.05	1.59	1.27	5.90	5.50	0.74	0.73	
HD 90839	36 UMa	F8 V	4.83	0.52	0.48	1.12	2.44	0.20	0.88	
HD 91232	46 Leo	M2 III	5.46	1.68		6.52	5.34	0.80	1.21:	0.12
HD 92055	U Hya	N2	4.82	2.69	1.77	9.09	9.00:	1.13	0.71	

TABLE II - Continued

Star		Spectrum	V	B-V	V-R	IUE-V	MgII-V	FeI	FeII	E(B-V)
HD 92626		Rp III	7.09	1.35	0.85	4.10	5.82	0.55	1.30	
HD 93497	Mu Vel	G5 III	2.69	0.90	0.69	2.61	3.57	0.33	0.73	
HD 94705	56 Leo	M5 III	5.81	1.45		6.48	4.96	0.85	0.37	
HD 95608	60 Leo	A0 V	4.42	0.05	0.08	0.19	0.39	0.03	0.18	0.03
HD 97603	Del Leo	A4 V	2.56	0.12	0.16	0.52	0.98	0.11	0.21	
HD 97778	72 Leo	M3 III	4.63	1.66	1.49	5.80	4.73:	0.70	0.30:	
HD 99028	Kap Leo	F2 IV	3.94	0.41	0.40	1.06	1.97	0.18	0.60	0.04
HD 99946	AW UMa	F2 V	7.05	0.36		0.76	1.49	0.08	0.52	
HD 99967		K2 III+cF	6.35	1.27		3.97	4.59	0.48	0.71	0.10
HD 99998	87 Leo	K4 III	4.77	1.54	1.20	6.06	5.74	1.11	0.71	0.13
HD 100764		R0	8.80	1.05	0.74	2.29	3.23	0.30	1.00	
HD 102212	Nu Vir	M1 IIIab	4.03	1.51	1.23	6.48	5.68	0.92	0.64	
HD 102365		G5 V	4.90	0.66	0.54	1.56	3.08		1.04	
HD 102870	Bet Vir	F8 V	3.61	0.55	0.46	1.29	2.63	0.14	0.91	
HD 103287	Gam UMa	A0 V	2.44	0.00	0.05	0.11	0.29	0.02	0.07	
HD 104035		A0 Ia	5.61	0.18		0.72	1.00	0.08	0.28	0.16
HD 104350	AG Vir	A8 V	8.51	0.25		0.57	1.08	0.05	0.32	
HD 107957	S Cen	R5 III	7.57	1.85	1.61	6.31	7.05	1.23		
HD 107966		A3 V	5.18	0.08		0.35	0.66	0.04	0.09	
HD 108105	SS Vir	C6,3e	6.60	4.19		10.60:	9.22			
HD 109358	Bet CVn	G0 V	4.26	0.59	0.54	1.30	2.75	0.18	0.98	
HD 109379	Bet Crv	G5 III	2.65	0.89	0.61	2.69	4.16	0.41	0.99	
HD 111775		A0 II	6.33	0.03		-0.02	0.11	0.00	0.10	
HD 111812	31 Com	G0 III	4.94	0.67	0.52	1.72	2.54	0.24	0.69	
HD 112142	Phi Vir	M3 IIIa+F	4.72	1.57	1.56	4.24	4.59	0.26	0.77	
HD 113139	78 UMa	F2 V	4.93	0.36	0.34	0.73	1.52	0.14	0.54	
HD 114710	Bet Com	G0 V	4.26	0.57	0.48	1.32	2.68	0.11	0.94	
HD 114961	SW Vir	M7 III	7.35	1.67		7.28	5.54	0.85	0.54:	
HD 115043		G1 V	6.83	0.60		1.50	2.68	0.17	0.86	
HD 115617	61 Vir	G6 V	4.74	0.71	0.58	1.96	3.34	0.37	1.00	0.02
HD 116842	80 UMa	A5 V	4.01	0.16	0.18	0.48	1.00	0.10	0.27	0.02
HD 117555	FK Com	G2 IIIa	8.12	0.87	0.70	2.25	1.11	0.20	0.12	
HD 119149	82 Vir	M2 IIIa	5.01	1.63		6.03	5.44	0.87	0.29	0.04
HD 120323	2 Cen	M5 III	4.19	1.50	2.14	5.99	5.41:	0.81	0.65	
HD 121370	Eta Boo	G0 IV	2.68	0.58	0.45	1.57	3.03	0.13	0.93	
HD 121447		M BaII	7.81	1.76		5.68	6.21	0.93	0.49:	
HD 122250	The Aps	M7 III	5.50	1.55		7.75	6.59	1.10		
HD 122408	Rho Vir	A3 III	4.26	0.10	0.14	0.42	0.79	0.02	0.17	
HD 123657	BY Boo	M4 III	5.27	1.59	1.85	6.07	5.18	0.74		
HD 123934	ET Vir	M2 III	4.87	1.66		6.53	5.49	0.78		0.07
HD 124689	RR Cen	F2 V	7.40	0.36		0.90	1.46	0.20	0.64	
HD 126327		M8 III	7.50			8.35	7.14	1.71:		
HD 126660	The Boo	F7 V	4.05	0.50	0.43	1.07	2.14	0.18	0.78	
HD 132813	RR UMi	M5 III	4.60	1.59	1.83	6.02	4.66	0.55	0.49	
HD 133774	Nu Lib	K5 III	5.20	1.58	1.24	6.04	6.43	1.06	0.71	0.04
HD 137613		R3 (HDC)	7.48	1.17	0.84	3.14	5.13	0.59	1.14	
HD 137759	Iot Dra	K2 III	3.29	1.16	0.78	4.48	5.56	0.92	0.91	
HD 142143		M6.5S	7.00			7.29:	6.13			

TABLE II - Continued

Star	Spectrum	V	B-V	V-R	IUE-V	MgII-V	FeI	FeII	E(B-V)
HD 144205 X Her	M8 III	6.58			7.77	6.86:	1.01	0.56	
HD 146051 Del Oph	M0.5 III	2.74	1.58	1.29	6.45	5.79	0.91	0.44	
HD 146233 18 Sco	G1 V	5.49	0.65	0.53	1.62	2.97	0.28	0.97	
HD 147547 Gam Her	A9 III	3.75	0.27	0.27	1.05	1.93	0.15	0.49	
HD 148783 30 g Her	M6 III	5.04	1.50	2.50	7.38	5.55	1.00	0.39:	
HD 150680 Zet Her	G0 IV	2.81	0.65	0.54	1.69	3.08	0.18	1.00	
HD 153751 Eps UMi	G5 III	4.23	0.89	0.70	2.54	3.27	0.36	0.69	
HD 156014 Alp Her	cM5+gG5	3.48	1.44	2.11	4.11	3.91	0.15	0.18	
HD 156074	R1	7.60	1.14	0.72	3.35	4.64	0.55	0.83	
HD 157244 Bet Ara	K3 Ib	2.85	1.46		4.96	5.02	0.78	0.58	0.04
HD 157792 44 Oph	A9 V	4.17	0.28	0.26	0.79	1.39	0.17	0.52	
HD 157950	F3 V	4.54	0.39	0.36	0.74	1.62	0.15	0.57	
HD 159181 Bet Dra	G2 II	2.79	0.98	0.68	2.82	3.44	0.36	0.64	0.11
HD 159441 V535 Ara	A8 V	7.30	0.20		0.70	1.26	0.08	0.30	
HD 159561 Alp Oph	A5 III	2.08	0.15	0.14	0.55	0.98	0.07	0.25	
HD 160346	K3 V	6.53	0.96	0.78	3.19	3.94	0.78	0.80	
HD 160365	F6 III	6.12	0.56		1.45	2.44	0.16	0.78	0.10
HD 161471 Iot1 Sco	F2 Ia	3.03	0.51		2.75	4.06	0.48:	0.63:	0.33
HD 161797 Mu Her	G5 V	3.43	0.75	0.54	2.17	3.49	0.35	0.90	0.05
HD 163506 89 Her	F2 Ib	5.46	0.34	0.31	2.16	3.18	0.44	0.84	0.16
HD 163611 V566 Oph	F0 V	7.65	0.45		0.63	1.52	0.13	0.72	0.14
HD 163990 OP Her	M5 IIS	6.02	1.64		6.76	5.46:	0.76		
HD 166205 Del Oph	A1 Vn	4.36	0.02	0.03	0.16	0.38	0.02	0.07	
HD 168227	R5	8.66	1.90	1.70	5.53	7.10			
HD 168574 V4028 Sgr	M5 III	6.25	1.84		6.76:	6.00	0.77		0.20
HD 172816 V3879 Sgr	M4 III	6.35	1.75		6.69:	5.91	0.89		0.15
HD 173667 110 Her	F6 V	4.19	0.46	0.39	1.08	2.22	0.19	0.72	
HD 175865 R Lyr	M5 III	4.04	1.59	2.03	6.84	5.48	0.67	0.36	
HD 176124	M4 III	6.40			6.21	5.77	0.49		
HD 182040	R0 (HDC)	7.00	1.09	0.81	2.71	4.66	0.50	0.85	
HD 185144 Sig Dra	K0 V	4.68	0.79	0.65	2.23	3.47	0.43	1.00	
HD 186408 16 Cyg	G2 V	5.96	0.64	0.45	1.69	3.13	0.25	0.90	
HD 186427	G2.5 V	6.20	0.66	0.44	1.74	3.19	0.26	1.02	
HD 186882 Del Cyg	B9.5 III+F	2.87	-0.03		-0.12	0.01	-0.02	0.02	
HD 188512 Bet Aql	G8 IV	3.71	0.86	0.66	2.59	4.03	0.46	0.93	0.04
HD 188665 23 Cyg	B5 V	5.14	-0.13	-0.06	-1.28	-1.30	-0.08	-0.05	
HD 189124 NU Pav	M6 III	5.10	1.53		6.22	5.60:	0.81	0.43	
HD 192876 Alp1 Cap	G3 Ib	4.24	1.07	0.79	3.19	4.03	0.38	0.60	0.15
HD 193793 V1687 Cyg	WC6/7+05/6	6.85	0.40		-0.06	-0.02	0.00	0.06	0.72
HD 196777 Nu Cap	M2 III	5.10	1.66		6.47	5.54:	0.69	0.13:	
HD 197345 Alp Cyg	A2 Ia	1.25	0.09	0.12	0.35	0.69	0.11	0.32	0.04
HD 197433 VW Cep	K0 V	7.42	0.84		2.38	2.56	0.39	0.51	
HD 198149 Eta Cep	K0 IV	3.43	0.92	0.67	2.87	4.25	0.51	1.02	
HD 198164	SC2/7.5	8.70			6.51:	7.21:			
HD 199178	G5 III-IV	7.30	0.92?		2.05	2.12	0.27	0.39	
HD 199629 Nu Cyg	A0 V	3.94	0.02	0.06	0.13	0.35	-0.02	0.08	
HD 201091 61 Cyg A	K5 V	5.21	1.18	1.02	4.37	4.30	0.91	0.43	0.03
HD 201092 61 Cyg B	K7 V	6.03	1.37	1.19	4.91	4.62	0.89	0.26	0.04

TABLE II - Concluded

Star		Spectrum	V	B-V	V-R	IUE-V	MgII-V	FeI	FeII	E(B-V)
HD 201626		R5-CH	8.16	1.10	0.82	2.42	3.74	-0.01	0.70	
HD 202874	T Ind	C7,2	6.00	2.33	1.55	8.57	8.31	1.01	0.90:	
HD 205372	GK Cep	A2 V	7.28	0.08		0.26	0.48	-0.02	0.16	
HD 205730	W Cyg	M5 III	5.53	1.58		6.73	5.83	0.28		
HD 206778	Eps Peg	K2 Ib	2.39	1.53	1.05	5.33	4.79	0.79	0.61	0.30
HD 207005	AG Cap	M3 III	6.00	1.66		6.49	5.53	0.88		0.06
HD 207076	EP Aqr	M7 III	6.57	1.50		7.86	7.12	1.32		
HD 209750	Alp Aqr	G2 Ib	2.96	0.98	0.66	3.00	4.03	0.41	0.86	0.10
HD 212087	Pi1 Gru	S5	6.00	2.01		5.49	5.80	0.31	0.79:	
HD 213080	Del2 Gru	M4 III	4.11	1.57	1.73	6.49	5.86	0.76	0.63:	
HD 214952	Bet Gru	M5 III	2.10	1.60	1.91	5.94	4.92	0.61	0.56	
HD 216386	Lam Aqr	M2 III+F?	3.74	1.64	1.40	5.04	5.19	0.37	0.66	
HD 216598	SW Lac	G5 V	8.69	0.75		1.96	2.37	0.31	0.62	
HD 216672	HR Peg	S4+/1+	6.30	1.80		6.74	5.77	0.71	0.44	
HD 216956	Alp PsA	A3 V	1.16	0.09	0.11	0.30	0.64	0.05	0.22	
HD 217906	Bet Peg	M2.5II-III	2.42	1.67	1.51	6.49	5.61	0.76	0.14:	0.07
HD 219134		K3 V	5.56	1.01		3.49	4.27	0.83	0.78	0.06
HD 219215	Phi Aqr	M1 III+B?	4.22	1.56	1.28	4.40	4.37	-0.09	-0.08	
HD 223075	TX Psc	N0	5.04	2.60	1.92	8.89	9.02	1.47		
HD 223392		R3	8.49	1.32		4.21	5.49	0.80	1.10	
HD 224062	XZ Psc	M5 III	5.61	1.59	1.90	5.96	4.72	0.79		
HD 234677	BY Dra	K7 Ve	8.20	1.24	1.13	3.74	2.59			
ER Del		S5,5	10.00			4.29:	4.22:			
UZ Lib		K0 III	9.20			3.14	2.40			
SU Mon		S5/7-S6/7	8.25	2.60	2.15	7.18:	7.57:			
SZ Sgr		N + ?	8.74	2.31	2.05	4.75	6.32v	0.22		
AL Vel		K0 III+B8	8.60			1.50:	1.81:		-0.66	
BD+2 3336		R2	9.40	1.91	1.28	5.29:	4.61:			
BD+17 3325		R0	8.72	1.18	0.83	3.91	5.53	0.76	0.30	
Sun (=Planets)		G2 Vp		0.63	0.52	(1.70)	2.96	0.18	0.97	

$$E(IUE-V)/E(B-V) = 2.63 - 0.06(IUE-V)$$
$$E(Mg\ II-V)/E(B-V) = 2.90$$
$$E(Fe\ I)/E(B-V) = 0.28$$
$$E(Fe\ II)/E(B-V) = 0.47$$

A more serious question of systematic error concerns interstellar and circumstellar line extinction in $Mg\ II\ \lambda\ 2800$. It is now widely accepted that almost all stars suffer such effects (Böhm–Vitense 1981), but as yet it is not possible to determine just how great a correction should be applied. We have applied none, but for some very cool giants this correction could make the $(Mg\ II-V)$ color bluer by about 0.7 mag or more.

Ground–based photometry has been taken from a variety of sources, but primarily from the *Bright Star Catalogue* and Iriarte *et al.* (1965). Photometry for the carbon stars comes from lists of Mendoza and Johnson (1965) and Walker (1979). A few stars have colors determined by me at Kitt Peak National Observatory. All $(V-R)$ colors are on the Johnson system, those determined originally on the Cape system having been transformed with the equations of Bessell (1979). Spectral types are likewise from a variety of sources, again primarily from the *Bright Star Catalogue*. The color excesses listed in the 11th column were found by comparing observed $(B-V)$ and $(V-R)$ colors with the standard relations of Johnson (1966), or were taken from Wu *et al.* (1983).

III. RELATIONS BETWEEN THE COLORS

We can study the data given in Table II best by plotting them in a group of color–color curves chosen to illustrate various properties of the colors. We will consider four of these in succession.

a) (B-V) versus (IUE-V)

Figure 2 gives this color–color curve. Both of these colors are sensitive to various properties of the stars. To reduce the clutter in the diagram, it is broken up into two panels. The top one gives the relation for A–M stars that are apparently not exceptional. These stars form a relatively sharp sequence which is roughly parallel to the reddening trajectory (line segment at right). For the A–G stars the sequence is really quite tight. There is a slight tendency for the giants to be redder in $(IUE-V)$ than the dwarfs, and this is especially true for the A–F supergiants. There is, however, a significant amount of scatter among the cooler giants which reflects the fact that $(B-V)$ is not a useful measure of effective temperature for M giants.

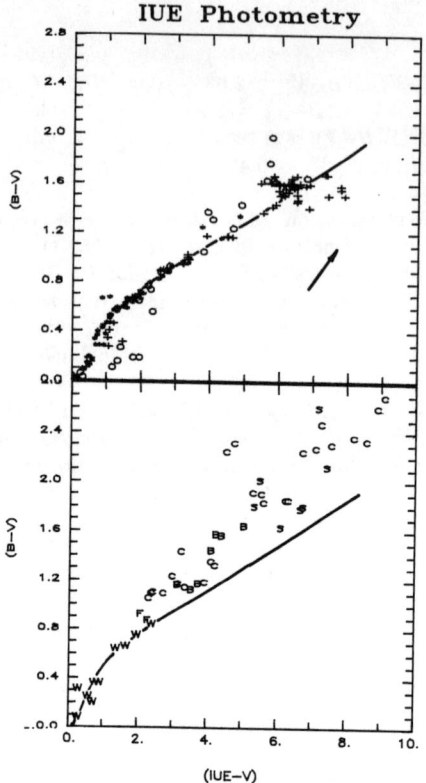

Figure 2. Color−color curve for $(B-V)$ vs. $(IUE-V)$. At the top are plotted normal stars, dwarfs given as stars, giants as plusses, and supergiants as open circles. In the lower graph are shown other, more interesting stars. Here we use symbols C for carbon stars, S for S stars, F for FK Com−type stars, W for W UMa−type binaries, and B for known or suspected binaries of various kinds. The line segment at the right in the upper panel indicates the reddening trajectory.

The peculiar stars in the lower panel are seen to deviate from the sequence in various ways and for a variety of reasons. Of special interest for studying the effect of chromospheric heating on the continuum are the W UMa binaries and FK Com giants. We see that both groups lie quite close to the locus of main−sequence stars. This is especially so for the warmer W UMa binaries, which have atmospheres so hot that they should not be changed much by such heating; however, it is also the case for the cooler objects—YY Eri, W UMa, SW Lac, and VW Cep—which lie within the photometric errors of the sequence. Such stars have significant opacity of neutral metals in their atmospheres that should be sensitive to chromospheric effects if they are indeed important. A similar conclusion applies for the FK Com stars, which may be slightly bluer in $(IUE-V)$ than the sequence, but not so by more than 0.2 mag. Thus, unless chromospheric effects simply move stars roughly parallel to the sequence, in this diagram they are not pronounced for rapidly rotating binary components and single stars.

Binary systems consisting of a red giant and faint blue companion tend to stand out from the standard sequence if the blue companion is bright enough. This is to be expected since IUE spectra have become the primary means of detecting faint blue companions (*e.g.*, Arellano Ferro and Madore 1986). We have identified seven stars in Table II with ultraviolet spectra apparently earlier than their spectral types which are likely binary systems. They are HD 35155, HD 59643, HD 99967, φ Vir, λ Aqr, and φ Aqr. As we shall see, all of them have colors inconsistent with other cool stars. In addition, in the ultraviolet the M3 giant η Gem looks more like a K star and may also be a binary.

This diagram sets apart the stars with peculiar abundances relatively well, especially the carbon stars and S stars. However, this is simply the result of the reddening of $(B-V)$ by the extra molecular features that fall in the B band. Of interest to note are several S stars that fall within the clump of normal M giants and two carbon stars much too blue in $(IUE-V)$ for their $(B-V)$ colors. The former group consists of HD 22649 (BD Cam = HR 1105), HD 49368, and HD 216672 (HR Peg = HR 8714). The latter comprises HD 59643 and SZ Sgr, both cool carbon stars with a hot companion of some sort.

b) $(MgII-V)$ versus $(IUE-V)$

Figure 3 gives our most interesting relationship, $(MgII-V)$ vs. $(IUE-V)$. This diagram is sensitive to some extent to metallicity, binarity, and other such complications, but is primarily a measurement of chromospheric line emission in terms of luminosity and effective temperature. Again for clarity the figure is divided into two panels. At the top we see that normal A–G stars form a sequence roughly parallel to the reddening trajectory. This is to be expected since most of the flux in the λ 2800 band is photospheric for such stars. Among the G stars, giants and supergiants again deviate from the sequence for dwarfs in the sense expected for stars with elevated chromospheric emission. Stars later than middle G show a spread of $(MgII-V)$ of about 1.0 mag which results from differences in the level of chromospheric activity. We can define a sequence of low Mg II emission in this diagram, given by the lines in the two panels, by excluding the stars such as the binaries with peculiar colors.

This diagram is especially telling for the peculiar stars. The FK Com giants and cooler W UMa binaries are all displaced downward in reflection of their enhanced chromospheric emission. Other binary systems show enhanced chromospheric emission as well. HD 352 = 5 Cet is seen at roughly $(IUE-V)$ = 3.1 mag. and $(Mg II-V)$ = 2.7 mag., giving it a very large deviation from the single stars. The chromospheric line contributes about 60% of the flux in the $Mg II$ band, so the displacement is caused mainly by elevated chromospheric emission ζ And, which is a long−period binary close enough to be an ellipsoidal variable of 0.09 mag amplitude, is in an intermediate position at $(IUE-V) = 3.5$ mag. and $(Mg\ II-V) = 3.6$ mag. The binary carbon and S stars HD 59643 and HD 35155 are both found in positions of very blue $(Mg\ II-V)$, but the binary carbon star SZ Sgr is not.

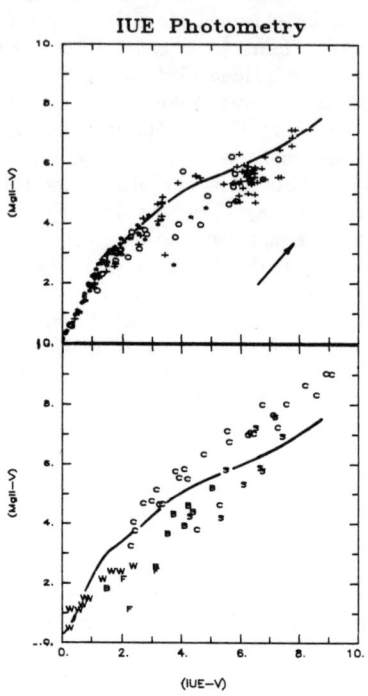

Figure 3. Color−color curve for $(MgII-V)$ vs. $(IUE-V)$. The division into two panels and the symbols are the same as in Figure 2.

Single carbon stars, in contrast, form roughly a separate sequence at redder $(Mg\ II-V)$ than the reddest oxygen−rich giants. The S stars appear to form two groups in this diagram, as in Figure 2. The three S stars with continuum colors like M giants' likewise have strong $Mg\ II$ emission like M giants.

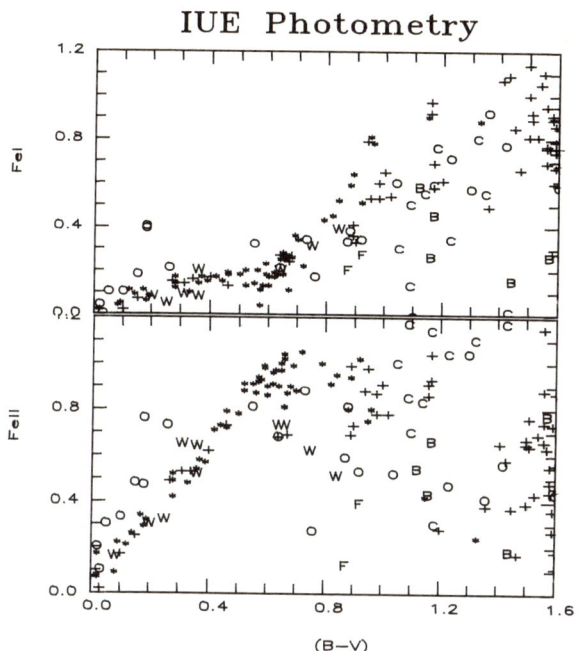

Figure 4. The variation of indices measuring the strength of Fe I and Fe II with $(B-V)$. Symbols are the same as in Figures 2 and 3. In the upper panel, we see that the Fe I feature is not detected for stars earlier than G. For the later types, most stars have Fe I strength determined by $(B-V)$, with those deviating from the relation being binary stars with peculiar spectra. Fe II strength, at bottom, is correlated with chromospheric activity, as if chromospheric emission is filling in the cores of absorption lines. The CH stars, which are quite metal poor, tend to have strong Fe II breaks for their $(B-V)$ colors and to be found near the top of the diagram.

c) Fe I and Fe II versus $(B-V)$

In Figure 4 we have plotted against $(B-V)$ the indices that measure strength of *Fe I* and *Fe II* features. *Fe I* is not especially interesting for the A−F stars; however, among the cooler stars, the giants and supergiants have a weaker *Fe I* feature as would be expected from their lower gravity and higher ionization. Chromospherically active stars show a slight tendency toward weaker *Fe I*. For the W UMa binaries, it is very slight, but for the FK Com giants the effect may be 0.2 mag in

Fe I. Other stars standing out in this diagram are the cool giants with blue companions which have the flat *UV* spectra of the hot companions. This group may contribute all the stars to the lower right in the diagram, and it certainly contains most of them.

The Fe II index gives a much better separation of the various classes. It is defined most reliably for stars with continuum extending to 2500 A, which limits its usefulness to $(B-V) < 1.1$ or so. Within this range, however, we see a strong separation of the A and F supergiants, which results from their bluer colors, and of the stars with enhanced chromospheric activity. All of the cooler W UMa binaries and the FK Com giants have weaker *Fe II* features than comparable stars with low activity. This result indicates that in this one instance we are indeed seeing chromospheric line emission in these stars photometrically.

IV. FURTHER RESULTS

a) Regularities and Colors

We may now ask ourselves just how much variation there is at a given spectral type in these ultraviolet colors and how well they characterize various properties of stellar energy distributions. Table III gives characteristic $(IUE-V)$ colors for K and M giants. Also listed are rough measures of the scatter about the mean, which is rather large in some cases. One thing immediately clear is that, for K giants, $(IUE-V)$ is well correlated with spectral type, hence with effective temperature. For the earlier M giants, on the other hand, $(IUE-V)$ is roughly constant before again becoming progressively redder for the very coolest M stars. This is the same effect seen in $(B-V)$, which is in the range 1.5–1.6 mag. throughout the range K5–M6. It suggests that in these stars the optical and ultraviolet continuum is always formed at a specific temperature, while the bulk of the radiation emitted by the star (in the infrared) comes from progressively cooler layers at the later spectral types.

The near constancy of $(IUE-V)$ for stars in the range K5–M5 gives us a chance to study the intrinsic variation of $(IUE-V)$. M giants that have especially blue colors tend to be binaries with approximately F-type companions making them bluer. These were not considered in forming Table III. However, there still remain several stars with spectra like those of M stars (not showing the characteristic absorption edges of F dwarfs and hot accreting components) with $(IUE-V)$ colors several tenths of a magnitude above the general trend. Examples are 57 Psc, η Gem, μ UMa, and 72 Leo. These objects indicate that there can be intrinsic variations at a given spectral type of up to a few tenths of a magnitude. Some of the variation is no doubt simply the result of our choice of *V* magnitude to use in forming the colors of cool variables. For the most part, we have used **mean** *V* magnitudes. Thus, for a star whose ultraviolet spectrum has the right shape for its spectral type, $(IUE-V)$ is probably defined to approximately 0.2–0.3 mag. At spectral types later than M4, $(IUE-V)$ shows much greater scatter, much of which must be due to uncertainties in the classification of the stars and to our choice of V magnitude. Indeed, most of this scatter would disappear with judicious changes in the spectral type of only 0.5–1 subclass.

TABLE III

$(IUE-V)$ and $(Mg\ II-V)$ Colors of Cool Giants

Spect Type	$(IUE - V)$	$(Mg\ II-V)$	Spect Type	$(IUE-V)$	$(MgII-V)$
K0	3.33 + 0.10	4.68	M2	6.29 + 0.16	5.44
K1	3.43 + 0.10	3.60	M3	6.38 + 0.10	5.26
K2	4.55 + 0.10	5.51	M4	6.24 + 0.31	5.50
K3	no stars in sample		M5	6.26 + 0.40	5.20
K4	5.81 + 0.10	5.62	M6	6.78 + 0.60	5.28
K5	6.06 + 0.10	5.80	M7	7.53 + 0.30	6.70
M0	6.33 + 0.10	5.75	M8	7.95 + 0.34	7.05
M1	6.40	5.60			

The $(Mg\ II-V)$ colors are somewhat less highly correlated with spectral type. There does appear to be a strong correlation between strength of $Mg\ II$ emission and blueness of the star in $(IUE-V)$, but this results from including stars we have identified as binaries on the basis of the shape of the ultraviolet spectrum. Among the M giants for which $(Mg\ II-V)$ is roughly constant, the individual stars scatter about the mean relation given in Table III by about 0.3–0.4 mag.

b) Chromospheric Losses in Mg II UV1

One of the intriguing mysteries of cool giants is the way chromospheric losses rapidly diminish with advancing spectral type. This has been studied previously in several ways, mainly with high–dispersion data (Basri and Linsky 1979; Linsky 1980; Steiman–Cameron et al. 1985). We now use the sample of K and M giants here for which photometric color temperatures are available from Wing (1978; see also Ridgway et al. 1980) to define this relationship for a large number of giants. The sample includes 44 stars. The importance of it lies in our ability to use color temperature T_c to find an effective temperature and the Barnes–

Evans relation (Barnes and Evans 1976, Eaton and Poe 1984) to derive the flux at the surface of the star in the V band. Thus we derive flux at the star in chromospheric $Mg\ II$ and, dividing by T^4_{eff}, find the fraction of the star's luminosity emitted in $Mg\ II\ \lambda\ 2800$. Logically, we have:

$$T_c \rightarrow T_{eff} \rightarrow \text{Flux(Bol)}$$
$$(V-R) \rightarrow \text{Flux(Vis)}$$
$$\text{Flux(Vis)} + (Mg\ II - V) \rightarrow \text{Flux}(Mg\ II)$$

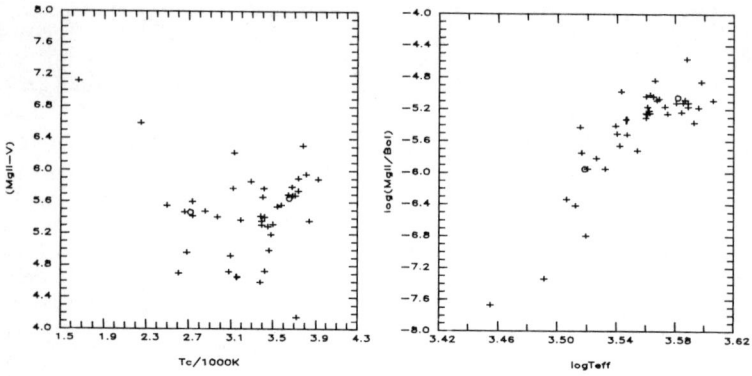

Figure 5. *Relative strength of chromospheric Mg II emission vs. temperature for K and M giants. On the left, the (Mg II–V) color is given as a function of color temperature. On the right we have transformed this to the flux of Mg II relative to the bolometric photospheric flux as a function of effective temperature. Here we see the rapid falloff of chromospheric heating with advancing spectral type. It is illustrated much more graphically here than in Steiman–Cameron et al. (1985) because of our larger sample of cool M giants.*

Figure 5 presents the results of this procedure. On the left we see how $(Mg\ II - V)$ color varies with color temperature. As expected from the discussion in Section IVa, it stays roughly constant for the higher temperatures, actually becoming somewhat bluer with decreasing T_c, before becoming quite red for the reddest stars. Although there is little temperature dependence, the bolometric correction—which, of course, is implicit in the assumptions we are making—increases markedly with decreasing temperature. This leads to the sharp drop in fractional luminosity of λ 2800 seen in the cooler stars in the right side of Figure 5. The fractional luminosity of Mg II drops by three orders of magnitude between M0 and M8. The surface flux drops as well, again by nearly three orders of magnitude over this range. What is not yet clear is the changes in the chromosphere that effect this drop. Total chromospheric heating becomes reduced by this great amount; however, the way that heating is transmitted to chromospheric gas has not yet become clear. There are several ways to reduce the emission of a chromosphere. One is to make it cooler, which is exponential and very effective. Another is to reduce the electron density so there are fewer particles floating about to excite each ion. A third is to reduce the extent of the chromosphere, to make it either less extended or to cover a smaller fraction of the surface of the star.

In order to choose among these possibilities and their various combinations, we must know fairly accurately the physical properties of the gas in the chromosphere. This can be done only by using high-dispersion spectra to study individual emission lines in detail. For cool giants there have been two approaches taken to such an analysis. In one (Ayres and Linsky 1975; Kelch et al. 1978) detailed models for chromospheric conditions are used to calculate the strength and profiles of individual strong lines, such as the Mg II doublet. In the other, diagnostic lines are used in conjunction with simple models to deduce conditions in the emitting gas. (e.g., Carpenter, Brown, and Stencel 1985; Judge 1986; Eaton and Johnson 1986). These analyses lead to estimates of such quantities as electron density, electron temperature, and mass of emitting gas. At the present time it appears that electron density decreases by an order of magnitude, or less, from K2 to M5. Electron temperature may decrease somewhat, as well, but this would require non-homogeneous chromospheres with different mixes of hotter and cooler material. The extent of the chromospheres in cool giants is likewise still controversial. Given a value of electron density and temperature, one can easily calculate the extent of a chromospheric shell sufficiently massive to emit all the carbon λ 2325 flux observed. Unfortunately, the size (thickness) derived depends critically on electron density used in the calculation. Indeed, the thickness could be anywhere between 1% and 100% of the radius of the star for different estimates of the electron density. Observations of atmospheric eclipses in ζ Aur–type binary stars (e.g., Stencel et al. 1984) indicate chromospheres can be highly extended. Thus, it is likely that the whole surface of a red giant is not covered with chromospheric gas, but whether the changes in the degree of coverage are required to explain the variation of Mg II strength with spectral type cannot be determined until more is known about the structure of red giants' chromospheres and how it changes with spectral type.

V. SUMMARY

We have defined a photometric system in the ultraviolet for use with cool stars and used it to study the physical properties of a range of such objects observed with IUE. Principal results are as follow:

Elevated chromospheric activity does not seem to change a star's ultraviolet continuum by very much, if any, in the wavelength range 2600–3200 A. FK Com giants and W UMa binaries both had essentially the same colors as other stars of the same spectral type. The *Fe II* UV1 multiplet, on the other hand, did seem to be filling in photospheric absorption lines to some extent in active stars. Changes in color caused by increases in photospheric ionization by chromospheric irradiation were not detected to within the errors of the photometry. One exception may be BY Dra, which was represented by a single spectrum and which seems to have $(IUE-V)$ color much bluer than expected.

Binary stars were found to have elevated *Mg II* emission for their spectral types, as might be expected from their rapid rotation. This was the case for binaries with either hot or cool components, and it seems not to result simply from heating of the chromosphere radiatively.

The $(IUE-V)$ color of a cool giant was found to be amazingly well correlated with spectral type when binaries containing hot secondaries were ignored. There were few stars in our sample with peculiar colors for which no cause of the peculiarity was evident.

VI. ACKNOWLEDGEMENTS

This research was carried out only with considerable help from NASA over the years. I am grateful to the staff of the IUE observatory for help with observing, especially to the encouragement of the directors, Drs. A. Boggess and Y. Kondo. Many of the spectra used were obtained from the National Space Science Data Center and from the IUE Condensed Data Archive. The considerable help of these two organizations is gratefully acknowledged. Over the years this research and its component parts have been supported by a variety of grants from NASA, including NAG 5–176, NAG 5–431, NAG 5–182, and NAG 5–599.

REFERENCES

Arellano Ferro, A. and Madore, B. F. 1986, *Astrophys. J.* 302, 767.
Ayres, T. F. and Linsky, J. L. 1975, *Astrophys. J.* 200, 660.
Barnes, T. G. and Evans, D. S. 1976, *Mon. Not. Roy. Astron. Soc.* 174, 489.
Basri, G. S. and Linsky, J. L. 1979, *Astrophys. J.* 234, 1023.
Bessell, M. S. 1979, *Publ. Astron. Soc. Pacific* 91, 589.
Böhm-Vitense, E. 1981, *Astrophys. J.* 244, 504.
Bopp, B. W. and Stencel, R. E. 1981, *Astrophys. J. Lett.*, 247, L131.
Carpenter, K. G., Brown, A., and Stencel, R. E. 1985, *Astrophys. J.* 289, 676.
Eaton, J. A. 1983, *Astrophys. J.* 268, 800.
Eaton, J. A. and Johnson, H. R. 1986, *Astrophys J.* submitted.
Eaton, J. A., Johnson, H. R., Baumert, J. H., and O'Brien, G. T. 1985, *Astrophys. J.* 290, 276.
Eaton, J. A. and Poe, C. H. 1984, *Acta Astron.*, 34, 97.
Hartmann, L. W., Dupree, A. K., and Raymond, J. C. 1982, *Astrophys. J.* 252, 214.
Iriarte, B., Johnson, H. L., Mitchell, R. I., Wisniewski, W. K. 1965, *Sky and Tel.* 30, 1.
Johnson, H. L. 1966, *Ann. Rev. Astron. Astrophys.* 4, 193.
Judge, P. G. 1986, *Mon. Not. Roy. Astron. Soc.* in press.
Kelch, W. L., Linsky, J. L., Basri, G. S., Chiu, H.-Y., Chang, S.-H., Maran, S. P., and Furenlid, I. 1978, *Astrophys. J.* 220, 962.
Linsky, J. L. 1980, *Ann. Rev. Astron. Astrophys.* 18, 439.
Linsky, J. L. and Ayres, T. R. 1978, *Astrophys. J.* 220, 619.
Mendoza, E. E. and Johnson, H. L. 1965, *Astrophys. J.* 141, 161.
Ridgway, S. T., Joyce, R. R., White, N. M., and Wing, R. F. 1980, *Astrophys. J.* 235, 126.
Savage, B. D. and Mathis, J. S. 1979, *Ann. Rev. Astron. Astrophys.* 17, 73.
Steiman-Cameron, T. Y., Johnson, H. R., and Honeycutt, R. K. 1985, *Astrophys. J. Lett.*, 291, L51.
Stencel, R. E., Hopkins, J. L., Hagen, W., Fried, R., Schmidtke, P.C., Kondo, Y., and Chapman, R. D. 1984, *Astrophys. J.* 281, 751.
Vilhu, O. and Rucinski, S. M. 1983, *Astron. Astrophys.*, 127, 5.
Walker, A. R. 1979, *South African Astron. Obs. Circ.* 1, No. 4, p. 112.
Wing, R. F. 1978, "Spectral Classifications and Color Temperatures for 280 Bright Stars in the Range K4-M8", private report of the author.
Wu, C.-C., Faber, S. M., Gallagher, J. S., Peck, M., and Tinsley, B. M. 1980, *Astrophys. J.* 237, 290.
Wu, C.-C., et al. 1983, *IUE-NASA Newsletter*, No. 22.

NAME INDEX

Adelman, S. J., 2, 118, 141, *157*

Baliunas, S. L., 2, *65*, 85, 87, *97*, 389
Barden, S. C., 229
Belserene, E. P., 91
Boggess, A., 430
Boyd, L. J., 5, *19*, *27*, *35*, *65*, 85, 90, *97*, 118
Boyd, P., 389
Boyle, R., 210
Brocious, D., 88, 90
Bruton, J., 229
Burke, E. W., Jr., 229

Chen, K.-Y., 342, *379*
Crawford, D. L., 1, 2, 5, 7, *19*, 39, 141, *145*, 149, 342, *345*
Criswell, S., 114

Doddington, H., 324
Donahue, R. A., *97*, 389
Dukes, R. J., Jr., 2, 3, 5, *117*, 118, 120, *125*
DuPuy, D. L., 91

Eaton, J. A., 342, *411*
Eichhorn, G., 4, *73*, *307*, *317*
Einstein, A., 394
Ely, D. W., *73*, *307*
Etzel, P. B., 352

Fernie, J. D., 91
Foote, J. L., 228, *233*
Francisco, M., 131
Frazer, J., 114
Fried, R. E., 227, *229*, 242
Fullerton, A., 245

Genet, D. R., *27*
Genet, R. M., 1, *19*, *27*, *35*, *65*, 85, 90, *97*, 163, *227*, 239, 245
Giovane, F. J., *73*, *307*, *317*
Grayzeck, E. J., Jr., 2, 118, 120
Griffin, R. F., 229
Groot, M. de, 90
Guinan, E. F., *97*, 118, 119, 342, *383*

Hall, D. S., 2, 5, 88, 90, 229

Harmanec, P., 239
Hayes, D. S., 1, *19*, *65*, 90, 118, 119, *139*, 141, 149, 159, 163, *185*, *193*, *341*
Herbst, W., 91, 229
Hillier, J., 114
Hopkins, J. L., 228, *295*
Howell, S. B., 229

Inman, C., 265

Johnson, A., 131
Johnson, H. R., 413

Karshner, G. B., *211*
Kennedy, H. D., 229
Kissell, K. E., 90
Kondo, Y., 430
Kubinec, W. R., 118, 131

Latham, D. L., 114, 199
Lee, J., 90
Limber, D. N., 239
Lines, H., 90, 396
Lines, R., 90, 396
Loeser, J. G., *97*, 389

Maloney, F. P., *383*
Manly, P. L., 228, *279*
Markworth, N. L., 5, *41*, 228, 247, *267*, 342, 343, *355*, *361*, 376
McCook, G. P., 2, 118, 119
McCrosky, R. E., 114
McDavid, D., 90, *237*
McKisson, J., *73*
Melsheimer, F. M., 239
Michaels, C., 376
Michaels, E. J., 255, 342, *369*
Misch, A., 114
Moffat, A. F. J., 91
Montes, B., 255
Mullikin, J., *41*

Nations, H. L., 90
Newton, I., 397
Nicastro, A., 210

Olson, E. C., 343, *351*, 352

Oswalt, T. D., 257

Pasachoff, J. M., 91
Pearce, E. C., 5, *51*
Percy, J. R., 91
Philip, A. G. D., 141, *165*, 342
Piercey, R. B., *73*

Rafert, J. B., 228, *247*, *257*, *267*, 342, 343, *355*, *361*, 362
Reber, G., 223
Richardson, T. R., 118
Rilum, J., *307*, 324

Scaligen, J., 380
Schmidtke, P. C., 141, *193*
Seeds, M. A., 90, *201*, *203*
Shapiro, I. I., 114
Smith, D. P., 2, 118
Smith, T., 131
Soderberg, B., *307*, 324
Stebbins, R. A., *217*

Trueblood, M., 228, *325*, 340

Upgren, A. R., 2, *133*, 342

Weisenberger, A., *307*
White, N. M., 141, *183*
White, R. E., 90
Wood, F. B., 342, *379*
Woodard, L., 114

Ziegler, K. W., 342, *403*

SUBJECT INDEX

AAVSO--American Association of Variable Star Observers, 381
acquisition, automatic, 35, 36, 41, 47, 51, 58
amateur scientists, 217
 and professionals, 221
 initial interest, 218
 types of, 219
APT--Automatic Photoelectric Telescope, 1, 2, 85
ASGT--Automatic Spectrographic Telescope, 158, 201, 203, 204
ASPT--Automatic Spectrophotometric Telescope, 141, 158
AT--Automatic Telescope, 28
atmosphere, extinction, 193

Be Stars, 108, 119
binaries, eclipsing, 383
 apsidal motion, 385
 circumbinary clouds, 391
 distant companion, 391
 orbit circularization, 390
 orbital and equatorial planes, 393
 times of minima, 379

centering, automatic, 35, 37
chromospheres, stellar, 66, 411
computers, telescope control, 11, 263, 267, 271
Consortium, Four-College APT, 117
 logistics, 122
Consortium, Pennsylvania Astronomical Research, 201, 203

DASS--Digitized Astronomical Supernova Search, 51
data storage, ASGT, 214
 image, 211
 methods, 211, 212
detectors
 CCDs, 139, 158, 166, 184, 188, 202, 211, 227, 282, 341
 CIDs, 282
 image-dissectors, 139
 intensified reticons, 139
 photomultipliers, 69, 71, 88, 158, 183, 186, 341, 359, 370

reticons, 139, 165, 184, 185, 188
vidicons, 282

extinction, atmospheric, 193

figures
 (B–V) vs. (IUE–V), 422
 (B–V) vs. (U–B), 151
 (Mg II–V) vs. (IUE–V), 424
 (u–b) vs. (b–y), 152
 89 Her: light curve; power spectrum, 105
 apsidal motion, 385
 APT computers, 92
 APT control system, 94
 APT Service facilities, 86
 APT Service Workshop: 1986 participants, 90
 APT Service: three APTs, 89
 APT weather sensors, 93
 bandwidth effects on Johnson U, 153
 BHB stars: energy distributions, 178
 Braeside Obs: observer's position, 230
 Braeside Obs: photometer and CCD camera, 231
 C–M diagram for 47 Tuc, 168
 C–M diagrams for three clusters, 168
 c1 and m1 vs. (b–y), 151
 chromospheric Mg II emission vs. temperature, K and M giants, 428
 commercial relay lenses, 285
 cool stars: IUE spectra, 412
 DASS luminosity fraction, 59, 62
 δ CrB: H and K line variation, 112, 113
 δ CrB: light curve, 112
 δ CrB: power spectrum, 112
 δ Scuti stars and all stars: number vs. apparent mag., 130
 δ Scuti stars: number vs. amplitude, 128
 δ Scuti stars: number vs. apparent mag., 127
 DI Her: light curve, 388
 DI Her: radial velocity curve, 388
 DI Her: relative orbit, 387
 early-type stars: energy distributions, 147

eyepiece as relay lens, 286
Fairborn 0.75−m telescope, 28
  base and RA drive, 31
  horseshoe/Dec drive, 32
  mirror cell/Dec. ring, 32
  superstructure, mirror cell, 32
Fairborn 10 APT, 87
Fe I and Fe II vs. (B−V), 425
FHB and BHB stars:
  spectrophotometry, 179
FHB stars: energy distributions, 178
FHB stars: IUE spectra, 175
FHB stars: Stromgren photometry, 173
FHB, BHB stars: log L vs. log T, 179
FIT 25−inch telescope, 258
FIT 25−inch, mount, 261
galaxy image profile, 55
Gaussian star image profile, 44
H−δ vs. (b−y), 153
HB stars: isochrones, 169
HB stars: spectra, 170
histogram, star field, 43
HPO low voltage power supply, 303
  MVL 100 pulse conditioner, 297
  photon counting block diagram, 296
  pulse conditioner layout, 300
  pulse conditioner PWB bottom, 301
  pulse conditioner PWB template, 302
  pulse conditioner PWB top, 300
  pulse conditioning circuit, 299
  pulse counter, 10 MHz, 304
image intensifier tubes, 281
Johnson U: effective wavelength, 154
later−type stars: energy distributions, 148
Limber Observatory, 238
  16−inch Telescope, 240
  Starlight−1 photometer, 241, 243
observed magnitudes vs. airmass, 197
ω Ori: light; color curves, 107
P Cyg: H−α profiles, 109
P cyg: light curve and H−α variation, 110
population II stars, meetings on, 167
RAAT system block diagram, 22
ρ Cas: light curve, 106
RW Tau: "Red" data, 357
SAL photometer configuration, 311
SAL software configuration, 322
SAL telescope drive gearing, 312
SAL telescope hardware config., 318
SAL telescope/photopolarimeter layout, 309
SAL telescope system timing, 321

SFA 41−inch drive disk, 250
SFA 41−inch mount, 251
SFA 41−inch telescope, 253
spectrophotometer: optical layout, 189
star image "boxing", 46, 47
star image position vectors, 48
SZ Her: light curve, 371
SZ Her: normalized minima, 373, 374
SZ Her: period analysis, 375
TAU system: sample screen, 271
TAU system: software flow, 270
TV camera, on telescope 287, 288
TV camera relay lenses, 284
TV camera support, 287
TV frames of star fields, 291
TV Psc: light curve, 102
UX Lyn: light curve, 103
Vanderbilt−16 APT, 88
W Ser stars: light curves, 364
W Ser stars: light curves, 365
WMO high−speed photometer, 327
  block diagram, 332
  counter circuit and device driver software, 337
  ECL/RS−422 schematic diagram, 340
  system concept, 326
  van computer block diagram, 335
  van computer rack, 334
Z Psc: light curve, 104
flux, absolute, 68
focal ratio, 11

H−line, Calcium, photometry, 66
horizontal−branch stars, 165

IAPPP−−International Amateur−Professional Photoelectric Photometry, 52, 239, 342, 380
image size, 11
image storage, methods, 211
images, data processing, 42, 51
images, galaxy, 53
images, stellar, 42, 51, 53
institutions, organizations
  American Association of Variable Star Observers, 100, 381
  Arizona State Univ., 193
  Arizona, Univ. of, 9, 10, 90
  Astronomisches Rechen−Institut, 134
  AutoScope, 27
  British−American Scientific Research Association, 223
  Bucknell Univ., 210
  Calgary, Univ. of, 217

SUBJECT INDEX 437

California Inst. of Technology, 9
California, Univ. of, 9
Cambridge Univ., England, 229
Case Institute of Technology, 171
Charleston, College of, 3, 117, 118, 125, 131
Citadel, The, 3, 117, 118, 157
DFM Engineering, Inc., 240
Dickinson College, 210
Florida Inst. of Technology, 247, 257, 267, 355, 361, 362
Florida, Univ. of, 73, 228, 307, 310, 316, 317, 324, 379
Franklin and Marshall College, 90, 201, 203, 210
Gettysburg College, 210, 211
Gila Astronomical Research Inst., 403, 407
Globe High School, 405
Harvard−Smithsonian Center for Astrophysics, 65, 86, 97, 389
Hawaii, Univ. of, 308, 316, 324
Illinois, Univ. of, Champaign−Urbana, 351
Indiana Univ., 411
Intel Corporation, 265
International Amateur−Professional Photoelectric Photometry, 52, 239, 342, 380
King College, 229
Lafayette College, 210
Montreal, Univ. of, 91
National Aeronautics and Space Administration, 80, 229, 307, 316, 324, 430
National Science Foundation, 39, 71, 87, 95, 114, 123, 137, 265, 363, 403
Nevada, Univ. of, Las Vegas, 3, 117, 118
New Mexico Inst. of Mining and Technology, 51
Ohio, Univ. of, 9
Pan American Univ., 249
Pennsylvania Astronomical Research Consortium, 201, 203
Pennsylvania, Central, Consortium of Colleges, 210
Royal Inst. of Technology, Stockholm, 307, 324
Saguaro Astronomy Club, Phoenix, 279
Smithsonian, 5, 71, 85, 95, 99, 114
Southern California, Univ. of, 307

Space Astronomy Lab., Univ. of Florida 74, 307, 316, 317, 324
Stephen F. Austin State Univ., 41, 247, 255, 257, 267, 268, 355, 361, 362, 369
Swiss Astronomical Society, Eclipsing Binary Observers of the (BBSAG), 381
Texas, Univ. of, at San Antonio, 237
Tinsley Laboratories, 249
Toronto, Univ. of, 91, 245
Union College, 165
Vanderbilt Univ., 87, 229
Villanova Univ., 3, 97, 117, 118, 383, 389
Virginia Military Inst., 91
Virginia, Univ. of, 237
Washington, Univ. of, 10
Wesleyan Univ., 91, 133, 229
Wiliams College, 91
instrumentation, space, 74
reliability, 73
instruments
   CCD cameras, 53, 186, 227, 229
   image intensifiers, 279, 280
   intensified CCD cameras, 35, 38, 70
   intensified reticon scanner (IRS), 190
   K−line photometer, 70
   objective prism, 171
   photometers, 317
      photon counting, 295
   photopolarimeters, 307, 310
      operations, 315
   SIT vidicon cameras, 53
   spectrophotometers, 157, 183, 185
   synthetic photometer, 183
   TV cameras, 264, 279, 282
      electronics, 288
      operations, 289
      optics, 283
   vidicon camera, 249, 282

K−line, Calcium, photometry, 65, 66

mass loss, stellar, 100
mirrors, light weight, 10
MOST−−Multiple−Object Spectroscopic Telescope, 13
motion, apsidal, 385

nearby stars, photometry, 133
NGIT−−New Generation Intermediate− sized Telescope, 2
NGLT−−New Generation Large Telescope, 2

NGST—New Generation Small Tel., 2, 7
NGT—New Generation Telescope, 2, 8
NNTT—National New Technology Tel., 8

objects, named
5 Cet, 423
57 Psc, 426
72 Leo, 426
89 Her, 99, 103
AG Per, 396
Alpha Vir, 396
AR Cas, 396
AS Cam, 389, 391, 393, 394, 395, 396, 397
BD Cam, 423
Beta Lyr, 362
Betelgeuse, 406
BW Vul, 229
BY Dra, 430
CW Cep, 396
CY Aqu, 125
δ CrB, 114
δ Sct, 125
DI Her, 386, 388, 389, 391, 393, 394–7
DY Peg, 125
Earth, 397
EK Cep, 396
ϵ Aur, 104
η Gem, 426
EZ Peg, 229
GD 428, 125
Halley's comet, 397
HD 26, 412
HD 2857, 176
HD 19058, 412
HD 35155, 423
HD 49368, 423
HD 57336, 175
HD 59643, 423
HD 86986, 169
HD 86986, 176
HD 99967, 423
HD 109995, 169, 176
HD 161817, 169, 176
HR 1952, 396
HR Peg, 423
ι Ori, 396
KR Per, 380
KU Cyg, 352
λ Aqr, 423
Mercury, 386, 389, 397
Moon, 397
Mu UMa, 426
NY Cep, 396
ω Ori, 108
P Cyg, 99, 108, 114
ϕ Aqr, 423
ϕ Vir, 423
PSR 1913+16, 394
QX Car, 396
R CrB, 106, 107
ρ Cas, 99, 103, 106, 108
RS CVn, 111
RW Tau, 357
S Cnc, 352
Sun, 66, 68, 119, 386, 397
SW Lac, 422
SX Phe, 125
SZ Her, 342, 369
SZ Sgr, 423
TV Psc, 101
Ursa Majoris System, 119
UU Her, 104
UX Lyn, 101
V CVn, 100
V1143 Cyg, 396
V356 Sgr, 363
V541 Cyg, 396
V889 Aql, 396
Vega, 68, 140, 160
VW Cep, 422
W Ser, 363
W UMa, 422
YY Eri, 422
YZ Cas, 381
Z Psc, 101
ζ And, 424

observatories
Anglo–Australian, 8
APT Service, 2, 5, 19, 23, 68, 85, 95, 117, 118, 122, 141, 186, 187
Armagh, 90
Astronomiczne Uniwersytetu Jagiellonskiego, 381
Big Cottonwood, 228, 233
Braeside, 227–9, 242
Canada–France–Hawaii, 8
Cerro Tololo Interamerican, 8, 169
Dyer, 87
European Southern, 8, 9
Fairborn, 1, 19, 23, 27, 35, 41, 65, 85, 97–99, 139, 141, 185, 186, 193, 227, 239, 274, 341, 342, 406
Fan Mountain, 239
Fred L. Whipple, 20, 85, 88, 90, 97–9
Gila, 406
Harvard College, 389, 396

SUBJECT INDEX 439

  Hopkins Phoenix, 295, 299, 301, 303, 305
  International Ultraviolet Explorer, 3, 175, 342, 391, 411, 430
  Kitt Peak National, 7, 8, 140, 145, 158, 159, 169, 180, 190, 229, 342, 345, 421
  Konkoly, 381
  Lick, 250
  Limber (D. Nelson Limber Memorial Observatory), 90, 228, 237, 238
  Lines, 90
  Lowell, 183, 326
  Maria Mitchell, 91
  McCormick, 133
  McDonald, 238, 362
  Mount Laguna, 352
  Mt. Haleakala, 74
  Mt. Wilson, 66–8, 111, 114, 158
  Multiple–Mirror Telescope, 20
  National Optical Astronomy, 7, 8, 145, 345
  Oak Ridge, 108, 114
  Palomar, 68, 158, 169
  Rosemary Hill, 379
  Smithsonian Astrophysical, 20
  Sonneberg, 381
  South Pole, 74
  Stephen F. Austin State Univ., 41, 362, 370
  U. S. Naval, 232
  Van Vleck, 133, 134, 165
  Villanova Univ., 390
  Winer Mobile, 228, 325
OGT––Old Generation Telescope, 8
optics, active, 11
orbital plane, 393
orbits, rapid circularization, 390

PARC––Pennsylvania Astronomical Research Consortium, 201, 203
pattern recognition, for acquisition, 47
photometers
  computer interface, 325
  computer–controlled, 317
  data logging, 267, 271
  GORT–1, 230
  high–speed, 325
  Hopkins Phoenix Obs., 406
  K–line, 70
  Optec SSP–3a, 99, 406
  photon counting, 295, 325
    power supply, 302
    conditing, 297
    universal counter, 303
  Starlight–1, 238, 240, 249, 258, 259, 357, 362, 370, 406
  Texas 362
  Vanderbilt 16, 88
  WMO high–speed, 325, 326, 340
    design, 328
photometric standards, 119
photometric systems
  Cousins UBVRI, 120
  H–α, 88, 121, 152
  H–β, 71, 88, 121, 146, 152, 346
  H–δ, 152
  H–γ, 152
  Johnson UBV (or subset), 71, 134, 146, 168, 190, 232, 240, 342, 345, 356, 397
  Johnson $UBVRI$ (or subset), 88, 99, 242, 359, 363, 406, 421
  Kron RI, 134
  modeling, 145
  Stromgren $uvby$, 71, 88, 121, 146, 165, 171, 195, 229, 232, 342, 345, 352, 397
    plus I, 352
    tutorial, 345
  ultraviolet, 414
photometry
  all–sky, 242
    reduction, 244
  automatic, 65, 165
  broadband, bright stars, 97
  differential, 244, 356, 361, 369, 379, 406
    data analysis, 364
    period study, 372
    reduction, 245, 351
    techniques, 352
  filter–defined, 183
  H–line (Calcium), 66, 87
  horizontal–branch stars, 165
  initial reduction, 358
  intermediate–band, 146, 345, 352
  K–line (Calcium), 65, 66, 87
  narrow–band, 147
    atmospheric extinction, 193
  nearby stars, 133
  photoelectric, 194, 237, 267, 341, 345, 355, 369, 403, 407
  quality–control check, 351
  reduction, 355
    algorithm, 196
    errors, 195
    transformation coefficients, 351
  synthetic, 145, 146, 161, 183
  ultraviolet, 411

color—color curves, 421
wide—band, 146, 345
photospheres, stellar, 411
places
  Brisbane, Australia, 229
  Carol Spring Mtn., Arizona, 408
  Czechoslovakia, 245
  Fairborn, Ohio, 85
  Flagstaff, Arizona, 232
  Globe, Arizona, 405, 406
  Harvard, Massachusetts, 108
  Kitt Peak, Arizona, 118
  La Silla, Chile, 9
  Mauna Kea, Hawaii, 10
  Mesa, Arizona, 20, 23
  Mt. Graham, Arizona, 9
  Mt. Haleakala, Hawaii, 308, 313
  Mt. Hopkins, Arizona, 20, 85, 89, 91, 99, 186
  Mt. Wilson, California, 87
  Nacogdoches, Texas, 248, 268, 370
  Nashville, Tennessee, 87
  Pipe Creek, Texas, 238
  San Antonio, Texas, 237
  Tucson, Arizona, 118
pulsation, stellar, 100

RAAT——Remote Access Automatic Telescope 3, 19
relativity, general, 390, 395, 397
  problems with, 394
  tests of, 383
reliability, automatic telescopes, 73
  concepts, 74
  construction procedures, 80
  economic considerations, 82
  hardware, 75
  role of procurement, 79
  software, 75
  space instrumentation concepts, 73
  testing for, 80

SAL——Space Astronomy Lab., Univ. of Florida, 307, 317
science, avocational, 220
  professional, and amateurs, 217
scientists, amateur, 217
  and professional, 221
SFA; SFASU——Stephen F. Austin State Univ., 42
spectrograph, automatic, design, 206
spectrophotometers
  automatic, 185
  design, 185
  design parameters, 187

HCO Scanner, 159, 169
  non—scanning, 183
  Oke Multichannel, 169
  scanning, 158, 169
  system performance, 190
spectrophotometry, 183
  atmospheric extinction, 193
  automatic, 139, 157, 162, 165
  horizontal—branch stars, 165
  low—resolution, 139, 185
spectroscopy, automatic, 201, 203
standards, photometric, 119
stars
  active chromosphere, 65
  Algol—type, 369
  Be, 99, 108
  binary, times of minima, 379
  bright, 98
  broadband photometry, 97
  chemically peculiar, 119
  chromospheres, 411
  cool, 87
  δ Scuti, 2, 5, 125
  early type, 171
  eclipsing binaries, 383
  FK Com, 413
  globular cluster, 174
  horizontal branch, 165, 166, 174
  nearby, 133
  photospheres, 411
  population II, 166
  RS CVn, 2, 87, 111, 406, 413
  solar—type, 66, 119
  W Ser, 342, 361, 362
  W UMa binaries, 413
  Wolf—Rayet, 108
starspots, 98, 111
Sun, absolute flux, 68
supernovae, detectability, 51

tables
  (IUE−V), (MgII−V) for cool giants, 427
  APT Consortium: participants, 118
  APT Consortium: projects, 119
  bandpasses for synthetic photometry, 148
  bright variable stars, 101
  candidate binary stars: orbital properties, 399
  check—comp star difference statistics, 353
  δ Sct stars: number vs. apparent magnitude, 131

SUBJECT INDEX

DI Her and AS Cam: orbital properties, 398
FHB stars: metallicities, 172
HB stars: temperatures and gravities, 177
IUE photometric bands, 414
late−type stars: visual and *UV* photometry, 416
McCormick stars: colors, 135
reliability concepts, 75
spectrophotometer: parameters, 188
telescope cost scaling law, 17
telescope drive system, 310
UBV transformation results, 150
TAU−−SFA telescope controller and data logger, 268

telescopes
Australian Nat. 2.5−m, 10
automatic, 27, 65, 74, 185, 187, 191, 203, 211, 307, 317
  design, 29
  mount, 30
  optics, 29
  reliability, 73
automatic photoelectric, 1, 20, 35, 38, 85, 97, 99, 117, 125, 165, 201, 239, 397
automatic spectrographic, 158, 201, 203
  design, 204
  management, 209
  operation, 207
automatic spectrophotometric, 141, 157, 158
Baker−Nunn Camera, 85
Big Cottonwood 0.5−m, 233
Case Schmidt, 134, 171
Cassegrain, 12−inch, 406
Celestron C−14, 238, 333, 406
computer−controlled, 227, 229, 233, 237, 247, 257, 267, 307, 317
  control computer, 263, 267, 271
  controller, 317
  design, 274
  drive, 262
  electronics, 253
  mount, 250, 260
  operations, 315
  optics, 249, 259
  positioning system, 314
  run scheduling, 269
  SAL system evaluation, 323
  system, 318
control computers, 234
cost scaling law, 16

DFM 16−inch, 239
drives, 234
ESO VLT 16−m, 9
Fairborn 0.75−m, 27, 85
Fairborn 10 APT, 85, 98, 99, 114
Fan Mountain 40−inch, 239
FIT 16−inch, 257, 268, 362
FIT 25−inch, 257, 268, 362
"Giant Binocular," 11.3−m, 9
global network of, 5
Hamburg Schmidt, 171, 180
Illinois 1−m reflector, 352
Keck 10−m, 1, 9, 10
Kitt Peak 36−inch, 190
Kitt Peak 4−m, 10
Lowell 24−inch, 406
Lowell 31−inch, 406
McDonald 76−cm telescope, 362
Mt. Wilson 100−inch, 111
Mt. Wilson 60−inch, 87, 111
Multiple Object Spectroscopic, 13
New Generation, 1, 2, 10
  Intermediate, 2
  Large, 2, 8
  Small, 2, 7, 14, 16, 65, 341
New Mexico DASS 0.75−m, 51
NNTT 15−m, 1, 8
Nordic 2.5−m, 10
Old Generation, 8
Phoenix 10 APT, 88, 90
photometric, 257, 295
Remote−Access Automatic, 3, 19, 21
SAL 32 cm, 310, 317
Schmidt−Cassegrain, 8−inch, 405
Schmidt−Cassegrain, 11−inch, 406
science per dollar ratio, 10, 13
SFA 18−inch, 257, 268, 362, 370
SFA 41−inch, 247, 257, 260, 268, 362
Soviet Six−Meter Reflector, 170
Vanderbilt 16−inch APT, 87, 95
Warner and Swasey Schmidt, 174, 180
WMO 30−inch, 326

variable stars, 98
irregular, 99
long−period Cepheids, 119
Mira, 100
multi−mode, 119
semi−regular, 99, 100
Vega, absolute flux, 68
VLT−−ESO Very Large Telescope, 9

winds, stellar, 100
WMO−Winer Mobile Observatory, 325
Wolf−Rayet stars, 108

ABOUT THE EDITORS

Donald S. Hayes

Donald S. Hayes holds a B. A. in Astronomy from Pomona College, and M. A. and PhD degrees in Astronomy from the University of California, Los Angeles. He has been on the faculty of Rensselaer Polytechnic Institute and Arizona State University, and on the staff of the Kitt Peak National Observatory. A member of the International Astronomical Union, the American Astronomical Society and the International Amateur–Professional Photoelectric Photometry (IAPPP), he is an author of many papers in photometry and spectrophotometry. He has been chief editor or a co–editor of the proceedings of six professional symposia, including *New Generation Small Telescopes*, and is a co–author of: *Supernova 1987A: Astronomy's Explosive Enigma*. He is with Fairborn Observatory and the Institute for Space Observations, and his research interests include the automation of photometric telescopes and several topics in stellar photometry. Currently, Donald and his wife Julie make their home in Scottsdale Arizona.

Russell M. Genet

Born in 1940, Russell Genet was raised in the small California town of Yucaipa in the San Bernardino mountains. He received his B.S. in Electrical Engineering from the University of Oklahoma, and an M.S. from the Air Force Institute of Technology. He has worked for the Air Force for over 30 years, primarily in research and development where he has been responsible for a new missile guidance system, space experiments, and new analytic techniques in operations research. For the past decade, he has conducted his research at the Air Force Human Resources Laboratory where he is currently working on low–cost simulators for training fighter pilots for combat. He is an instrument–rated pilot and flight instructor. Russell and his wife Ann have five children and make their home in Mesa, Arizona.

In 1979 he founded the Fairborn Observatory which was devoted to making photoelectric observations of variable stars. In 1980, with Dr. Douglas S. Hall, he founded the International Amateur–Professional Photoelectric Photometry (IAPPP) Association, and its quarterly journal, the *IAPPP Communications*. The IAPPP has grown to some 750 members from 45 countries. With Louis J. Boyd, Genet has developed automatic photoelectric telescopes (APTs), and the Fairborn Observatory in conjunction with the Smithsonian Institution now operates a number of APTs on Mt. Hopkins in southern Arizona. In the area of astronomy and use of microcomputers in astronomy, he is author or editor of a dozen books. Russell Genet is a recipient of the Astronomical Society of the Pacific's *Amateur Achievement Award* and the Astronomical League's *Leslie C. Peltier Award*. He is a member of the AAS, AAVSO, ASP, and IAPPP.

David R. Genet

David R. Genet received a B. S. in Industrial Engineering from Southern Illinois University in 1985. Since then he has been employed in the Simulations Department of a major aerospace firm. He was a co-editor of *New Generation Small Telescopes* and chief editor of the *Photoelectric Photometry Handbook* as well as co-author of *Supernova 1987A: Astronomy's Explosive Enigma*. He is a member of the IAPPP and a publisher associate member of the AAS. In addition to writing and publishing books in the area of astronomy, he enjoys flying airplanes, firing Scandanavian pottery on his back porch during the summer months, and complaining about the heat with Debbie, the chief pilot for Dave's Lear Jet. David, his Siberian Huskie Sasha, and his cats, Cheapet and Punkinhead, make their home in Gilbert, Arizona.